LONDON MATHEMATICAL SOCIETY STUDENT TEXTS

Managing editor: Dr C.M. Series, Mathematics Institute
University of Warwick, Coventry CV4 7AL, United Kingdom

1 Introduction to combinators and λ-calculus, J.R. HINDLEY & J.P. SELDIN
2 Building models by games, WILFRID HODGES
3 Local fields, J.W.S. CASSELS
4 An introduction to twistor theory: Second edition, S.A. HUGGETT & K.P. TOD
5 Introduction to general relativity, L.P. HUGHSTON & K.P. TOD
6 Lectures on stochastic analysis: diffusion theory, DANIEL W. STROOCK
7 The theory of evolution and dynamical systems, J. HOFBAUER & K. SIGMUND
8 Summing and nuclear norms in Banach space theory, G.J.O. JAMESON
9 Automorphisms of surfaces after Nielsen and Thurston, A. CASSON & S. BLEILER
10 Nonstandard analysis and its applications, N. CUTLAND (ed)
11 Spacetime and singularities, G. NABER
12 Undergraduate algebraic geometry, MILES REID
13 An introduction to Hankel operators, J.R. PARTINGTON
14 Combinatorial group theory: a topological approach, DANIEL E. COHEN
15 Presentations of groups, D.L. JOHNSON
16 An introduction to noncommutative Noetherian rings, K.R. GOODEARL & R.B. WARFIELD, JR.
17 Aspects of quantum field theory in curved spacetime, S.A. FULLING
18 Braids and coverings: selected topics, VAGN LUNDSGAARD HANSEN
19 Steps in commutative algebra, R.Y. SHARP
20 Communication theory, C.M. GOLDIE & R.G.E. PINCH
21 Representations of finite groups of Lie type, FRANÇOIS DIGNE & JEAN MICHEL
22 Designs, graphs, codes, and their links, P.J. CAMERON & J.H. VAN LINT
23 Complex algebraic curves, FRANCES KIRWAN
24 Lectures on elliptic curves, J.W.S. CASSELS
25 Hyperbolic geometry, BIRGER IVERSEN
26 An introduction to the theory of L-functions and Eisenstein series, H. HIDA
27 Hilbert Space: compact operators and the trace theorem, J.R. RETHERFORD
28 Potential theory in the complex plane, T. RANSFORD
29 Undergraduate commutative algebra, M. REID
31 The Laplacian on a Riemannian manifold, S. ROSENBERG
32 Lectures on Lie Groups and Lie Algebras, R. CARTER, G. SEGAL & I. MACDONALD
33 A primer of algebraic *D*-modules, S.C. COUTINHO
34 Complex algebraic surfaces, A. BEAUVILLE
37 A mathematical introduction to wavelets, P. WOJTASZCZYK

London Mathematical Society Student Texts 37

A Mathematical Introduction to Wavelets

P. Wojtaszczyk
Warsaw University

PUBLISHED BY THE PRESS SYNDICATE OF THE UNIVERSITY OF CAMBRIDGE
The Pitt Building, Trumpington Street, Cambridge, United Kingdom

CAMBRIDGE UNIVERSITY PRESS
The Edinburgh Building, Cambridge CB2 2RU, UK www.cup.cam.ac.uk
40 West 20th Street, New York, NY 10011-4211, USA www.cup.org
10 Stamford Road, Oakleigh, Melbourne 3166, Australia
Ruiz de Alarcón 13, 28014 Madrid, Spain

First published 1997
Reprinted 1999

Printed in the United Kingdom at the University Press, Cambridge

A catalogue record for this book is available from the British Library

ISBN 0 521 57020 4 hardback
ISBN 0 521 57894 9 paperback

Contents

Preface

In many places in engineering or science or mathematics we are faced with a version of the following archetypical problem. We are given a function $f(t)$ defined for $t \in \mathbb{R}$. Let us imagine that this function describes some real-life phenomenon. To make things mathematically simple let us assume that $f \in L_2(\mathbb{R})$. Our aim (admittedly vague) is to transmit (or store or analyze) this function using some 'finite' device. A good illustration might be that f represents a voice signal and we want to transmit it over the telephone lines or put it on a compact disk. The whole function f is given by the totality of its values at points of $\mathbb{R}$, and this makes it a continuum of points – we can not do much with 'finite' device. So let us suppose that as our background knowledge we have some orthonormal basis $(f_n)_{n\in N}$ in $L_2(\mathbb{R})$. Then we know that we can write

$$f = \sum_{n\in N} a_n f_n$$

where the series converges in $L_2(\mathbb{R})$ and the coefficients are uniquely determined by the formulas $a_n = \langle f, f_n \rangle$ for $n \in N$. Thus instead of transmitting the function f it suffices to transmit the sequence of coefficients $(a_n)_{n\in N}$ and let the recipient sum the series himself. It is still not a finite procedure but it looks better. To make it really finite we have to choose a finite set $A \subset N$ such that $\sum_{n\in A} a_n f_n$ will be 'almost equal' to $\sum_{n\in N} a_n f_n$. But life is not perfect, so we cannot expect to have perfect tranmission every time. This means that the recipient is really forming the sum $\sum_{n\in A} \tilde{a}_n f_n$, where $\tilde{a}_n$ is 'close' to a_n, and has to hope that the result is still 'almost equal' to $\sum_{a\in N} a_n f_n$. But what does 'almost equal' mean? If we are in any real situation we have to decide and base this decision on our experience. Mathematically speaking we need a distance between functions, which very often is provided by some norm, usually different from the hilbertian norm.

This is a very general story and there were and are many ways to deal with various special instances of different aspects of this archetypical problem. Wavelets are just one new tool to deal with this type of problem.

The subject of wavelets appeared in the mid 1980s influenced by ideas from both pure mathematics (harmonic analysis, functional analysis, approximation theory, fractal sets etc.) and applied mathematics (signal processing, mathematical physics etc.). Almost instantanously it became a success story with thousands of papers written by now and wide ranging applications. The reader may learn some history of this subject from [87] or [86] or from introductions and comments in [24], [84] or [85]. I do not want to discuss the history in any detail. Let me simply state that by now wavelets find applications in many areas of mathematics, science or technology. Just to show how diverse are applications of wavelets let me say that *Wavelet Literature Survey* [92] divides its entries into the following categories: acoustics, astronomy, atomic decompositions, $ax+b$ group, Bernoulli–Gaussian processes, chord–arc curves, fractal and Cantor sets, frames, Franklin wavelets, Gabor representations, image processing, irregular sampling, non-orthogonal expansions, numerical algorithms, partial differential equations, seismology, signal processing, splines, theory, wavelet bases, and wavelet transform.

In order to make this discussion a bit more precise let us state here (in the case of one variable only) the definition of a wavelet which (naturally enough) is the main concept discussed in this book.

Definition 2.1 *A wavelet is a function $\Psi(t) \in L_2(\mathbb{R})$ such that the family of functions*

$$\Psi_{j,k} =: 2^{j/2}\Psi(2^j t - k)$$

where j and k are arbitrary integers, is an orthonormal basis in the Hilbert space $L_2(\mathbb{R})$.

If we look at this definition we can guess how wavelets fit into the general scheme depicted above. They provide the orthonormal system that we wanted to have as our 'background knowledge'.

When one looks at the above definition for the first time, one question arises immediately: *do wavelets exist at all?* To this question we will give an affirmative answer many times in this book. We start with some examples in Chapter 1 and will give more in subsequent chapters. Let us assume for the time being that wavelets exist. Then the next question springs to mind: why bother with them?

Well, the first and really the most important answer is that wavelets fit very well into many concrete cases of our archetipical story. In short: *they are useful.* Naturally this answer can be justified only after working with wavelets on some concrete problems.

The second answer is: simply for the fun of it. To be more serious; it is not clear that such a function exists, so when it turns out that it indeed exists we have a mild surprise. Thus we may wish to investigate such functions in some detail, to know how many such strange functions exist, what additional properties they may have etc. Also each such function gives us an orthonormal basis in $L_2(\mathbb{R})$. So it may be interesting to investigate such bases. There is also a natural question of generalizations, e.g. what happens in $\mathbb{R}^d$. All this will be studied in this book and I hope that the reader will be convinced that wavelets are interesting.

Since I am a pure mathematician by profession, education and character this second answer is close to my heart. Thus I study wavelets as a beautiful mathematical idea. It seems to me that it is this beauty and simplicity of the wavelet concept that has attracted so many people. This intrinsic simplicity makes wavelets a convenient framework unifying various earlier methods.

Even from this Preface it should be clear that the literature on wavelets in enormous and is growing at a tremendous rate. Nevertheless when in 1993 I began preparing a Part III (beginning graduate students in mathematics) course on wavelets at Cambridge University, I had difficulties in finding a text appropriate for such students. Out of [24], [85], [27] and many expository and research papers, and out of my own experience I put together a course which, with many additions, I repeated in 1994/95 for senior undergraduates in mathematics at Warsaw University. The present book is an outgrowth of those efforts. My idea was to present the essential mathematical core of theory of wavelets. I decided to concentrate on orthonormal wavelets as the most complete and 'cleanest' part of the theory. My aim was to give detailed constructions of the most important wavelets and present the usefulness of wavelet bases in decomposing functions. All this is done in the framework of function spaces.

So this is a purely mathematical book, although constantly I try to make my calculations as explicit as possible and I concentrate on theoretical questions that should have relevance to applications. But regrettably I discuss no *real* applications.

In other words, I take it for granted that wavelets are useful, and this belief is one of the motivations for studying wavelets, but I explain only their basic mathematical theory. I hope that each reader will find a favorite application for wavelets. I also believe that the knowledge of orthogonal wavelets discussed in this book should make the study of various generalizations much easier.

This book is basically a course of lectures aimed at students of mathematics (maybe even pure mathematics). Thus I start rather slowly but as the book progresses the pace quickens a bit. There are also more than a hundred exercises for the reader to solve.

Let me explain the content and organization of this book in more detail.

Chapters 1–4 discuss one-variable orthonormal wavelets (as defined above). This is the backbone of the mathematical theory. Chapter 1 in a sense presents an overview of the book. Without any general theory we discuss two wavelets: the Haar wavelet and the Strömberg wavelet. They were invented and investigated well before the emergence of the general theory or indeed the notion of wavelet. We present the construction and properties of those wavelets and show some sample theorems about convergence of wavelet expansions. As its title suggests, the aim of this chapter is to convey the general spirit of the book by presenting important but relatively easy and explicit examples. Formally its results are not used later except as a motivation or in examples and exercises. Chapter 2 discusses the general theory. We present and discuss here the concept of multiresolution analysis and the scaling function. Then we describe all wavelets associated with a given multiresolution analysis. We also show how the scaling function can be used to build the multiresolution analysis. Here we also discuss in general terms periodic wavelets. In Chapter 3 we show how the above general theory can be applied in concrete cases. We construct Meyer's wavelets and spline wavelets and discuss in detail their smoothness and decay. This chapter also includes a self-contained introduction to spline functions. We also discuss in this chapter examples of wavelets not associated with any multiresolution analysis. Chapter 4 discusses wavelets with compact support. We present a general approach to constructing compactly supported wavelets and apply this to a construction of smooth, compactly supported wavelets. We also present an 'elementary' construction of a continuous wavelet whose support is $[0,3]$. These Chapters use only the rudiments of Hilbert spaces and the Fourier transform on the real line. Chapter 5 discusses multivariable generalizations. We discuss briefly a tensor product technique and next generalize to $\mathbb{R}^d$ the concept of multiresolution analysis. Then we discuss the general procedures leading from multiresolution analysis to wavelets in this context. It turns out that in general we need a finite *wavelet set* instead of one wavelet. We exhibit many examples of Haar-like wavelets on $\mathbb{R}^d$, i.e. wavelets Ψ such that $|\Psi(x)|$ is the characteristic function of a set. Then we construct more smooth examples.

These five chapters constitute an introduction to constructions of orthogonal wavelets. The rest of the book deals with wavelet expansions. In Chapter 6 we give a self-contained presentation of the basic theory of L_p spaces and H_1 and BMO. Our main results are the necessary interpolation theorems. In Chapter 7 we introduce unconditional convergence of series in Banach spaces and discuss the concept of unconditional basis. In Chapter 8 we prove that good wavelets provide unconditional bases in $L_p(\mathbb{R}^d)$ and in $H_1(\mathbb{R}^d)$ and we show the equivalence of various wavelet bases. We also give a characterization of those function spaces in terms of wavelet expansions. We also discuss periodized Meyer wavelets since they lead to systems of trigonometric polynomials. We conclude with Chapter 9 where, on $\mathbb{R}$ only, we discuss moduli of continuity and Besov norms and their connections with wavelets.

The obvious prerequisite to reading this book is a sound understanding of real analysis, functions, series of functions etc. A familiarity with Lebesgue integration is very useful, in particular the Lebesgue dominated convergence theorem is used several times. Additionally some elementary knowledge of the Fourier transform and Hilbert spaces are needed. All the facts needed are summarized in Sections A.1 and A.2. To read Chapters 1–4 (with the exception of some portions of Chapter 1) one needs only to know the Fourier transform on $\mathbb{R}$ and be familiar with the Hilbert space $L_2(\mathbb{R})$. To read Chapter 5 one needs to know the Fourier transform on $\mathbb{R}^d$. To read the remaining chapters one needs the concept and very basic properties of Banach spaces. They are summarized in Section A.3. The more advanced facts are carefully presented when needed. In particular Chapter 6 contains a detailed and selfcontained presentation of the necessary parts of the theory of L_p spaces and the theory of H_1 and BMO. The concept of unconditional basis is introduced and presented in some detail in Chapter 7.

As the reader could have guessed anyway, the book is organized into chapters, which are divided into sections. The numbering is decimal by chapter. Displayed formulas (if numbered) are numbered consecutively. There is a single numbering sequence for theorems, propositions, lemmas, corollaries and definitions. Each chapter ends with a section 'Sources and comments' and with exercises. Exercises range from quite easy and routine to rather difficult. The last chapter of the book is an Appendix which contains sections listing results used about Hilbert spaces, Fourier transforms and Banach spaces. It also contains list of symbols and list of spaces. References to a result from the Appendix are easily recognizable by the presence of the letter A and Roman numerals, e.g. A.2–II

Since this is clearly a textbook the Bibliography lists only works actually referred to in the text, mostly in 'Sources and comments'. I have not tried to give full bibliography of the subject and it would be impossible anyway. For readers looking for more information on the mathematical part of wavelet literature I can only offer the following suggestions:

- To check recent volumes of *Mathematical Reviews* or *Zentralblatt für Mathematik*. If possible this should be done using the corresponding computer data base. The search for the keyword *wavelet* will immediately exhibit hundreds of items.
- The book *Wavelet Literature Survey* [92] lists more than 1000 items of the wavelet literature. They cover both the theory and many diverse applications.
- There is a whole series *Wavelet analysis and its applications* published by Academic Press, which contains books on various aspects of wavelets. The book [15] was the first publication in this series.
- There are mathematical journals which regularly publish papers connected with wavelets. Browsing through volumes of *Applied and Computational Harmonic Analysis, The Journal of Fourier Analysis and Applications, Constructive Approximation* or *Studia Mathematica* should easily yield many interesting papers.

Aknowledgements. Part of the work on this book was done while I was a Central and East European Fellow under the Human Capital and Mobility Programme of the Commision of the European Communities and while I was a Visiting Scholar of St. John's College, Cambridge. I would like to thank St. John's College, for the hospitality that was extended to me and my family. During work on this book I was partially supported by KBN grant no 2P301004.06. I would like to thank Dr K. Nowiński of the Interdisciplinary Center for Mathematical Modelling, Warsaw University, for his help in preparing the computer graphics appearing in this book.

1
A small sample

The main point of this introductory chapter is to present two wavelets: the Haar wavelet and the Strömberg wavelet. We do so without using any general theory and without even giving the definition of a wavelet. We present the construction of these wavelets and indicate how they can be used to represent functions from some simple, natural classes. This (as the chapter title indicates) represents a sample of this book. Such an approach is also justified historically. Both these wavelets were well known (without the use of the word 'wavelet') before the emergence of the general theory.

1.1 The Haar wavelet

In this section we will discuss in some detail the most elementary wavelet, called the Haar wavelet.

DEFINITION 1.1 *The Haar wavelet is the function defined on the real line* $\mathbb{R}$ *as*

$$H(t) = \begin{cases} 1 & \textit{for } t \in [0, \frac{1}{2}) \\ -1 & \textit{for } t \in [\frac{1}{2}, 1] \\ 0 & \textit{otherwise.} \end{cases} \tag{1.1}$$

We are interested in the family $\{2^{j/2}H(2^j t - k)\}_{j\in\mathbb{Z},k\in\mathbb{Z}}$. To simplify the notation let us denote $H_{jk}(t) =: 2^{j/2}H(2^j t - k)$. Observe that

$$\operatorname{supp} H_{j,k} = [k2^{-j}, (k+1)2^{-j}]. \tag{1.2}$$

The intervals $[k2^j, (k+1)2^j]$ for $k, j \in \mathbb{Z}$ form the family of dyadic intervals. This family of intervals splits naturally into levels; the j-th level consists of intervals whose length is 2^{-j}. Inside each level distinct dyadic intervals are non-overlapping. The following two properties of dyadic intervals will be useful in further considerations:

(i) either two dyadic intervals do not overlap or one is contained in the other

(ii) if one dyadic interval is strictly contained in the other, then it is contained either in the left half or in the right half of it.

Those observations easily give the following proposition.

Proposition 1.2 *The system* $\{2^{j/2}H(2^jt-k)\}_{j\in\mathbb{Z},k\in\mathbb{Z}}$ *is orthonormal in* $L_2(\mathbb{R})$.

Proof Let us look at the scalar products $\langle H_{j,k}, H_{j',k'}\rangle$. We can assume that $j \le j'$. Using the substitution $u = 2^j - k$ we see that

$$\langle H_{j,k}, H_{j',k'}\rangle = \int_{-\infty}^{\infty} 2^{s/2}H(t)H(2^st-r)dt \tag{1.3}$$

where $s = j'-j$ and $r = 2^{j'-j}k - k'$. If $j = j'$ and $k = k'$ then the integral in 1.3 clearly equals 1. If $j = j'$ but $k \neq k'$ then $r \neq 0$ so $\mathrm{supp}H(2^st-r)\cap\mathrm{supp}H(t) = \emptyset$ and the integral in 1.3 is 0. When $j' > j$ then we have either $\mathrm{supp}H(2^st-r)\cap\mathrm{supp}H(t) = \emptyset$ so the integral is 0 or $\mathrm{supp}H(2^st-r) \not\subseteq \mathrm{supp}H(t)$. But in this case $H(t)$ is constant on $\mathrm{supp}H(2^st-r)$, so the integral is also 0 because $\int_{-\infty}^{\infty} H(t)dt = 0$. □

In order to show that $\{2^{j/2}H(2^jt-k)\}_{j\in\mathbb{Z},k\in\mathbb{Z}}$ is an orthonormal *basis* in $L_2(\mathbb{R})$ let us consider two families of closed subspaces of $L_2(\mathbb{R})$:

$$S_n = \mathrm{span}\,\{H_{j,k}\}_{j<n,k\in\mathbb{Z}} \tag{1.4}$$

and

$$L_n = \left\{\begin{matrix}\text{all functions in } L_2(\mathbb{R}) \text{ constant on all}\\ \text{intervals } [k2^{-n},(k+1)2^{-n}], \text{ for } k\in\mathbb{Z}\end{matrix}\right\}. \tag{1.5}$$

Both these families have the following properties (we formulate them for the S_n's only, but the same holds for the L_n's).

$$\ldots \subset S_{-1} \subset S_0 \subset S_1 \subset \ldots \tag{1.6}$$

$$f(t)\in S_n \iff f(2t)\in S_{n+1} \tag{1.7}$$

$$f(t)\in S_0 \iff f(t+k)\in S_0 \quad \text{for } k\in\mathbb{Z}. \tag{1.8}$$

Lemma 1.3 *For all* $n\in\mathbb{Z}$ *we have* $L_n = S_n$.

Proof From 1.7 above we see that it suffices to show that $S_0 = L_0$. Since each $H_{j,k}$ for $j < 0$ is constant on any interval $[r, r+1]$ we see that $S_0 \subset L_0$. On the other hand each function from L_0 can be written as $\sum_{r\in\mathbb{Z}} a_r \mathbf{1}_{[r,r+1]}$ so by 1.8 it suffices to show that $\mathbf{1}_{[0,1]} \in S_0$.

To show this let us consider the series

$$\sum_{j<0} 2^{j/2} H_{j,0} = \sum_{j<0} 2^j H(2^j t).$$

Since $\|2^j H(2^j t)\|_2 = 2^{j/2}$ and $j < 0$ this series is absolutely convergent in $L_2(\mathbb{R})$. One can easily see from the definition of $H(t)$ that

$$\sum_{j<0} 2^{j/2} H_{j,0}(t) = 0 \quad \text{for } t \leq 0,$$

$$\sum_{j<0} 2^{j/2} H_{j,0}(t) = \sum_{j<0} 2^j = 1 \text{ for } 0 < t < 1$$

and for $2^r < t < 2^{r+1}$ where $r = 0, 1, 2, \ldots$ one has

$$\sum_{j<0} 2^{j/2} H_{j,0}(t) = -2^{-r-1} + \sum_{j=r+2}^{\infty} 2^{-j} = 0.$$

This shows that $S_0 = L_0$ so $L_n = S_n$ for all $n \in \mathbb{Z}$. □

From Proposition 1.2 and Lemma 1.3 and the fact that $\bigcup_{n=-\infty}^{\infty} L_n$ is dense in $L_2(\mathbb{R})$ we get immediately:

Theorem 1.4 *The system* $\{2^{j/2} H(2^j t - k)\}_{j\in\mathbb{Z},k\in\mathbb{Z}}$ *is an orthonormal basis in* $L_2(\mathbb{R})$.

This means that for a function $f \in L_2(\mathbb{R})$ we have a decomposition

$$f = \sum_{j\in\mathbb{Z}} \sum_{k\in\mathbb{Z}} \langle f, H_{jk}\rangle H_{jk}. \tag{1.9}$$

Since $H \in L_p(\mathbb{R})$ for all p, $1 \leq p \leq \infty$, we can write the right hand side for any $f \in L_p(\mathbb{R})$, $1 \leq p \leq \infty$. For the rest of this section we will investigate the convergence of this series for $f \in L_p(\mathbb{R})$. Later (Section 8.2) we will show that for $1 < p < \infty$ this series converges when arranged in any order. For the time being we will discuss only the most natural order. Thus we will study operators Q_j^μ and P_r defined as

$$Q_j^\mu(f) = \sum_{k\leq\mu} \langle f, H_{jk}\rangle H_{jk} \tag{1.10}$$

and

$$P_r(f) = \sum_{j<r} \sum_{k \in \mathbb{Z}} \langle f, H_{jk} \rangle H_{jk}. \tag{1.11}$$

Theorem 1.5 *If $f \in L_p(\mathbb{R})$ with $1 < p < \infty$ or $f \in C_0(\mathbb{R})$, then $\lim_{r\to\infty} P_r(f) = f$ and for each $r \in \mathbb{Z}$ $\lim_{\mu\to\infty} P_r(f) + Q_r^\mu(f) = P_{r+1}(f)$. The convergence is in the norm of the space.*

Proof For $1 < p < \infty$ let us denote by S_n^p and L_n^p the spaces defined by 1.4 and 1.5 where the closure is taken in $L_p(\mathbb{R})$ not in $L_2(\mathbb{R})$. The proof of Lemma 1.3 can be easily modified to show that $L_n^p = S_n^p$ for all $n \in \mathbb{Z}$. Since $P_r(f)$ is an orthogonal projection onto S_r and $S_r = L_r$ we can write a different representation of the operator P_r, namely we have

$$P_r(f) = \sum_{k\in\mathbb{Z}} 2^r \int_{k2^{-r}}^{(k+1)2^{-r}} f(t)dt \cdot \mathbf{1}_{[k2^{-r},(k+1)2^{-r}]}.$$

The validity of this representation follows immediately from the fact that the right hand side of the above equation defines an orthogonal projection onto L_r. Let us also note that $\bigcup_{n\in\mathbb{Z}} L_n$ is dense in $L_p(\mathbb{R})$ for $1 < p < \infty$.

Thus the first claim for $1 < p < \infty$ follows from the fact that the norm of P_r as an operator on $L_p(\mathbb{R})$ equals 1. This is a simple consequence of Hölder's inequality as follows:

$$\begin{aligned} \|P_r f\|_p &= \left(\sum_{k\in\mathbb{Z}} 2^{rp} \left| \int_{k2^{-r}}^{(k+1)2^{-r}} f(t)dt \right|^p 2^{-r} \right)^{1/p} \\ &\leq \left(\sum_{k\in\mathbb{Z}} 2^{rp} \left(\int_{k2^{-r}}^{(k+1)2^{-r}} |f(t)|^p dt \right) 2^{-r} 2^{-rp/q} \right)^{1/p} \\ &= \left(\int_{-\infty}^{\infty} |f(t)|^p dt \right)^{1/p}. \end{aligned}$$

If $f \in C_0(\mathbb{R})$ then it is uniformly continuous. Thus given $\varepsilon > 0$ we can find an N such that for $r > N$ and for each $k \in \mathbb{Z}$

$$\sup \{ |f(x) - f(y)| \; : \; x, y \in [k2^{-r}, (k+1)2^{-r}] \} < \varepsilon.$$

For a given $r > N$ and each $t \in \mathbb{R}$ we fix an integer k such that $t \in [k2^{-r}, (k+1)2^{-r}]$. Then

$$\begin{aligned}
|P_r f(t) - f(t)| &= \left|2^r \int_{k2^{-r}}^{(k+1)2^{-r}} f(s)\,ds - f(t)\right| \\
&= \left|2^r \int_{k2^{-r}}^{(k+1)2^{-r}} (f(s) - f(t))\,ds\right| < \varepsilon.
\end{aligned}$$

This implies that

$$\sup_{t\in\mathbb{R}} |P_r(f)(t) - f(t)| \to 0 \quad \text{as } r \to \infty.$$

This proves the first claim of the theorem.

Since for a fixed j the functions $\{H_{jk}\}_{k\in\mathbb{Z}}$ have disjoint supports, we have

$$\begin{aligned}
\Big\|\sum_{k\le\mu} \langle f, H_{jk}\rangle H_{jk}\Big\|_p &= \left(\textstyle\sum_{k\le\mu} |\langle f, H_{jk}\rangle|^p \, \|H_{jk}\|_p\right)^{1/p} \\
&= \left(\textstyle\sum_{k\le\mu} 2^{j(\frac{1}{2}-\frac{1}{p})} \, |\langle f, H_{jk}\rangle|^p\right)^{1/p}.
\end{aligned}$$

This shows that $\lim_{\mu\to\infty} Q_j^\mu(f)$ exists in the norm of the space. Clearly it equals $P_{j+1}(f) - P_j(f)$. This completes the proof of the theorem. □

The Haar wavelet is very well localized. The supports of the functions $\{2^{j/2}H(2^j t - k)\}_{j\in\mathbb{Z},k\in\mathbb{Z}}$ are dyadic intervals and are easily understood. The chief drawback is that H is not continuous. This implies that even for $f \in C_0(\mathbb{R})$ the functions $P_r f$ are not continuous. They are step functions.

1.2 The Strömberg wavelet

In this section we discuss the Strömberg wavelet, which is a continuous, piecewise linear function. Its definition is much more involved than the definition of the Haar wavelet.

Let us start by defining some subsets of $\mathbb{R}$. Let us put

$$\begin{aligned}
\mathbb{Z}_+ &= \{1, 2, \ldots\} \\
\mathbb{Z}_- &= -\mathbb{Z}_+ \\
A_0 &= \mathbb{Z}_+ \cup \{0\} \cup \tfrac{1}{2}\mathbb{Z}_- \\
A_1 &= A_0 \cup \{\tfrac{1}{2}\}.
\end{aligned}$$

Note that in the above definitions we use our standard notation that for a number a and a subset $A \subset \mathbb{R}$ we have $aA = \{ax : x \in A\}$ and $a + A = \{a + x : x \in A\}$.

Given a discrete subset $V \subset \mathbb{R}$ let $\mathcal{S}(V)$ be the space of all functions $f \in L_2(\mathbb{R})$ which are continuous on $\mathbb{R}$ and linear on every interval $I \subset \mathbb{R}$ such that $I \cap V = \emptyset$. It is clear that $\mathcal{S}(V)$ is non-empty if $|V| \geq 3$. Also if $V_1 \subset V_2$ are discrete subsets of $\mathbb{R}$ then $\mathcal{S}(V_1) \subset \mathcal{S}(V_2)$. In particular $\mathcal{S}(A_0) \subset \mathcal{S}(A_1)$ are non-trivial closed subspaces of $L_2(\mathbb{R})$, and it is easy to see that $\mathcal{S}(A_0)$ has codimension 1 in $\mathcal{S}(A_1)$. One can simply write each function $f \in \mathcal{S}(A_1)$ as a sum $f = g + \alpha\Lambda$ where $g \in \mathcal{S}(A_0)$ is defined as $g(r) = f(r)$ for $r \in A_0$ and $\Lambda \in (A_1)$ is defined as

$$\Lambda(r) = \begin{cases} 0 & \text{if } r \in A_0 \\ 1 & \text{if } r = \frac{1}{2}. \end{cases} \tag{1.12}$$

DEFINITION 1.6 *The Strömberg wavelet is a function $S \in \mathcal{S}(A_1)$ such that $\|S\|_2 = 1$ and S is orthogonal to $\mathcal{S}(A_0)$.*

REMARK 1.1. Such a function S is actually defined only up to a unimodular multiplicative constant, but this will not matter.

We will proceed analogously to Section 1.1 and show that the system $\{2^{j/2}S(2^j t - k)\}_{j\in\mathbb{Z},k\in\mathbb{Z}}$ is an orthonormal basis in $L_2(\mathbb{R})$. First let us check that this system is orthonormal.

The argument for this is similar to the argument given in the previous section for the orthogonality of $\{2^{j/2}H(2^j t - k)\}_{j\in\mathbb{Z},k\in\mathbb{Z}}$. Let us take two pairs (j, k) and (j', k'). Let us assume that $j \leq j'$. Using the substitution $u = 2^j t - k$ we see that

$$\langle S_{jk}, S_{j'k'}\rangle = 2^{s/2} \int_{-\infty}^{\infty} S(t) S(2^s t - r)\, dt \tag{1.13}$$

where $s = j' - j$ and $r = 2^{j'-j}k - k'$.

If $s \geq 1$ (i.e. $j \neq j'$) then for each $r \in \mathbb{Z}$

$$S(2^s t - r) \in \mathcal{S}(\tfrac{1}{2}\mathbb{Z}) \subset \mathcal{S}(A_0)$$

so the integral in 1.13 equals zero. If $s = 0$ (i.e. $j = j'$) and $k \neq k'$ then we can assume that $k' > k$. In this situation $r = k - k' < 0$ so $S(t - r) \in \mathcal{S}(A_1 - r) \subset \mathcal{S}(A_0)$. This means that the integral in 1.13 equals zero, so $\{2^{j/2}S(2^j t - k)\}_{j\in\mathbb{Z},k\in\mathbb{Z}}$ is an orthonormal system.

The fact that $\{2^{j/2}S(2^j x - k)\}_{j,k\in\mathbb{Z}}$ is an orthonormal basis follows immediately from the following three facts:

$$\bigcup_{n\in\mathbb{Z}} \mathcal{S}(2^{-n}\mathbb{Z}) \quad \text{is dense in} \quad L_2(\mathbb{R}) \tag{1.14}$$

$$\bigcap_{n\in\mathbb{Z}} \mathcal{S}(2^{-n}\mathbb{Z}) = \{0\} \tag{1.15}$$

and

$$\mathcal{S}(2^{-j-1}\mathbb{Z}) = \mathcal{S}(2^{-j}\mathbb{Z}) \oplus \operatorname{span}\{S_{jk}\}_{k\in\mathbb{Z}}. \tag{1.16}$$

The proof of 1.14 is a routine approximation argument, e.g. we know that continuous functions with compact support are dense in $L_2(\mathbb{R})$ and each such function can be approximated by functions from $\mathcal{S}(2^{-n}\mathbb{Z})$.

To see 1.15 note that if $f \in \mathcal{S}(2^n\mathbb{Z})$ for an integer n then f is linear on the intervals $[-2^n, 0]$ and $[0, 2^n]$, so if $f \in \bigcap_{n\in\mathbb{Z}} \mathcal{S}(2^n\mathbb{Z})$ then f is linear on both half-lines $[0,\infty)$ and $(-\infty, 0]$. For a function in $L_2(\mathbb{R})$ this means that $f = 0$.

One easily checks, using an obvious change of variables, that the map $f \mapsto 2^{j/2} f(2^j x)$ is a unitary map of $L_2(\mathbb{R})$. One also checks that it maps $\mathcal{S}(\frac{1}{2}\mathbb{Z})$ onto $\mathcal{S}(2^{-j-1}\mathbb{Z})$ and $\mathcal{S}(\mathbb{Z})$ onto $\mathcal{S}(2^{-j}\mathbb{Z})$ and $\operatorname{span}\{S_{0,k}\}_{k\in\mathbb{Z}}$ onto $\operatorname{span}\{S_{jk}\}_{k\in\mathbb{Z}}$. Thus 1.16 will follow once we establish

$$\mathcal{S}(\tfrac{1}{2}\mathbb{Z}) = \mathcal{S}(\mathbb{Z}) \oplus \operatorname{span}\{S_{0,k}\}_{k\in\mathbb{Z}}. \tag{1.17}$$

To show 1.17 we will use the translation operator T_N defined by the formula $(T_N f)(x) = f(x - N)$. This is clearly a unitary operator on $L_2(\mathbb{R})$. Applying the translation operator T_N to the definition of S we see that for each $N \in \mathbb{N}$

$$\mathcal{S}(A_1 - N) = \mathcal{S}(A_0 - N) \oplus \operatorname{span}\{S_{0,-N}\}. \tag{1.18}$$

Since $A_0 = A_1 - 1$, we can repeatedly apply 1.18 and obtain that for each N and $r \geq 0$

$$\begin{aligned}
&\mathcal{S}(A_1 - (N - r)) \\
&\quad = \mathcal{S}(A_0 - (N - r)) \oplus \operatorname{span}\{S_{0,-N+r}\} \\
&\quad = \mathcal{S}(A_1 - (N - r + 1)) \oplus \operatorname{span}\{S_{0,-N+r}\} \\
&\quad = \mathcal{S}(A_0 - (N - r + 1)) \oplus \operatorname{span}\{S_{0,-N+r}, S_{0,-N+r-1}\} \\
&\quad \vdots \\
&\quad = \mathcal{S}(A_0 - N) \oplus \operatorname{span}\{S_{0,k}\}_{k=-N}^{-N+r}.
\end{aligned}$$

If we take $N > 0$ and $r = 2N$ we get

$$\mathcal{S}(A_1 + N) = \mathcal{S}(A_0 - N) \oplus \operatorname{span}\{S_{0,k}\}_{k=-N}^{N}.$$

Letting $N \to \infty$ we obtain 1.17.

So we know that $\{2^{j/2}S(2^jt-k)\}_{j\in\mathbb{Z},k\in\mathbb{Z}}$ *is an orthonormal basis in* $L_2(\mathbb{R})$.

The definition of S gives no direct clue to what it looks like. Our next task is to compute the function S explicitly. It turns out that although S is supported on the whole real line $\mathbb{R}$, it decays very fast at infinity. We have the following:

Theorem 1.7 *At points of the set* A_1 *the values of the Strömberg wavelet* S *are given by*

$$\begin{aligned} S(k) &= S(1)(\sqrt{3}-2)^{k-1} \quad \textit{for } k=1,2,3,\ldots \\ S(\tfrac{1}{2}) &= -S(1)(\sqrt{3}+\tfrac{1}{2}) \\ S(0) &= S(1)(2\sqrt{3}-2) \\ S(-\tfrac{k}{2}) &= S(1)(2\sqrt{3}-2)(\sqrt{3}-2)^k \quad \textit{for } k=1,2,3,\ldots \end{aligned}$$

where $S(1)$ *has to be fixed so that* $\|S\|_2=1$. *Note that since* $S\in\mathcal{S}(A_1)$ *the above values determine* S *completely.*

Let us point out some facts which follow immediately from this Theorem.

(a) The Strömberg wavelet S oscillates; because $\sqrt{3}-2<0$ it changes sign between any two consecutive points of A_1.

(b) The Strömberg wavelet S has exponential decay, i.e. there exist constants C and $\alpha>0$ such that

$$|S(t)|\le C\exp(-\alpha|t|) \tag{1.19}$$

for all $t\in\mathbb{R}$. Clearly $\alpha=-\ln(2-\sqrt{3})\sim 1.316$ works.

(c) The Strömberg wavelet S is in $L_p(\mathbb{R})$ for all $1\le p\le\infty$.

(d) We have $S(-\frac{k}{2})=(10-6\sqrt{3})S(k)$ for $k=1,2,3,\ldots$. This shows that S admits a certain symmetry. This symmetry mirrors the fact that A_1 is twice as dense in $(-\infty,0]$ as in $[0,\infty)$.

Proof of Theorem 1.7. For $\sigma\in A_0$ let the function $\Lambda_\sigma\in\mathcal{S}(A_0)$ be defined as

$$\Lambda_\sigma(t)=\begin{cases}1 & \text{if } t=\sigma\\ 0 & \text{if } t\in A_0 \text{ and } t\neq\sigma\\ \text{linear} & \text{otherwise}\end{cases} \tag{1.20}$$

Such functions are called simple tents in $\mathcal{S}(A_0)$. Clearly $\langle S,\Lambda_\sigma\rangle=0$ for all $\sigma\in A_0$. Since S is piecewise linear, these scalar products can easily be computed explicitly in terms of values of S on A_1. From these calculations we obtain

$$0 = \langle S, \Lambda_\sigma \rangle = S(\sigma - 1) + 4S(\sigma) + S(\sigma + 1) \tag{1.21}$$

for $\sigma = 2, 3, 4, \ldots$, and

$$0 = \langle S, \Lambda_\sigma \rangle = S(\sigma - \tfrac{1}{2}) + 4S(\sigma) + S(\sigma + \tfrac{1}{2}) \tag{1.22}$$

for $\sigma = -\frac{1}{2}, -1, -\frac{3}{2}, -2, \ldots$.

The remaining two tents Λ_0 and Λ_1 are a bit more involved. We obtain

$$0 = \langle S, \Lambda_0 \rangle = 2S(-\tfrac{1}{2}) + 9S(0) + 6S(\tfrac{1}{2}) + S(1) \tag{1.23}$$

and

$$0 = \langle S, \Lambda_1 \rangle = S(0) + 6S(\tfrac{1}{2}) + 13S(1) + 4S(2). \tag{1.24}$$

It is natural to look for the solution of the system of equations 1.21 in the form $S(\sigma) = S(1)q^{\sigma-1}$. When we substitute this into 1.21, all equations give for q the same quadratic equation $1 + 4q + q^2 = 0$. It has two solutions $q = -2 \pm \sqrt{3}$. However, S should be in $L_2(\mathbb{R})$ so we need $|q| < 1$; thus we must take $q = \sqrt{3} - 2$.

Now we try to find the solution of 1.22 in the form $S(-\frac{k}{2}) = S(0)q^k$ for $k = 1, 2, 3, \ldots$ and in the same way we obtain the same $q = \sqrt{3} - 2$. If we substitute the values obtained so far for $S(-\frac{1}{2})$ and $S(2)$ into 1.23 and 1.24 we obtain the system

$$\begin{aligned} 0 &= 2(\sqrt{3} - 2)S(0) + 9S(0) + 6S(\tfrac{1}{2}) + S(1) \\ 0 &= S(0) + 6S(\tfrac{1}{2}) + 13S(1) + (\sqrt{3} - 2)S(1). \end{aligned}$$

Solving this system we obtain

$$\begin{aligned} S(0) &= (2\sqrt{3} - 2)S(1) \\ S(\tfrac{1}{2}) &= -(\sqrt{3} + \tfrac{1}{2})S(1). \end{aligned}$$

Thus we obtain a sequence $(S(\sigma))_{\sigma \in A_1}$ satisfying all the equations 1.21–1.24. This sequence yields a function $S \in \mathcal{S}(A_1)$ which is orthogonal to all simple tents in $\mathcal{S}(A_0)$. Since, as is easily seen, all simple tents in $\mathcal{S}(A_0)$ are linearly dense in $\mathcal{S}(A_0)$, we obtain that S is orthogonal to $\mathcal{S}(A_0)$. This S when properly normalized is the Strömberg wavelet. $\square$

Because $S \in L_p(\mathbb{R})$ for all p, $1 \leq p \leq \infty$, the coefficients with respect to the Strömberg basis, namely $\langle f, S_{jk} \rangle = \int_{-\infty}^{\infty} f(t) S_{jk}(t)\, dt$, can be computed for any function f which is in some $L_p(\mathbb{R})$ for $1 \le p \le \infty$. So we can consider the convergence of the series

$$\sum_{j\in\mathbb{Z}}\sum_{k\in\mathbb{Z}} \langle f, S_{jk}\rangle \cdot S_{jk} \tag{1.25}$$

for f in various function spaces. We will discuss here only the case $f \in C_0(\mathbb{R})$. Analogously to what we did with the Haar wavelet we start by considering operators P_r defined as

$$P_r f = \sum_{j<r}\sum_{k\in\mathbb{Z}} \langle f, S_{jk}\rangle\, S_{jk} \tag{1.26}$$

which in this case are orthogonal projections onto $\mathcal{S}(2^{-s}\mathbb{Z})$. We have

Theorem 1.8 *There exists a constant C such that for any $f \in C_0(\mathbb{R}) \cap L_2(\mathbb{R})$ the operator P_r defined by 1.26 satisfies*

$$\|P_r f\|_\infty \le C\,\|f\|_\infty$$

for all $r \in \mathbb{Z}$.

The technicalities of the proof of Theorem 1.8 are contained in the following two lemmas.

Lemma 1.9 *If g_{abc} is a continuous, piecewise linear function on $[-1, 1]$ given by*

$$g_{abc}(x) = \begin{cases} a & x = -1 \\ c & x = 0 \\ b & x = 1 \\ \textit{linear} & \textit{otherwise} \end{cases} \tag{1.27}$$

then

$$I =: \int_{-1}^{1} |g_{abc}(x)|^2 dx = \frac{a^2 + b^2 + 2c^2 + ac + bc}{3}.$$

Proof Simply compute

$$\begin{aligned} I &= \int_{-1}^{0} \left[(c-a)x + c\right]^2 dx + \int_{0}^{1} \left[(b-c)x + c\right]^2 dx \\ &= \frac{(a-c)^2}{3} + c(a-c) + c^2 + \frac{(b-c)^2}{3} + c(b-c) + c^2 \\ &= \frac{a^2 + b^2 + 2c^2 + ac + bc}{3}. \end{aligned}$$

□

Lemma 1.10 *Suppose we have a positive real number c and real numbers a, b with $|a|, |b| \leq c$. Then*

$$\inf_{-c\leq\alpha\leq c}\left(\int_{-1}^{1}|g_{ab\alpha}(x)|^2dx\right)^{1/2} \leq \left(\int_{-1}^{1}|g_{abc}(x)|^2dx\right)^{1/2} - \frac{c}{6\sqrt{2}+2\sqrt{6}} \tag{1.28}$$

where the function $g_{ab\alpha}$ is defined by 1.27 above.

Proof We divide 1.28 by c and from Lemma 1.9 infer that we need to prove that

$$\left(\frac{a^2+b^2+2+a+b}{3}\right)^{1/2} - \inf_{-1\leq\alpha\leq 1}\left(\frac{a^2+b^2+2\alpha^2+\alpha a+\alpha b}{3}\right)^{1/2} \geq \frac{1}{6\sqrt{2}+2\sqrt{6}}$$

for all numbers a, b with $|a|, |b| \leq 1$. Minimizing the quadratic polynomial in α on the left hand side, we see that the minimum is attained for $\alpha = -\frac{a+b}{4}$, so we need to prove

$$\begin{aligned} L &= \left(\frac{a^2+b^2+2+a+b}{3}\right)^{1/2} - \left(\frac{a^2+b^2-\frac{1}{8}(a+b)^2}{3}\right)^{1/2} \\ &\geq \frac{1}{6\sqrt{2}+2\sqrt{6}}. \end{aligned}$$

But for $|a|, |b| \leq 1$ we have

$$\begin{aligned} L &= \frac{1}{\sqrt{3}}\frac{2+a+b+\frac{1}{8}(a+b)^2}{(a^2+b^2+2+a+b)^{1/2}+(a^2+b^2-\frac{1}{8}(a+b)^2)^{1/2}} \\ &\geq \frac{1}{\sqrt{3}}\frac{2+a+b+\frac{1}{8}(a+b)^2}{\sqrt{6}+\sqrt{2}} \\ &\geq \frac{1}{\sqrt{3}}\frac{1}{2}\frac{1}{\sqrt{6}+\sqrt{2}} = \frac{1}{6\sqrt{2}+2\sqrt{6}}. \end{aligned}$$

□

Proof of the Theorem 1.8 Let us start with $r = 0$ and let us take $f \in C_0(\mathbb{R})$ with compact support and $\|f\|_\infty = 1$ and let $c = \|P_0 f\|_\infty$. Replacing f by an appropriate translation of it and multiplying by ± 1

we can assume that $P_0(f)(0) = c$. We also know that P_0 is an orthogonal projection onto $\mathcal{S}(\mathbb{Z})$, so

$$\begin{aligned}\int_{-\infty}^{\infty} |f - P_0 f|^2 &= \inf\left\{\int_{-\infty}^{\infty} |f-g|^2 \,:\, g \in \mathcal{S}(\mathbb{Z})\right\} \\ &\leq \inf\left\{\int_{-\infty}^{\infty} |f-g|^2 \,:\, g \in \mathcal{S}(\mathbb{Z}) \text{ and} \right. \\ &\qquad \left. g(k) = P_0(f)(k) \text{ for } k \in \mathbb{Z} \setminus \{0\}\right\}. \qquad (1.29)\end{aligned}$$

Let us denote $P_0(f)(-1) = a$ and $P_0(f)(1) = b$. From 1.29 we get

$$\left(\int_{-1}^{1} |f - g_{abc}|^2\right)^{1/2} \leq \inf_{-c \leq \alpha \leq c} \left(\int_{-1}^{1} |f - g_{ab\alpha}|^2\right)^{1/2}.$$

Applying the triangle inequality we obtain

$$\begin{aligned}&\left(\int_{-1}^{1} |g_{abc}|^2\right)^{1/2} - \left(\int_{-1}^{1} |f|^2\right)^{1/2} \\ &\qquad \leq \left(\int_{-1}^{1} |f|^2\right)^{1/2} + \inf_{-c\leq\alpha\leq c} \left(\int_{-1}^{1} |g_{ab\alpha}|^2\right)^{1/2}.\end{aligned}$$

Since $(\int_{-1}^{1} |f|^2)^{1/2} \leq \sqrt{2}$ (because $\|f\|_\infty = 1$) we get

$$\left(\int_{-1}^{1} |g_{abc}|^2\right)^{1/2} \leq 2\sqrt{2} + \inf_{-c\leq\alpha\leq c} \left(\int_{-1}^{1} |g_{ab\alpha}|^2\right)^{1/2}.$$

From Lemma 1.10 we obtain

$$\left(\int_{-1}^{1} |g_{abc}|^2\right)^{1/2} \leq 2\sqrt{2} + \left(\int_{-1}^{1} |g_{abc}|^2\right)^{1/2} - \frac{c}{2\sqrt{3} + 2\sqrt{6}}.$$

This implies that $c \leq 2\sqrt{2} \cdot (2\sqrt{3} + 2\sqrt{6})$. Since functions with compact support are dense in $C_0(\mathbb{R})$ we conclude that $\|P_0\| \leq 2\sqrt{2} \cdot (2\sqrt{3} + 2\sqrt{6})$.

To prove the estimate for $r \neq 0$ let us observe that

$$P_r f(x) = \sum_{j<r} \sum_{k\in\mathbb{Z}} \int_{-\infty}^{\infty} f(t) 2^{j/2} S(2^j t - k)\, dt\, 2^{j/2} S(2^j x - k).$$

Substituting $2^{-r} u = t$ and $2^{-r} y = x$ we get

$$\begin{aligned}&P_r f(x) \\ &\quad = \sum_{j<r} \sum_{k\in\mathbb{Z}} 2^{j-r} \int_{-\infty}^{\infty} f(2^{-r} u) S(2^{j-r} u - k)\, du\, S(2^{j-r} y - k) \\ &\quad = (P_0 g)(2^r x)\end{aligned}$$

where $g(x) = f(2^{-r}x)$. This implies

$$\begin{aligned}\|P_r f\|_\infty &= \|P_0 g\|_\infty \leq \|P_0\| \, \|g\|_\infty \\ &= \|P_0\| \, \|f\|_\infty .\end{aligned}$$

□

REMARK 1.2. This argument is quite wasteful and gives $\|P_0\|$ much too big. One can improve Lemma 1.10 and get a better constant – for some improvements see Exercise 1.6.

Since $\bigcup_{j\in\mathbb{Z}} S(2^j\mathbb{Z})$ is dense also in $C_0(\mathbb{R})$ we get

Corollary 1.11 *If $f \in C_0(\mathbb{R})$ then $P_r(f)$ converges uniformly to f when r tends to ∞.*

Now we will analyze the operators Q_j^μ given by

$$Q_j^\mu(f) = \sum_{k\leq\mu} \langle f, S_{jk}\rangle \, S_{jk}.$$

Our aim is to show that $Q_j^\mu(f) \to P_{j+1}(f) - P_j(f)$ as $r \to \infty$. First let us look at the operators Q_0^μ. Observe that for every $f \in C_0(\mathbb{R})$

$$\lim_{|k|\to\infty} \langle f, S_{0,k}\rangle \, \|S_{0,k}\|_\infty = 0. \tag{1.30}$$

To see this note that

$$\begin{aligned}|\langle f, S_{0,k}\rangle| \, \|S_{0,k}\| &= C\left|\int_{-\infty}^{\infty} f(x)S(x-k)dx\right| \\ &= C\left|\int_{-\infty}^{\infty} f(u+k)S(u)du\right|\end{aligned} \tag{1.31}$$

and this expression tends to zero by the Lebesgue Dominated Convergence Theorem because $S(u)$ is integrable (see remark (c) after Theorem 1.7) and $|f(u+k)| \leq \|f\|_\infty$ and for each $u \in \mathbb{R}$ $\lim_{|k|\to\infty} |f(u+k)| = 0$ because $f \in C_0(\mathbb{R})$. It also follows from 1.31 that

$$|\langle f, S_{0,k}\rangle| \, \|S_{0,k}\| \leq C \, \|f\|_\infty \tag{1.32}$$

for all $k \in \mathbb{Z}$.

Also for any sequence of scalars $(\alpha_k)_{k\in\mathbb{Z}}$ we have

$$\left\| \sum_{k\in\mathbb{Z}} \alpha_k \frac{S_{0,k}}{\|S_{0,k}\|_\infty} \right\| \leq C \sup |\alpha_k|. \tag{1.33}$$

This follows from the fact that S decays exponentially (see remark (b) after Theorem 1.7), i.e.

$$|S_{0,k}(u)| \leq C \|S_{0,k}\|_\infty \exp(-\alpha|u-k|)$$

so for every $u \in \mathbb{R}$ we have

$$\left|\sum_{k\in\mathbb{Z}} \alpha_k \|S_{0,k}\|_\infty^{-1} S_{0,k}(u)\right| \leq C \sup_{k\in\mathbb{Z}} |\alpha_k| \sum_{k\in\mathbb{Z}} \exp(-\alpha|u-k|) \leq C_1 \sup_{k\in\mathbb{Z}} |\alpha_k|.$$

Putting together 1.30, 1.32 and 1.33 we obtain that $\|Q_0^\mu\| \leq C$ for all $\mu \in \mathbb{Z}$ and the series $\sum_{k\leq\mu} \langle f, S_{0k}\rangle S_{0k}$ converges in $C_0(\mathbb{R})$ and

$$\lim_{r\to\infty} \sum_{k\leq r} \langle f, S_{0k}\rangle S_{0k} = P_1 f - P_0 f.$$

The passage to general j's is accomplished exactly as at the end of the proof of Theorem 1.8.

Sources and comments

As was noted in the text, both wavelets discussed in this chapter were really introduced well before the emergence of the general theory. In his paper [52] A. Haar introduced an orthonormal system of functions on the interval $[0,1]$, which is now generally called the Haar system. It is the system $2^{j/2}H(2^j t-k) \mid [0,1]$ with $j = 0,1,\ldots$ and $k = 0,1,\ldots,2^j-1$ supplemented by the constant function. In the terminology of our book it is a periodic Haar wavelet as discussed in Section 2.5. Haar's main aim was to construct an orthonormal system such that each continuous function on $[0,1]$ has a uniformly convergent Fourier series with respect to this system. Thus he showed a version for continuous functions on the interval $[0,1]$ of our Theorem 1.5, with basically the same proof (cf. Exercise 1.9). The convergence in $L_p[0,1]$ for $1 \leq p < \infty$ was shown later by M. J. Schauder [99]. An orthogonal system of continuous, piecewise linear functions on $[0,1]$ was constructed by Ph. Franklin [41]. He also showed the uniform convergence of the Franklin–Fourier series of any continuous function, i.e. our Theorem 1.8 and Corollary 1.11. In fact our proofs closely follow ideas of Franklin [41]. Although in the Haar system the passage from $[0,1]$ to $\mathbb{R}$ is quite obvious, it is not so for the Franklin system. This step (and much more) was done by J.-O. Strömberg [107]. His real aim was to construct unconditional bases in $H_p(\mathbb{R}^d)$ for $0 < p \leq 1$, so he considered not only piecewise linear continuous functions but also smoother splines (see Section 3.3). Hardy spaces $H_p(\mathbb{R}^d)$, but only for $p = 1$, are discussed later in this book, in Sections 6.2, 8.2 and 8.3. Both the Haar system and the Franklin system on $[0,1]$ are among the most important orthogonal systems and have been investigated in great detail, cf. [56]. In particular the fact that the classical Franklin system on $[0,1]$ has exponential decay was shown by Z. Ciesielski [16], the proof being somewhat similar to the Strömberg's proof of Theorem 1.7, which we give. Since the Franklin system is not a wavelet, each function has a slightly different shape, so one can not attain the precision and simplicity of Theorem 1.7. I would also like to mention the paper [82] where Strömberg wavelets are computed in terms of the Fourier transform and equation 2.45.

About the exercises. Exercise 1.8 is a part of Strömberg's [107] construction of wavelets on $\mathbb{R}^d$. Exercise 1.9 is Haar's [52] original formulation and result. Exercise 1.6 is a step towards the exact computation of $\|P_0\|$ and $\|P_0 + Q_0^0\|$ on $C_0(\mathbb{R})$. These values are formally unknown, but in the (probably similar) case of the classical Franklin system such calculations were done in [17].

Exercises

1.1 Suppose $\Phi \in L_0$ (cf. 1.5) is such that the system $\{\Phi(x-k)\}_{k\in\mathbb{Z}}$ is orthonormal and suppΦ is compact. Show that for some $s \in \mathbb{Z}$ we have $\Phi(x) = \pm\mathbf{1}_{[0,1]}(x-s)$.

1.2 Let X be the closed span of the set $\left\{2^{j/2}H(2^j t - k)\right\}_{j\in\mathbb{Z},k\in\mathbb{Z}}$ in the space $L_1(\mathbb{R})$. Show that

- If $f \in X$ then $\int_{-\infty}^{\infty} f(x)\,dx = 0$, whence $\mathbf{1}_{[0,1]} \notin X$.
- The function $H(x - \frac{1}{2})$ is not in X.

1.3 Suppose that the function f on $\mathbb{R}$ satisfies Hölder's condition with exponent α, $0 < \alpha \leq 1$, i.e. there exists a constant C such that

$$|f(x) - f(y)| \leq C|x-y|^{\alpha}$$

for all $x, y \in \mathbb{R}$. Show that $\langle f, H_{jk}\rangle = \int_{-\infty}^{\infty} f(t)H_{jk}(t)\,dt$ exists for all $j, k \in \mathbb{Z}$ and satisfies

$$|\langle f, H_{jk}\rangle| \leq C \cdot 2^{-j(\alpha+1/2)}.$$

1.4 Suppose that $f \in L_2(\mathbb{R})$. Show that f is even if and only if $\langle f, H_{jk}\rangle = -\langle f, H_{j,-k-1}\rangle$ for all $j, k \in \mathbb{Z}$. Show that f is odd if and only if $\langle f, H_{jk}\rangle = \langle f, H_{j,-k-1}\rangle$ for all $j,k \in \mathbb{Z}$. Let $g = f \cdot \mathbf{1}_{[0,\infty)}$. Show that

$$g = \sum_{j\in\mathbb{Z}}\sum_{k\geq 0} \langle f, H_{jk}\rangle\, H_{jk}.$$

1.5 Let S denote the Strömberg wavelet.

- Show that

$$\int_{-\infty}^{\infty} S(t)\,dt = \int_{-\infty}^{\infty} tS(t)\,dt = 0.$$

- Show that for all $x \in \mathbb{R}$ with $|x| > 1$ we have $S(-x/2) = (10 - 6\sqrt{3})S(x)$

1.6 Let P_0 be an orthonormal projection onto $\mathcal{S}(\mathbb{Z})$.

- Improve Lemma 1.10 and show e.g. that $(6\sqrt{2}+2\sqrt{6})^{-1}$ can be replaced by $\frac{\sqrt{5}-2}{\sqrt{6}}$, which is a bit better but still not optimal.
- How does this improve the estimate for $\|P_0\|$?
- Show that $\|P_0\|_\infty > 1$.

1.7 Suppose that $f \in \mathcal{S}(\mathbb{Z})$ has compact support. Show that the system $\{f(t-m)\}_{m\in\mathbb{Z}}$ is not an orthonormal basis in $\mathcal{S}(\mathbb{Z})$.

1.8 Let $\psi \in \mathcal{S}(\mathbb{Z}_+ \cup \{0\})$ be such that $\|\psi\|_2 = 1$ and $\psi \perp \mathcal{S}(\mathbb{Z}_+)$.

- Show that $\{\psi(t-m)\}_{m\in\mathbb{Z}}$ is an orthonormal basis in $\mathcal{S}(\mathbb{Z})$.
- Show that ψ has exponential decay.

1.9 For $n = 1,2,\ldots$ write $n = 2^j + k$ with $j = 0,1,\ldots$ and $k = 0,1,\ldots,2^j-1$ and define $h_n(t) =: 2^{j/2}H(2^j t - k) \mid [0,1]$. Define also $h_0(t) =: 1$. Show that $(h_n)_{n=0}^\infty$ is a complete orthonormal system in $L_2[0,1]$. Show that for $f \in C[0,1]$ the series $\sum_{n=0}^\infty \langle f, h_n\rangle\, h_n$ converges uniformly to f.

2

General constructions

2.1 Basic concepts

Let us start this chapter with the definition of a wavelet – the main concept discussed in this book.

DEFINITION 2.1 *A wavelet is a function $\Psi(t) \in L_2(\mathbb{R})$ such that the family of functions*

$$\Psi_{j,k} =: 2^{j/2}\Psi(2^j t - k),$$

where j and k are arbitrary integers, is an orthonormal basis in the Hilbert space $L_2(\mathbb{R})$.

Let me stress that for a function Ψ the notation Ψ_{jk} to denote the function $2^{j/2}\Psi(2^j t - k)$ will be used throughout this book. According to the above definition, with each wavelet we have associated an orthonormal system $\left\{2^{j/2}\Psi(2^j t - k)\right\}_{j\in\mathbb{Z},k\in\mathbb{Z}}$ which will be called a *wavelet basis*. As we remarked in the Preface, the very existence of wavelets is not clear. Two examples of wavelets were given in Sections 1.1 and 1.2 and many more will be constructed in subsequent chapters. Good wavelets are usually constructed starting from a multiresolution analysis. Let us give the definition of this concept.

DEFINITION 2.2 *A multiresolution analysis is a sequence $(V_j)_{j\in\mathbb{Z}}$ of subspaces of $L_2(\mathbb{R})$ such that*

(i) $\ldots \subset V_{-1} \subset V_0 \subset V_1 \subset \ldots$
(ii) $\operatorname{span} \bigcup_{j\in\mathbb{Z}} V_j = L_2(\mathbb{R})$
(iii) $\bigcap_{j\in\mathbb{Z}} V_j = \{0\}$
(iv) $f(x) \in V_j$ *if and only if* $f(2^{-j}x) \in V_0$
(v) $f \in V_0$ *if and only if* $f(x - m) \in V_0$ *for all* $m \in \mathbb{Z}$

(vi) *there exists a function $\Phi \in V_0$, called a scaling function, such that the system $\{\Phi(t-m)\}_{m\in\mathbb{Z}}$ is an orthonormal basis in V_0.*

It would be good to give at this point some interesting examples of multiresolution analyses. I will not do so, because the next two chapters will be devoted to this task. So I will only point out that the spaces $(L_n)_{n=-\infty}^{\infty}$ defined by 1.5 form a multiresolution analysis. In this case the function $\mathbf{1}_{[0,1]}$ can be taken to be the scaling function. Also, the spaces $\{\mathcal{S}(2^j\mathbb{Z})\}_{j\in\mathbb{Z}}$ considered in Section 1.2 clearly satisfy conditions (i)–(v) of Definition 2.2. In this case there is a problem with the choice of scaling function. We will return to this problem at the end of Section 2.2 and we will show that they form a multiresolution analysis. It is clear from Definitions 2.1 and 2.2 that operators of dilation and translation of functions should be essential to studying wavelets. In this section we will define those operators. The notations established in the process will be used throughout the whole book. Let us formally define translation and dilation operators.

DEFINITION 2.3 *Given a real number h we define the translation operator T_h acting on functions defined on $\mathbb{R}$ by the formula*

$$T_h(f)(x) = f(x-h).$$

DEFINITION 2.4 *Given an integer s we define the dyadic dilation operator J_s acting on functions defined on $\mathbb{R}$ by the formula*

$$J_s(f)(x) = f(2^s x).$$

REMARK 2.1. Clearly we can define a dilation by any real number, not only by 2^s. We restrict our definition to the dyadic case simply because in this and the next two chapters we will use only such dilations.

Clearly all the above operators are formally invertible; we have $T_h^{-1} = T_{-h}$ and $J_s^{-1} = J_{-s}$. Let us also note how they act on $L_2(\mathbb{R})$. A simple change of variables gives the proof of the following lemma.

Lemma 2.5 *(a) For every number $h \in \mathbb{R}$ the operator T_h is an isometry on $L_2(\mathbb{R})$.*

(b) For every integer $s \in \mathbb{Z}$ the operator $2^{s/2}J_s$ is an isometry on $L_2(\mathbb{R})$.

Equipped with the above definitions, let us make some comments about multiresolution analyses. Let us note that conditions (i)–(iii) mean that every function in $L_2(\mathbb{R})$ can be approximated by elements of the subspaces V_j, and as j increases to ∞ the precision of approximation increases. Conditions (iv) and (v) express the invariance of the system of subspaces $(V_j)_{j\in\mathbb{Z}}$ with respect to translation and dilation operators. Using the notation introduced in Definitions 2.3 and 2.4 we can rephrase conditions (iv) and (v) of Definition 2.2 as conditions

(iv′) $V_j = J_{-j}(V_0)$ *for all* $j \in \mathbb{Z}$
(v′) $V_0 = T_n(V_0)$ *for all* $n \in \mathbb{Z}$

From Lemma 2.5 or simply using the change of variables we see that condition (vi) can be rephrased as

(vi′) *for each* $j \in \mathbb{Z}$ *the system* $\{2^{j/2}\Phi(2^j x - k)\}_{k\in\mathbb{Z}}$ *is an orthonormal basis in* V_j.

The concept of multiresolution analysis as defined above forms a very natural and transparent scheme. However, the definition given above is not 'minimal' in the sense that some of the conditions (i)–(vi) follow from the remaining conditions. It is clear at the first glance that (v) follows from (vi). We will see later in Proposition 2.15 that the condition (iii) is superfluous. The interdependence of the other conditions is partially discussed in exercises at the end of this chapter.

It is immediately clear that we can adopt two ways of looking at a multiresolution analysis.

- We take the subspaces $(V_j)_{j\in\mathbb{Z}}$ as our basic, given objects. They have to satisfy conditions (i)–(v) which usually are rather easy to check. Then we need to find a scaling function satisfying (vi). This is usually not so obvious.
- We start with the function Φ. We define V_0 as span $\{\Phi(t-m)\}_{m\in\mathbb{Z}}$ and the other spaces V_j are defined by condition (iv) or (iv′). Condition (v) is automatically satisfied and we need to check conditions (i)–(iii) and (vi).

There is also a third way, not so clear directly from the definition. It is to start with the scaling equation (which is the expression of the fact that $V_0 \subset V_1$) and to build the scaling function from there. This is discussed in detail in Remark 2.6. Whichever point of view we adopt, it is clear that we will need to know something about the system of translates of a given function and the subspaces spanned by such systems. This topic will be discussed in Section 2.2.

2.2 Multiresolution analyses

Let us start with the following concept from the geometry of Hilbert spaces.

DEFINITION 2.6 *A sequence of vectors $(x_n)_{n\in A}$ in a Hilbert space H is called a Riesz sequence if there exist constants $0 < c \leq C$ such that*

$$c\left(\sum_{n\in A}|a_n|^2\right)^{1/2} \leq \left\|\sum_{n\in A} a_n x_n\right\| \leq C\left(\sum_{n\in A}|a_n|^2\right)^{1/2} \tag{2.1}$$

for all sequences of scalars $(a_n)_{n\in A}$. A Riesz sequence is called a Riesz basis if additionally $\operatorname{span}(x_n)_{n\in A} = H$.

Since any orthonormal system satisfies 2.1 with $c = C = 1$, the concept of a Riesz system is a generalization of the notion of orthonormal system, and a Riesz basis is a generalization of an orthonormal basis.

REMARK 2.2. Let us also note a reformulation of Definition 2.6. *A Riesz sequence in H is the image of the unit vector basis in ℓ_2 under an isomorphic embedding.* Using this language we can also say that an orthonormal sequence is the image of the unit vector basis in ℓ_2 under an isometric operator.

Lemma 2.7 *(a) If we have a Riesz basis $(x_n)_{n\in\mathbb{Z}}$ in H then there exist biorthogonal functionals $(x_n^*)_{n\in\mathbb{Z}}$, i.e. vectors in H such that $\langle x_n^*, x_m\rangle = \delta_{n,m}$. The sequence $(x_n^*)_{n\in\mathbb{Z}}$ is also a Riesz basis in H.*

(b) If $(x_n)_{n\in\mathbb{Z}}$ is a Riesz basis in H then there exist constants $0 < c \leq C$ such that

$$c\,\|x\| \leq \left(\sum_{n\in\mathbb{Z}} |\langle x, x_n\rangle|^2\right)^{1/2} \leq C\,\|x\| \tag{2.2}$$

for all $x \in H$.

REMARK 2.3. A system of vectors $(x_n)_{n\in A}$ in a Hilbert space H is called a *frame* if there exist constants $0 < c \leq C$ such that 2.2 holds for each $x \in H$. If $c = C$ the frame is called a *tight frame.* Sometimes frames are used as a substitute for a Riesz basis. The important difference between frames and Riesz bases is that if $(x_n)_{n\in A}$ is a frame, the vectors $(x_n)_{n\in A}$ need not be linearly independent; as a trivial example take $(x_n)_{n\in A}$ to be the disjoint union of two orthonormal systems. We will not discuss this concept in this book, except in the Exercises (cf. Exercises 2.15 and 4.12).

Proof Let $\ell_2(\mathbb{Z})$ denote as usual the Hilbert space of all sequences $(a_n)_{n\in\mathbb{Z}}$ such that $\sum_{n\in\mathbb{Z}} |a_n|^2 < \infty$, and let e_n denote the n-th unit vector in $\ell_2(\mathbb{Z})$, i.e. the sequence which has all coordinates equal to 0 except the n-th coordinate which equals 1. Let us consider the operator $I : \ell_2(\mathbb{Z}) \to H$ defined by the condition $I(e_n) = x_n$ for all $n \in \mathbb{Z}$. Since $(x_n)_{n\in\mathbb{Z}}$ is a Riesz basis, we see that I is an isomorphism of $\ell_2(\mathbb{Z})$ onto H. Thus the operator $(I^{-1})^*$ mapping $\ell_2(\mathbb{Z})$ onto H is also an isomorphism. We define $x_n^* = (I^{-1})^*(e_n)$. Then we have

$$\langle x_n^*, x_m\rangle = \langle (I^{-1})^*(e_n), I(e_m)\rangle = \langle e_n, e_m\rangle$$

so $(x_n^*)_{n\in\mathbb{Z}}$ is the sequence of biorthogonal functionals to $(x_n)_{n\in\mathbb{Z}}$. It is a Riesz basis because the operator $(I^{-1})^*$ is an isomorphism. This gives part (a) of the Lemma. To get (b) we use (a) and write $x = \sum_{n\in\mathbb{Z}} a_n x_n^*$. Then we have

$$\sum_{n\in\mathbb{Z}} |\langle x, x_n\rangle|^2 = \sum_{n\in\mathbb{Z}} |a_n|^2$$

and since we already know from part (a) that $(x_n^*)_{n\in\mathbb{Z}}$ is a Riesz basis we get 2.2. □

In order to understand and efficiently apply condition (vi) of Definition 2.2 we need to understand sets of the form $\{\Phi(t-m)\}_{m\in\mathbb{Z}}$ and the spaces they span. Let us take a finite sequence of scalars (a_n). From A1.2–III we see that

$$\mathcal{F}\left(\sum_n a_n \Phi(x-n)\right) = \left(\sum_n a_n e^{-in\xi}\right)\hat{\Phi}(\xi). \tag{2.3}$$

Since the Fourier transform $\mathcal{F}$ is a unitary operator on $L_2(\mathbb{R})$ and the function $\sum_n a_n e^{-in\xi}$ is 2π-periodic, we infer that

$$\left\|\sum_n a_n\Phi(x-n)\right\|_2 = \left(\int_{-\infty}^{\infty}\left|\sum_n a_n e^{-in\xi}\right|^2 |\hat{\Phi}(\xi)|^2\, d\xi\right)^{1/2}$$

$$= \left(\int_0^{2\pi}\left|\sum_n a_n e^{-in\xi}\right|^2 \sum_{l\in\mathbb{Z}} |\hat{\Phi}(\xi+2\pi l)|^2\, d\xi\right)^{1/2}. \tag{2.4}$$

Note that the series $\sum_{l\in\mathbb{Z}} |\hat{\Phi}(\xi+2\pi l)|^2$ converges almost everywhere to a finite value, because

$$\int_0^{2\pi} \sum_{l\in\mathbb{Z}} |\hat{\Phi}(\xi+2\pi l)|^2 = \int_{-\infty}^{\infty} |\hat{\Phi}(\xi)|^2 \, d\xi < \infty.$$

These observations allow us to characterize functions $\Phi \in L_2(\mathbb{R})$ such that $\{\Phi(t-m)\}_{m\in\mathbb{Z}}$ is a Riesz sequence, namely we have:

Proposition 2.8 *Let Φ be a function in $L_2(\mathbb{R})$ and let $0 < a \leq A$ be two constants. The following two conditions are equivalent:*

(i) *for every sequence of scalars $(a_n)_{n\in\mathbb{Z}}$ we have*

$$a\left(\sum_{n\in\mathbb{Z}} |a_n|^2\right)^{1/2} \leq \left(\int_{-\infty}^{\infty} \Big|\sum_{n\in\mathbb{Z}} a_n \Phi(x-n)\Big|^2 dx\right)^{1/2} \leq A\left(\sum_{n\in\mathbb{Z}} |a_n|^2\right)^{1/2} \tag{2.5}$$

(ii) *for almost all $\xi \in \mathbb{R}$*

$$\frac{a^2}{2\pi} \leq \sum_{l\in\mathbb{Z}} |\hat{\Phi}(\xi+2\pi l)|^2 \leq \frac{A^2}{2\pi}. \tag{2.6}$$

Proof If 2.6 holds then we infer from 2.4 that for each finite sequence of scalars we have

$$\frac{a}{2\pi}\left(\int_0^{2\pi} |\sum_n a_n e^{-in\xi}|^2 \, d\xi\right)^{1/2} \leq \Big\|\sum_n a_n \Phi(x-n)\Big\| \leq \frac{A}{2\pi}\left(\int_0^{2\pi} |\sum_n a_n e^{in\xi}|^2 \, d\xi\right)^{1/2}.$$

Since

$$\left(\int_0^{2\pi} |\sum_n a_n e^{-in\xi}|^2 \, d\xi\right)^{1/2} = \sqrt{2\pi}\left(\sum_n |a_n|^2\right)^{1/2}$$

we obtain 2.5 for all finite sequences of scalars. But this easily implies that 2.5 holds for all infinite sequences of scalars as well. To prove the other implication, consider the set

$$A_\alpha = \{\xi \in [0,2\pi] \, : \, \sum_{l\in\mathbb{Z}} |\hat{\Phi}(\xi+2\pi l)|^2 > \alpha\}.$$

If for some α this set has positive measure we take a sequence of scalars $(a_n)_{n\in\mathbb{Z}} \in \ell_2(\mathbb{Z})$ such that $\mathbf{1}_{A_\alpha} = \sum_{n\in\mathbb{Z}} a_n e^{-in\xi}$ for $\xi \in [0, 2\pi]$. From 2.4 we obtain

$$\left\| \sum_{n\in\mathbb{Z}} a_n \Phi(x-n) \right\|_2 = \left(\int_{A_\alpha} \sum_{l\in\mathbb{Z}} |\hat{\Phi}(\xi + 2\pi l)|^2 \right)^{1/2} \geq \sqrt{\alpha}|A_\alpha|^{1/2}. \tag{2.7}$$

Since

$$\left(\sum_{n\in\mathbb{Z}} |a_n|^2 \right)^{1/2} = \frac{1}{\sqrt{2\pi}} \|\mathbf{1}_{A_\alpha}\|_2 = \frac{1}{\sqrt{2\pi}} |A_\alpha|^{1/2}$$

we infer that

$$\left\| \sum_{n\in\mathbb{Z}} a_n \Phi(x-n) \right\|_2 \geq \sqrt{2\pi}\sqrt{\alpha} \left(\sum_{n\in\mathbb{Z}} |a_n|^2 \right)^{1/2}.$$

Comparing this with 2.5 we obtain that $2\pi\alpha \leq A^2$. This gives the right hand inequality in 2.6. Analogously, considering the set

$$B_\alpha = \{\xi \in [0, 2\pi] : \sum_{l\in\mathbb{Z}} |\hat{\Phi}(\xi + 2\pi l)|^2 < \alpha\}$$

we obtain the left hand inequality in 2.6. □

Corollary 2.9 *The system $\{\Phi(t-m)\}_{m\in\mathbb{Z}}$ is an orthonormal system if and only if*

$$\sum_{l\in\mathbb{Z}} |\hat{\Phi}(\xi + 2\pi l)|^2 = \frac{1}{2\pi} \tag{2.8}$$

for almost all $\xi \in \mathbb{R}$.

Proof From Proposition 2.8 we infer that condition 2.8 is equivalent to the fact that $\left\| \sum_{n\in\mathbb{Z}} a_n \Phi(x-n) \right\|_2 = (\sum_{n\in\mathbb{Z}} |a_n|^2)^{1/2}$ for all sequences of scalars. This is clearly equivalent to the system $\{\Phi(t-m)\}_{m\in\mathbb{Z}}$ being orthonormal. □

It will be essential in our future arguments that we understand clearly which functions are in the span of $\{\Phi(t-m)\}_{m\in\mathbb{Z}}$ when the system $\{\Phi(t-m)\}_{m\in\mathbb{Z}}$ is a Riesz sequence. For future reference we will formulate it in the form of the following, practically self-evident proposition.

Proposition 2.10 *Suppose that* $\{\Phi(t-m)\}_{m\in\mathbb{Z}}$ *is a Riesz sequence satisfying condition 2.5. Then*

(i) *Any function* $g \in \operatorname{span}\{\Phi(t-m)\}_{m\in\mathbb{Z}}$ *can be written as*

$$g(x) = \sum_{n\in\mathbb{Z}} a_n \Phi(x-n)$$

and the series converges in $L_2(\mathbb{R})$.

(ii) $g \in \operatorname{span}\{\Phi(t-m)\}_{m\in\mathbb{Z}}$ *if and only if*

$$\hat{g}(\xi) = \phi(\xi)\hat{\Phi}(\xi) \tag{2.9}$$

for some 2π*-periodic function* $\phi(\xi)$ *with* $\int_0^{2\pi} |\phi(\xi)|^2\, d\xi < \infty$.

(iii) *If* $g(x) = \sum_{n\in\mathbb{Z}} a_n \Phi(x-n)$ *and* $\phi(\xi)$ *is given by 2.9 above then* $\phi(\xi) = \sum_{n\in\mathbb{Z}} a_n e^{-in\xi}$ *and conversely.*

(iv) *The norms of* g *and* ϕ *are related as follows:*

$$A^{-1}\|g\|_2 \leq \left(\frac{1}{2\pi}\int_0^{2\pi} |\phi(\xi)|^2\, d\xi\right)^{1/2} \leq a^{-1}\|g\|_2 \tag{2.10}$$

where a *and* A *are the constants appearing in 2.5. Conversely, if 2.10 holds for some constants* a *and* A *and all* $g \in \operatorname{span}\{\Phi(t-m)\}_{m\in\mathbb{Z}}$ *and corresponding* ϕ *given by (iii) above, then 2.5 holds with constants* a *and* A.

Using the above description we can prove:

Proposition 2.11 *Suppose that* $\Phi \in L_2(\mathbb{R})$ *is a function such that* $\{\Phi(t-m)\}_{m\in\mathbb{Z}}$ *is a Riesz sequence. Then there exists a function* $\Phi_1 \in \operatorname{span}\{\Phi(t-m)\}_{m\in\mathbb{Z}}$ *such that* $\{\Phi_1(t-m)\}_{m\in\mathbb{Z}}$ *is an orthonormal system. For each such* Φ_1 *the system* $\{\Phi_1(t-m)\}_{m\in\mathbb{Z}}$ *is an orthonormal basis in the space* $\operatorname{span}\{\Phi(t-m)\}_{m\in\mathbb{Z}}$.

Proof From Corollary 2.9 we see that in order to have $\{\Phi_1(t-m)\}_{m\in\mathbb{Z}}$ orthonormal we have to define the Fourier transform of Φ_1 as

$$\hat{\Phi}_1(\xi) = h(\xi)\hat{\Phi}(\xi)$$

with $h(\xi)$ a 2π-periodic function such that

$$|h(\xi)| = \frac{1}{\sqrt{2\pi}}\left(\sum_{l\in\mathbb{Z}} |\hat{\Phi}(\xi+2\pi l)|^2\right)^{-1/2}.$$

It is clear from Proposition 2.8 that any such $h(\xi)$ satisfies

$$0 < b \le |h(\xi)| \le B \tag{2.11}$$

for some constants b and B. From Proposition 2.10 and 2.11 we obtain that $\Phi_1 \in \text{span}\,\{\Phi(t-m)\}_{m\in\mathbb{Z}}$. Also we easily see that such a choice ensures that $\sum_{l\in\mathbb{Z}} |\Phi_1(\xi+2\pi l)|^2 = \frac{1}{\sqrt{2\pi}}$, so by Corollary 2.9 the system $\{\Phi_1(t-m)\}_{m\in\mathbb{Z}}$ is orthonormal. From 2.11 we infer that the set of functions $\hat{g}$ which can be written as $\hat{g}(\xi) = f(\xi)\Phi(\xi)$ with $f(\xi)$ 2π-periodic and $\int_0^{2\pi} |f(\xi)|^2 < \infty$ is equal to the set of functions which can be written as $\hat{g}(\xi) = f(\xi)\Phi_1(\xi)$ with f as above. Thus Proposition 2.10 tells us that $\text{span}\,\{\Phi_1(t-m)\}_{m\in\mathbb{Z}} = \text{span}\,\{\Phi(t-m)\}_{m\in\mathbb{Z}}$. □

REMARK 2.4. Clearly the most natural choice for Φ_1 giving an orthonormal basis in $\text{span}\,\{\Phi(t-m)\}_{m\in\mathbb{Z}}$ is given by

$$\hat{\Phi}_1(\xi) = \frac{1}{\sqrt{2\pi}}\left(\sum_{l\in\mathbb{Z}} |\hat{\Phi}(\xi+2\pi l)|^2\right)^{-1/2} \hat{\Phi}(\xi). \tag{2.12}$$

We will use this choice repeatedly in our constructions of wavelets.

REMARK 2.5. It follows from the above arguments that if $\Phi(x)$ is a scaling function of a multiresolution analysis then this multiresolution analysis has many other scaling functions. A function $\Phi_1(x)$ is a scaling function of this multiresolution analysis if and only if $\hat{\Phi}_1(\xi) = m(\xi)\hat{\Phi}(\xi)$ for some 2π-periodic function $m(\xi)$ such that $|m(\xi)| = 1$ a.e.

EXAMPLE 2.1. Let us conclude this section with one more example of a multiresolution analysis. Namely we want to show that the spaces $\mathcal{S}(2^{-j}\mathbb{Z})$ discussed in Section 1.2 form a multiresolution analysis. We have already remarked, and it is almost obvious, that they satisfy conditions (i)–(v) of Definition 2.2. We only have to ensure that (vi) holds also. From Proposition 2.11 we see that it suffices to find a function $\Phi \in \mathcal{S}(\mathbb{Z})$ such that $\{\Phi(t-m)\}_{m\in\mathbb{Z}}$ is a Riesz basis in $\mathcal{S}(\mathbb{Z})$. The very natural candidate for such Φ is a simple tent Λ defined by the conditions

$$\Lambda \in \mathcal{S}(\mathbb{Z}),\ \Lambda(0) = 1,\ \Lambda(n) = 0 \text{ for } n \in \mathbb{Z}\setminus\{0\}.$$

We already know from Section 1.2 that $\mathcal{S}(\mathbb{Z}) = \text{span}\,\{\Lambda(t-m)\}_{m\in\mathbb{Z}}$. There are at least two ways to check that $\{\Lambda(t-m)\}_{m\in\mathbb{Z}}$ is a Riesz system. One is to calculate directly $\int_{-\infty}^{\infty} |\sum_n a_n\Lambda(x-n)|^2\,dx$ for a finite sequence (a_n). Since $\sum_n a_n\Lambda(x-n)$ is a continuous, piecewise linear function having the value a_n for $x = n$, a direct calculation (like the one in Lemma 1.9) gives

$$\int_{-\infty}^{\infty} |\sum_n a_n \Lambda(x-n)|^2\,dx = \frac{1}{3}\sum_n (a_n^2 + a_{n+1}^2 + a_n a_{n+1}).$$

From Hölder's inequality we obtain $|\sum_n a_n a_{n+1}| \leq \sum_n |a_n|^2$ so

$$\int_{-\infty}^{\infty} |\sum_n a_n \Lambda(x-n)|^2\,dx \leq \sum_n |a_n|^2.$$

On the other hand we have

$$\begin{aligned} \frac{1}{3}\sum_n (a_n^2 + a_{n+1}^2 + a_n a_{n+1}) &\geq \frac{1}{3}\sum_n (a_n^2 + a_{n+1}^2 - |a_n||a_{n+1}|) \\ &\geq \frac{1}{3}\sum_n a_n^2. \end{aligned}$$

So we obtain that $\{\Lambda(t-m)\}_{m\in\mathbb{Z}}$ is a Riesz basis in $\mathcal{S}(\mathbb{Z})$. Another argument can be obtained using the Fourier transform. One can easily compute directly $\hat{\Lambda}(\xi)$ and check condition 2.6. I will not present details here because they will be presented in greater generality in the next chapter. •

It should be also noted that we exhibited a scaling function for this multiresolution analysis in Exercise 1.8.

2.3 From scaling function to multiresolution analysis

An inspection of Definition 2.2 shows that a scaling function Φ determines the multiresolution analysis completely. Also, as we will see later 2.47 gives an explicit formula for a wavelet in terms of the scaling function. Thus, if we are interested in constructing wavelets it may be essential to be able to recognize a scaling function of an multiresolution analysis. Let us concentrate now on the scaling function Φ appearing in Definition 2.2. From conditions (i) and (vi) of Definition 2.2 we see that the scaling function Φ is in V_1, so (iv) gives $\Phi(x/2) \in V_0$. From (vi) we obtain

$$\Phi(x/2) = \sum_{n\in\mathbb{Z}} a_n \Phi(x-n) \qquad (2.1$$

or equivalently

$$\Phi(x) = \sum_{n\in\mathbb{Z}} a_n \Phi(2x-n). \qquad (2.1$$

Using Proposition 2.10 the above equations can be rewritten equivalently as

$$\hat{\Phi}(\xi) = m_\Phi(\xi/2)\hat{\Phi}(\xi/2) \tag{2.15}$$

or equivalently as

$$\hat{\Phi}(2\xi) = m_\Phi(\xi)\hat{\Phi}(\xi) \tag{2.16}$$

where m_Φ is a 2π-periodic function given by

$$m_\Phi(\xi) = \frac{1}{2}\sum_{n\in\mathbb{Z}} a_n e^{-in\xi}. \tag{2.17}$$

Naturally, the coefficients a_n in 2.17 are the same as in 2.13 and 2.14. Since $\|\Phi(x/2)\| = \sqrt{2}$ we infer from 2.13 that $\sum_{n\in\mathbb{Z}} |a_n|^2 = 2$ so

$$\left(\frac{1}{2\pi}\int_0^{2\pi} |m_\Phi(\xi)|^2 d\xi\right)^{1/2} = \frac{1}{\sqrt{2}}.$$

Each of the equivalent equations 2.13–2.16 is called a *scaling equation.* They play a fundamental role in the theory of wavelets. Before we proceed, we will prove a lemma about the function $m_\Phi(\xi)$ appearing in the scaling equations 2.15 and 2.16.

Lemma 2.12 $|m_\Phi(\xi)|^2 + |m_\Phi(\xi+\pi)|^2 = 1$ *for almost all* $\xi \in \mathbb{R}$.

Proof From Corollary 2.9 we know that

$$\sum_{l\in\mathbb{Z}} |\hat{\Phi}(\xi + 2\pi l)^2 = \frac{1}{2\pi} \tag{2.18}$$

for almost all ξ. Using 2.15 and the fact that $m_\Phi(\xi)$ is 2π-periodic, we get

$$\begin{aligned}
\frac{1}{2\pi} &= \sum_{l\in\mathbb{Z}} |m_\Phi(\tfrac{\xi}{2} + \pi l)|^2 \cdot |\hat{\Phi}(\tfrac{\xi}{2} + \pi l)|^2 \\
&= \sum_{k\in\mathbb{Z}} |m_\Phi(\tfrac{\xi}{2} + 2k\pi)|^2 |\hat{\Phi}(\tfrac{\xi}{2} + 2k\pi)|^2 \\
&\quad + \sum_{k\in\mathbb{Z}} |m_\Phi(\tfrac{\xi}{2} + \pi + 2k\pi)|^2 |\hat{\Phi}(\tfrac{\xi}{2} + \pi + 2k\pi)|^2 \\
&= |m_\Phi(\tfrac{\xi}{2})|^2 \cdot \sum_{k\in\mathbb{Z}} |\hat{\Phi}(\tfrac{\xi}{2} + 2k\pi)|^2 \\
&\quad + |m_\Phi(\tfrac{\xi}{2} + \pi)|^2 \cdot \sum_{k\in\mathbb{Z}} |\hat{\Phi}(\tfrac{\xi}{2} + \pi + 2k\pi)|^2
\end{aligned}$$

Using 2.18 once more we obtain the lemma. □

Each of equations 2.13–2.16 means that $\Phi \in V_1$, so if we want condition (i) of Definition 2.2 to hold we need to assume 2.14 (or any other of the equivalent conditions 2.13–2.16). The other condition that a scaling function has to satisfy is that $\{\Phi(t-m)\}_{m\in\mathbb{Z}}$ is an orthonormal system. But we saw in Proposition 2.11 that we can easily pass from a function Φ such that $\{\Phi(t-m)\}_{m\in\mathbb{Z}}$ is a Riesz sequence to a function Φ_1 such that $\{\Phi_1(t-m)\}_{m\in\mathbb{Z}}$ is an orthonormal system spanning the same space. These considerations set the stage for the following theorem, which is the main result of this section.

Theorem 2.13 *Suppose we have a function Φ in $L_2(\mathbb{R})$ such that*

(i) $\{\Phi(t-m)\}_{m\in\mathbb{Z}}$ *is a Riesz sequence in* $L_2(\mathbb{R})$

(ii) $\Phi(x/2) = \sum_{k\in\mathbb{Z}} a_k \Phi(x-k)$ *with the convergence of the series understood as the norm convergence in* $L_2(\mathbb{R})$

(iii) $\hat{\Phi}(\xi)$ *is continuous at 0 and* $\hat{\Phi}(0) \neq 0$.

Then the spaces

$$V_j = \operatorname{span}\{\Phi(2^j x - k)\}_{k\in\mathbb{Z}} \tag{2.19}$$

with $j \in \mathbb{Z}$ form a multiresolution analysis.

Condition (iii) can be relaxed a bit, see Exercise 2.13. The main part of the proof of this theorem is contained in the following two propositions.

Proposition 2.14 *Suppose we have a function $\Phi \in L_2(\mathbb{R})$ satisfying condition* (i) *of Theorem 2.13. Let the spaces V_j be defined by 2.19 and let P_j be an orthonormal projection onto V_j. Then for each $f \in L_2(\mathbb{R})$ we have* $\lim_{j\to-\infty} P_j f = 0$, *in particular* $\bigcap_{j\in\mathbb{Z}} V_j = \{0\}$.

Proposition 2.15 *Suppose we have a function $\Phi \in L_2(\mathbb{R})$ satisfying conditions* (i) *and* (iii) *of Theorem 2.13. Then $\bigcup_{j\in\mathbb{Z}} V_j$ is dense in $L_2(\mathbb{R})$, where the spaces V_j are defined by 2.19.*

Assume for the moment that both these propositions are true. The proof of Theorem 2.13 is then simply a matter of checking the conditions of Definition 2.2. The very definitions of V_0 and V_j show that (iv) of Definition 2.2 holds. Condition (ii) of Theorem 2.13 ensures that condition (i) of Definition 2.2 holds. Proposition 2.14 shows that condition (iii) of Definition 2.2 holds and Proposition 2.15 shows that condition (ii) of

Definition 2.2 holds. From Proposition 2.11 we infer that condition (vi) of Definition 2.2 holds.

Proof of Proposition 2.14 Since functions with compact support are dense in $L_2(\mathbb{R})$ it suffices to show that for such a function g we have $\lim_{j\to\infty} \|P_j g\| = 0$. Let us say that supp $g \subset [-R, R]$. From Lemma 2.5(b) we infer that for each $j \in \mathbb{Z}$ the system

$$\{\Phi_{jk}(x)\}_{k\in\mathbb{Z}} = \left\{2^{j/2}\Phi(2^j x - k)\right\}_{k\in\mathbb{Z}}$$

is a Riesz basis in V_j with constants independent of j. Thus from Lemma 2.7 we get for arbitrary $j \in \mathbb{Z}$

$$\begin{aligned}
\|P_j g\|^2 &\leq C\sum_{k\in\mathbb{Z}} |\langle P_j g, \Phi_{jk}\rangle|^2 \\
&= C\sum_{k\in\mathbb{Z}} |\langle g, \Phi_{jk}\rangle|^2 = C\sum_{k\in\mathbb{Z}} \left|\int_{-R}^{R} g(s)\overline{\Phi_{jk}(s)}\, ds\right|^2 \\
&\leq C\sum_{k\in\mathbb{Z}} \int_{-R}^{R} |g(s)|^2 ds \cdot \int_{-R}^{R} |\Phi_{jk}(s)|^2 ds \qquad (2.20) \\
&= C\|g\|^2 \sum_{k\in\mathbb{Z}} 2^j \int_{-R}^{R} |\Phi(2^j s - k)|^2 ds \\
&= C\|g\|^2 \sum_{k\in\mathbb{Z}} \int_{-2^j R - k}^{2^j R - k} |\Phi(u)|^2 du.
\end{aligned}$$

Observe that there exists a $j_0 \in \mathbb{Z}$ such that for $j < j_0$ we have $2^j R < \frac{1}{2}$. For such j's the integrals in the last line of 2.20 are taken over disjoint sets, so from 2.20 we obtain

$$\|P_j g\|^2 \leq C\|g\|^2 \int_{U_j} |\Phi(u)|^2 du \qquad (2.21)$$

where $U_j = \bigcup_{k\in\mathbb{Z}}[-2^j R - k, 2^j R - k]$. Since $\int_{-\infty}^{\infty} |\Phi(u)|^2 du < \infty$ from the Lebesgue dominated convergence theorem we easily infer that $\int_{U_j} |\Phi(u)|^2 du \to 0$ as $j \to -\infty$. This means that $\|P_j g\| \to 0$ as $j \to -\infty$. For $f \in \bigcap_{j\in\mathbb{Z}} V_j$ we have $P_j f = f$ for all $j \in \mathbb{Z}$ so $f = 0$. □

Proof of Proposition 2.15 Take $f \perp \bigcup_{j\in\mathbb{Z}} V_j$. Let $g \in L_2(\mathbb{R})$ be such that $\hat{g} = \hat{f} \cdot \mathbf{1}_{[-R,R]}$ for some $R > 0$ and $\|f - g\| < \varepsilon$. This is possible because, by Plancherel's theorem, the Fourier transform is a unitary operator. As in the previous proof let P_j denote the orthogonal projection onto V_j. Clearly $P_j f = 0$ for all $j \in \mathbb{Z}$, so

$$\|P_j g\| < \varepsilon \text{ for all } j \in \mathbb{Z}. \tag{2.22}$$

From Lemma 2.5(b) we infer that for each $j \in \mathbb{Z}$ the system

$$\{\Phi_{jk}(x)\}_{k\in\mathbb{Z}} = \left\{2^{j/2}\Phi(2^j x - k)\right\}_{k\in\mathbb{Z}}$$

is a Riesz basis in V_j with constants independent of j. Thus from Lemma 2.7 using A1.2–III and A1.2–IV we get for arbitrary $j \in \mathbb{Z}$

$$\begin{aligned}
\|P_j g\|^2 &\geq C\sum_{k\in\mathbb{Z}} |\langle P_j g, \Phi_{jk}\rangle|^2 = C\sum_{k\in\mathbb{Z}} |\langle g, \Phi_{jk}\rangle|^2 \\
&= C\sum_{k\in\mathbb{Z}} \left|\int_{-\infty}^{\infty} \hat{g}(\xi)\overline{\hat{\Phi}_{jk}(\xi)}\, d\xi\right|^2 \qquad (2.23) \\
&= C\sum_{k\in\mathbb{Z}} \left|\int_{-\infty}^{\infty} \hat{g}(\xi)2^{-j/2}\overline{e^{ik2^{-j}\xi}\hat{\Phi}(2^{-j}\xi)}\, d\xi\right|^2.
\end{aligned}$$

Now let us assume that j is such that $2^j\pi > R$. Then the integrals in the last line of 2.23 can be written as

$$\int_{-2^j\pi}^{2^j\pi} \hat{g}(\xi)\overline{\hat{\Phi}(2^{-j}\xi)}2^{-j/2}e^{ik2^{-j}\xi}\, d\xi.$$

We interpret this integral as the $(-k)$-th Fourier coefficient of the function $\sqrt{2\pi}\hat{g}(\xi)\overline{\hat{\Phi}(2^{-j}\xi)}$ on the interval $[-2^j\pi, 2^j\pi]$ with respect to the system

$$\left\{\frac{1}{\sqrt{2\pi}}2^{-j/2}e^{ik2^{-j}\xi}\right\}_{k\in\mathbb{Z}}$$

which is a complete orthonormal system in $L_2(-2^j\pi, 2^j\pi)$ (cf. A1.2–XII). Thus from 2.23 and our assumption about supp $\hat{g}$ we get

$$\begin{aligned}
\|P_j g\|^2 &\geq C2\pi\int_{-2^j\pi}^{2^j\pi} |\hat{g}(\xi)\overline{\hat{\Phi}(2^{-j}\xi)}|^2 d\xi \\
&= C2\pi\int_{-R}^{R} |\hat{g}(\xi)\overline{\hat{\Phi}(2^{-j}\xi)}|^2 d\xi. \qquad (2.24)
\end{aligned}$$

Since $\hat{\Phi}$ is continuous at 0 we see that $\hat{\Phi}(2^{-j}\xi)$ tends uniformly on $[-R, R]$ to $\hat{\Phi}(0)$ as $j \to \infty$. So from 2.22 and 2.24 we get

$$\varepsilon^2 \geq C2\pi|\hat{\Phi}(0)|\|g\|^2.$$

Since ε is arbitrary and $\hat{\Phi}(0) \neq 0$ we get $\|g\| = 0$. This implies that $f = 0$, so $\bigcup_{j\in\mathbb{Z}} V_j$ is dense in $L_2(\mathbb{R})$. □

Using arguments from the above proof we can easily establish the following fact, which is interesting in itself and will be used later.

Proposition 2.16 *Let $\Phi(x)$ be a scaling function of a multiresolution analysis and assume additionally that $\hat{\Phi}(\xi)$ is continuous at 0. Then $|\hat{\Phi}(0)| = \frac{1}{\sqrt{2\pi}}$. In particular, if $\Phi \in L_1(\mathbb{R})$ then*

$$\left| \int_{-\infty}^{\infty} \Phi(x)\, dx \right| = 1.$$

Proof Let P_j denote the orthogonal projection onto V_j and let g be a fixed non-zero function in $L_2(\mathbb{R})$ such that supp $\hat{g} \subset [-1, 1]$. Repeating arguments used in the proof of Proposition 2.15 and using the orthonormality of each of the systems

$$\{\Phi_{jk}(x)\}_{k\in\mathbb{Z}} = \left\{2^{j/2}\Phi(2^j x - k)\right\}_{k\in\mathbb{Z}}$$

we get equality in 2.23 with $C = 1$, so for $j > 0$ we get (exactly as we got 2.24, only using $R = 1$)

$$\|P_j g\|^2 = 2\pi \int_{-1}^{1} |\hat{g}(\xi)\overline{\hat{\Phi}(2^{-j}\xi)}|^2 d\xi. \tag{2.25}$$

As $j \to \infty$ the left hand side of 2.25 tends to $\|g\|^2$ (the V_j's are dense) while the right hand side tends to $2\pi \int_{-1}^{1} |\hat{g}(\xi)|^2\, d\xi \cdot |\hat{\Phi}(0)|^2$ (since $\hat{\Phi}$ is continuous at 0, $\hat{\Phi}(2^{-j}\xi)$ tends to $\hat{\Phi}(0)$ uniformly on $[-1, 1]$). Since by Plancherel's theorem A1.2–IV

$$\|g\|^2 = \int_{-1}^{1} |\hat{g}(\xi)|^2\, d\xi$$

the claim is established. □

Now let us return once more to equations 2.13–2.17, this time under the additional assumption that $\hat{\Phi}$ is continuous at 0. In particular this covers the case $\Phi \in L_1(\mathbb{R}) \cap L_2(\mathbb{R})$. From the above Proposition 2.16 we know that $|\hat{\Phi}(0)| \neq 0$, so from 2.16 we infer that $m_\Phi(\xi)$ is continuous at 0 and $m_\Phi(0) = 1$. From Lemma 2.12 we conclude that $|m_\Phi(\xi)|$ is continuous at π and

$$m_\Phi(\pi) = 0. \tag{2.26}$$

Note that this implies that $m_\Phi(\xi)$ is continuous at π.

The following proposition lists some interesting and very useful properties of a scaling function $\Phi \in L_1(\mathbb{R}) \cap L_2(\mathbb{R})$.

Proposition 2.17 *Suppose that $\Phi \in L_1(\mathbb{R}) \cap L_2(\mathbb{R})$ is a scaling function of a multiresolution analysis. Then*

(i) $|\hat{\Phi}(0)| = \frac{1}{\sqrt{2\pi}}$ *and* $\hat{\Phi}(2\pi k) = 0$ *for* $k \neq 0$, $k \in \mathbb{Z}$

(ii) $\sum_{k\in\mathbb{Z}} \Phi(x+k) = \alpha$, *where* α *is a constant of absolute value* 1.

Proof From Corollary 2.9 we know that $\sum_{k\in\mathbb{Z}} |\hat{\Phi}(\xi + 2\pi k)|^2 = \frac{1}{2\pi}$ almost everywhere. Since $\hat{\Phi}(\xi)$ is continuous (recall that $\Phi \in L_1(\mathbb{R})$) and we know from Proposition 2.16 that $|\hat{\Phi}(0)| = \frac{1}{\sqrt{2\pi}}$ we infer that $|\hat{\Phi}(2\pi k)| = 0$ for $k \neq 0$, $k \in \mathbb{Z}$, which gives (i). Condition (ii) actually follows from (i). From the Poisson summation formula A1.2–XIV we infer that the Fourier coefficients of the function $\sum_{k\in\mathbb{Z}} \Phi(x+k) \mid [0,1]$ are all zero except 0-th which has modulus 1. This gives (ii). □

REMARK 2.6. Let us observe also that equation 2.15 offers the possibility of building a scaling function Φ given a function $m_\Phi(\xi)$, or equivalently the sequence of scalars $(a_n)_{n\in\mathbb{Z}}$ appearing in 2.13 and 2.17. Simply observe that substituting 2.15 into itself N times we get

$$\hat{\Phi}(\xi) = \prod_{j=1}^{N} m_\Phi(2^{-j}\xi)\hat{\Phi}(2^{-N}\xi). \tag{2.27}$$

Thus if Φ is to be a scaling function such that $\hat{\Phi}$ is continuous at 0 we must have

$$\hat{\Phi}(\xi) = \frac{1}{\sqrt{2\pi}} \prod_{j=1}^{\infty} m_\Phi(2^{-j}\xi). \tag{2.28}$$

Clearly only very special sequences $(a_n)_{n\in\mathbb{Z}}$ yield $m_\Phi(\xi)$ (via 2.17) such that 2.28 actually is a scaling function of a multiresolution analysis. Nevertheless this approach will be used succesfully in Chapter 4 to construct smooth compactly supported wavelets.

2.4 Construction of wavelets

Our main aim in this section is to construct wavelets, given a multiresolution analysis. Of course we want our wavelet to be closely connected with the given multiresolution analysis. Suppose we are given a multiresolution analysis $(V_j)_{j\in\mathbb{Z}}$ in $L_2(\mathbb{R})$ with the scaling function $\Phi(x)$.

We want to find a wavelet $\Psi \in V_1$ such that $\operatorname{span}\{\Psi_{jk}\}_{k\in\mathbb{Z}, j<s} = V_s$ for all $s \in \mathbb{Z}$. Let us introduce subspaces W_j of $L_2(\mathbb{R})$ defined by the condition

$$V_j \oplus W_j = V_{j+1}. \tag{2.29}$$

Let J_j denote the dilation operator defined in Definition 2.4. From Lemma 2.5 we know that $2^{j/2}J_j$ is a unitary map, and we know from Definition 2.2 that $J_j(V_1) = V_{j+1}$. Thus we have

$$V_{j+1} = J_j(V_0 \oplus W_0) = J_j(V_0) \oplus J_j(W_0) = V_j \oplus J_j(W_0).$$

This gives

$$W_j = J_j(W_0) \quad \text{for all } \ j \in \mathbb{Z} \tag{2.30}$$

From conditions (i)–(iii) of Definition 2.2 we see that we obtain an orthogonal decomposition

$$L_2(\mathbb{R}) = \oplus \sum_{j\in\mathbb{Z}} W_j. \tag{2.31}$$

Thus we need to find a function $\Psi \in W_0$ such that $\{\Psi(t-m)\}_{m\in\mathbb{Z}}$ is an orthonormal basis in W_0. Any such function Ψ is a wavelet; this follows directly from 2.30 and 2.31. If a wavelet is obtained from a multiresolution analysis in the way described above we will say that it is associated with this multiresolution analysis.

Our construction of Ψ will be given in terms of the Fourier transform. We start with an explicit description of the subspaces V_0 and V_1 in terms of the Fourier transform. From the description given in Proposition 2.10 we know that $g \in V_0$ if and only if

$$\hat{g}(\xi) = \phi(\xi)\hat{\Phi}(\xi) \tag{2.32}$$

for some 2π-periodic function $\phi(\xi)$, and

$$\|g\|_2 = \left(\frac{1}{2\pi}\int_0^{2\pi} |\phi(\xi)|^2\, d\xi\right)^{1/2}. \tag{2.33}$$

From condition (iv) of Definition 2.2 we know that $f \in V_1$ if and only if

$$f(x) = \sqrt{2}g(2x) \tag{2.34}$$

for some $g \in V_0$; clearly we have $\|f\| = \|g\|$. From 2.32 and 2.34 and elementary properties of the Fourier transform (see A1.2–VII) we obtain immediately that $f \in V_1$ if and only if

$$\hat{f}(\xi) = m_f(\xi/2) \cdot \hat{\Phi}(\xi/2) \tag{2.35}$$

where $m_f(\xi)$ is a 2π-periodic function and

$$\left(\frac{1}{2\pi}\int_0^{2\pi} |m_f(\xi)|^2\right)^{1/2} = \frac{1}{\sqrt{2}}\|f\|_2. \tag{2.36}$$

Now we want to describe the space $W_0 = V_1 \ominus V_0$ in terms of the Fourier transform.

Proposition 2.18 *A function f belongs to W_0 if and only if*

$$\hat{f}(\xi) = e^{i\xi/2}v(\xi)\overline{m_\Phi(\tfrac{\xi}{2}+\pi)}\hat{\Phi}(\xi/2) \tag{2.37}$$

where m_Φ is the function defined by 2.15 and $v(\xi)$ is a 2π-periodic function. We have $\|f\| = \left(\frac{1}{2\pi}\int_0^{2\pi}|v(\xi)|^2\,d\xi\right)^{1/2}$.

REMARK 2.7. The reader should compare 2.37 with 2.35 and note that $m_f(\xi/2)$ is actually a 4π-periodic function of ξ, while in 2.37 we have a 2π-periodic function $v(\xi)$. This reflects the fact that W_0 is only a 'half' of V_1.

Proof Clearly $f \in W_0$ if and only if $f \in V_1$ and $f \perp V_0$. This in turn is equivalent to $f \in V_1$ and $\langle f, \Phi(x-k)\rangle = 0$ for all $k \in \mathbb{Z}$. We have

$$\begin{aligned}
\langle f, \Phi(x-k)\rangle &= \int_{-\infty}^{\infty} \hat{f}(\xi)\overline{\mathcal{F}(\Phi(x-k))}\,d\xi \\
&= \int_{-\infty}^{\infty} \hat{f}(\xi)e^{ik\xi}\overline{\hat{\Phi}(\xi)}\,d\xi \qquad (2.38)\\
&= \int_0^{2\pi} e^{ik\xi}\left(\sum_{l\in\mathbb{Z}}\hat{f}(\xi+2\pi l)\cdot\overline{\hat{\Phi}(\xi+2\pi l)}\right)d\xi.
\end{aligned}$$

Note that

$$\begin{aligned}
&\sum_{l\in\mathbb{Z}}\int_0^{2\pi} |\hat{f}(\xi+2\pi l)||\hat{\Phi}(\xi+2\pi l)| \\
&\qquad \le \int_{-\infty}^{\infty} |\hat{f}(\xi)||\hat{\Phi}(\xi)| \le \|f\|_2\|\Phi\|_2
\end{aligned}$$

so the series $\sum_{l\in\mathbb{Z}}\hat{f}(\xi+2\pi l)\overline{\hat{\Phi}(\xi+2\pi l)}$ represents an integrable function on $[0,2\pi]$, call it $F(\xi)$. Looking at 2.38 we realize that $\langle f, \Phi(x-k)\rangle$ equals the $(-k)$-th Fourier coefficient (multiplied by 2π) of the function F. So we can say that $f \in W_0$ if and only if $f \in V_1$ and

$$\sum_{l\in\mathbb{Z}}\hat{f}(\xi+2\pi l)\overline{\hat{\Phi}(\xi+2\pi l)} = 0 \quad \text{a.e.} \tag{2.39}$$

because when all Fourier coefficients of F are zero, then the function F itself has to be zero (cf. A1.2–XII). Since $f \in V_1$ equation 2.35 holds, so we can substitute 2.35 and 2.15 into 2.39 to obtain

$$0 = \sum_{l \in \mathbb{Z}} m_f(\xi/2 + \pi l)\hat{\Phi}(\xi/2 + \pi l)\overline{m_\Phi(\xi/2 + \pi l)\hat{\Phi}(\xi/2 + \pi l)}.$$

Let us call $\eta = \xi/2$ and split this sum into two; odd numbered summands and even numbered summands. We get

$$\begin{aligned} 0 &= \sum_{k \in \mathbb{Z}} |\hat{\Phi}(\eta + 2\pi k)|^2 m_f(\eta + 2\pi k)\overline{m_\Phi(\eta + 2\pi k)} \\ &\quad + \sum_{k \in \mathbb{Z}} |\hat{\Phi}(\eta + \pi + 2\pi k)|^2 m_f(\eta + \pi + 2\pi k)\overline{m_\Phi(\eta + \pi + 2\pi k)}. \end{aligned}$$

Since m_f and m_Φ are 2π-periodic, and Φ as a scaling function satisfies 2.8, we obtain

$$\begin{aligned} 0 &= m_f(\eta)\overline{m_\Phi(\eta)} \sum_{k \in \mathbb{Z}} |\hat{\Phi}(\eta + 2\pi k)|^2 \\ &\quad + m_f(\eta + \pi)\overline{m_\Phi(\eta + \pi)} \sum_{k \in \mathbb{Z}} |\hat{\Phi}(\eta + \pi + 2\pi k)|^2 \qquad (2.40) \\ &= \frac{1}{2\pi}\left(m_f(\eta)\overline{m_\Phi(\eta)} + m_f(\eta + \pi)\overline{m_\Phi(\eta + \pi)}\right). \end{aligned}$$

The above argument is reversible, so $f \in W_0$ if and only if $f \in V_1$ and 2.40 holds. But 2.40 means that the 2-dimensional vector

$$(m_f(\eta), m_f(\eta + \pi))$$

is orthogonal to the 2-dimensional vector

$$((m_\Phi(\eta), m_\Phi(\eta + \pi))$$

for almost all $\eta \in \mathbb{R}$. Note that Lemma 2.12 shows that the vector $(m_\Phi(\eta), m_\Phi(\eta + \pi))$ is not zero. This means that

$$(m_f(\eta), m_f(\eta + \pi)) = \alpha(\eta)\left(\overline{m_\Phi(\eta + \pi)}, -\overline{m_\Phi(\eta)}\right) \qquad (2.41)$$

for some 2π-periodic, complex valued function $\alpha(\eta)$. Replacing in 2.41 η by $\eta + \pi$ and using the fact that all functions involved are 2π-periodic we obtain

$$(m_f(\eta + \pi), m_f(\eta)) = \alpha(\eta + \pi)\left(\overline{m_\Phi(\eta)}, -\overline{m_\Phi(\eta + \pi)}\right). \qquad (2.42)$$

Comparing 2.41 and 2.42 we obtain that $m_f(\eta) = \alpha(\eta)\overline{m_\Phi(\eta + \pi)}$ and

$\alpha(\eta) = -\alpha(\eta+\pi)$. Once again the above reasoning is reversible and we can summarize our considerations as follows: $f \in W_0$ if and only if

$$\hat{f}(\xi) = m_f(\xi/2)\hat{\Phi}(\xi/2) \tag{2.43}$$

with

$$m_f(\eta) = \alpha(\eta)\overline{m_\Phi(\eta+\pi)} \tag{2.44}$$

where $\alpha(\eta)$ is a 2π-periodic function satisfying $\alpha(\eta) = -\alpha(\eta+\pi)$. Observe that this last condition is equivalent to $h(\eta) =: e^{-i\eta}\alpha(\eta)$ being π-periodic. Writing $v(\xi) = h(\xi/2)$ we see that 2.37 is equivalent to 2.43 and 2.44. If 2.37 holds, then from 2.36 we obtain

$$\begin{aligned}\|f\|_2 &= \sqrt{2}\left(\frac{1}{2\pi}\int_0^{2\pi} |v(2u)m_\Phi(u+\pi)|^2\,du\right)^{1/2}\\ &= \left(\frac{1}{2\pi}\int_0^{\pi} |v(2u)|^2\left[|m_\Phi(u)|^2 + |m_\Phi(u+\pi)|^2\right]du\right)^{1/2}.\end{aligned}$$

From Lemma 2.12 we get

$$\|f\|_2 = \sqrt{2}\left(\frac{1}{2\pi}\int_0^{\pi} |v(2u)|^2\,du\right)^{1/2} = \left(\frac{1}{2\pi}\int_0^{2\pi} |v(u)|^2\,du\right)^{1/2}.$$

□

Now that we have a description of W_0, we need to identify those functions $\Psi \in W_0$ such that $\{\Psi(t-m)\}_{m\in\mathbb{Z}}$ is an orthonormal system. This we can do with the help of Corollary 2.9.

Lemma 2.19 *Suppose $f \in W_0$ and $\hat{f}(\xi)$ is written as in 2.37. The system $\{f(t-m)\}_{m\in\mathbb{Z}}$ is an orthonormal basis in W_0 if and only if $|v| = 1$ a.e.*

Proof Clearly we want to check condition 2.8 of Corollary 2.9. Writing $\hat{f}$ in the form 2.37 and splitting the sum into even and odd summands we get

$$\begin{aligned}&\sum_{k\in\mathbb{Z}} |\hat{f}(\xi+2\pi k)|^2\\ &\quad= |v(\xi)|\Bigg(|m_\Phi(\xi/2+\pi)|^2\sum_{k\in\mathbb{Z}}|\hat{\Phi}(\xi/2+2\pi k)|^2\\ &\qquad\quad+|m_\Phi(\xi/2)|^2\sum_{k\in\mathbb{Z}}|\hat{\Phi}(\xi/2+\pi+2\pi k)|^2\Bigg).\end{aligned}$$

Since Φ is a scaling function of a multiresolution analysis we get from Corollary 2.9 and next from Lemma 2.12 that

$$\begin{aligned}
&\sum_{k\in\mathbb{Z}} |\hat{f}(\xi+2\pi k)|^2 \\
&\quad = |v(\xi)|^2 \frac{1}{2\pi}\left(|m_\Phi(\xi/2+\pi)|^2 + |m_\Phi(\xi/2)|^2\right) \\
&\quad = \frac{1}{2\pi}|v(\xi)|^2.
\end{aligned}$$

Now Corollary 2.9 tells us that $\{f(t-m)\}_{m\in\mathbb{Z}}$ is an orthonormal system if and only if $|v|=1$ a.e. We still need to show that for each such f the system $\{f(t-m)\}_{m\in\mathbb{Z}}$ is actually an orthonormal basis in W_0. But this follows directly from Proposition 2.10 and the form of $\hat{f}$; if we compare 2.9 with 2.37 we infer that $W_0 = \text{span}\,\{f(t-m)\}_{m\in\mathbb{Z}}$. □

Now we will summarize our considerations in this Section in the following Theorem.

Theorem 2.20 *Suppose that $\ldots \subset V_{-1} \subset V_0 \subset V_1 \subset \ldots$ is a multiresolution analysis with the scaling function $\Phi \in V_0$. The function $\Psi \in W_0 = V_1 \ominus V_0$ is a wavelet if and only if*

$$\hat{\Psi}(\xi) = e^{i\xi/2} v(\xi) \overline{m_\Phi(\xi/2+\pi)}\hat{\Phi}(\xi/2) \tag{2.45}$$

for some 2π-periodic function $v(\xi)$ such that $|v(\xi)| = 1$ a.e. Each such wavelet Ψ has the property that $\text{span}\{\Psi_{jk}\}_{k\in\mathbb{Z}, j<s} = V_s$ for every $s \in \mathbb{Z}$.

REMARK 2.8. It follows from our discussion at the beginning of this section that the set of wavelets we get using Theorem 2.20 does not depend on the choice of the scaling function Φ, it depends only on the multiresolution analysis. It is clear that different functions $v(\xi)$ give different wavelets. However, when $v_1(\xi) = e^{ir\xi}v_2(\xi)$ for some $r \in \mathbb{Z}$, the resulting wavelets are just translations of each other. This implies that the resulting wavelet bases are the same (only labeled differently), so the wavelets are essentially the same.

If we want to construct a wavelet using Theorem 2.20 we need to fix the function v. Clearly the most obvious choice is $v = 1$. This gives a wavelet Ψ such that

$$\hat{\Psi}(\xi) = e^{i\xi/2}\overline{m_\Phi(\xi/2+\pi)}\hat{\Phi}(\xi/2). \tag{2.46}$$

Using 2.17 we obtain

$$\hat{\Psi}(\xi) \;=\; \tfrac{1}{2}e^{i\xi/2}\sum_{n\in\mathbb{Z}} \overline{a_n} e^{in(\xi/2+\pi)}\hat{\Phi}(\xi/2)$$

$$= \sum_{n\in\mathbb{Z}} \overline{a_n}(-1)^n \tfrac{1}{2} e^{i(n+1)\xi/2} \hat{\Phi}(\xi/2)$$

so

$$\Psi(x) = \sum_{n\in\mathbb{Z}} \overline{a_n}(-1)^n \Phi(2x+n+1) \tag{2.47}$$

where $a_n = \int_{-\infty}^{\infty} \Phi(x/2)\overline{\Phi(x-n)}\,dx$. It is worth noting that 2.47 gives a formula for a wavelet directly in terms of the scaling function.

Assume now that we have a multiresolution analysis with a scaling function $\Phi \in L_1(\mathbb{R}) \cap L_2(\mathbb{R})$. Let Ψ be any wavelet associated with this multiresolution analysis. From 2.26 we know that $m_\Phi(\xi)$ is continuous at π and $m_\Phi(\pi) = 0$. Using this and 2.45 (remember that $v(\xi)$ and $m_\Phi(\xi)$ are bounded, see Lemma 2.12) we see that $\hat{\Psi}(\xi)$ is continuous at 0 and $\hat{\Psi}(0) = 0$. We also know that $\hat{\Phi}(\xi)$ is continuous and $\hat{\Phi}(2\pi k) = 0$ for $k \in \mathbb{Z}$, $k \neq 0$ (see Proposition 2.17(i)) so analogously $\hat{\Psi}(\xi)$ is continuous at points $4k\pi$ with $k \neq 0$ and $\hat{\Psi}(4k\pi) = 0$. Thus we have

$$\hat{\Psi}(\xi) \text{ is continuous at points } 4k\pi, \quad \text{and } \hat{\Psi}(4k\pi) = 0, \; k \in \mathbb{Z}. \tag{2.48}$$

2.5 Periodic wavelets

The aim of this section is to present 'periodic wavelets', i.e. 'wavelets' on the circle $\mathbb{T}$. Although the circle $\mathbb{T}$ is a group, so we can define translations, it does not admit dilations. Thus we cannot repeat our general procedure. Instead, we will use the fact (used already many times) that the real line $\mathbb{R}$ can be 'wrapped around' the circle $\mathbb{T}$; the map is $t \mapsto e^{2\pi i t}$. This provides the identification between functions on the circle $\mathbb{T}$ and 1-periodic functions on $\mathbb{R}$. Our main point in this section is that a good wavelet basis on $\mathbb{R}$ yields a nicely structured complete orthonormal system on $\mathbb{T}$ (identified quite often with $[0, 1)$). There is a very natural procedure, which given a function on $\mathbb{R}$ gives a function on $\mathbb{T}$. We have encountered this procedure many times already. If f is a function on $\mathbb{R}$ we define a 1-periodic function on $\mathbb{R}$, that is a function on $\mathbb{T}$, as

$$\mathcal{P}f(t) = \sum_{k\in\mathbb{Z}} f(t+k). \tag{2.49}$$

It is not defined for every function on $\mathbb{R}$, but it is obviously well defined when $f \in L_1(\mathbb{R})$. Now let us assume that we have a multiresolution analysis on $\mathbb{R}$, call it $\ldots \subset V_{-1} \subset V_0 \subset V_1 \subset \ldots$, with a scaling function Φ and an associated wavelet Ψ.

In order to ensure that $\mathcal{P}\Phi$ and $\mathcal{P}\Psi$ are well defined let us assume throughout this section that $\Phi, \Psi \in L_1(\mathbb{R})$.

Now we want to know what the operator $\mathcal{P}$ does to the function Φ and Ψ, or more precisely to the wavelet basis

$$\left\{2^{j/2}\Psi(2^j t - k)\right\}_{j\in\mathbb{Z},k\in\mathbb{Z}}$$

and to bases in the spaces V_j given as

$$\Phi_{jk}(t) = 2^{j/2}\Phi(2^j t - k) \quad \text{with} \quad k \in \mathbb{Z}.$$

First let us observe that for $j \geq 0$ we have

$$\begin{aligned} \mathcal{P}\Psi_{jk}(x) &= \sum_{s\in\mathbb{Z}} 2^{j/2}\Psi(2^j(x+s)-k) \\ &= \sum_{s\in\mathbb{Z}} 2^{j/2}\Psi(2^j x + 2^j s - k). \end{aligned} \tag{2.50}$$

Since the same formulas as 2.50 also hold for Φ_{jk} we see immediately that

$$\mathcal{P}\Psi_{jk}(t) = \mathcal{P}\Psi_{j,k+2^j}(t) \tag{2.51}$$

$$\mathcal{P}\Psi_{j,k+1}(t) = \mathcal{P}\Psi_{jk}(t+2^{-j}) \tag{2.52}$$

$$\mathcal{P}\Phi_{jk}(t) = \mathcal{P}\Phi_{j,k+2^j}(t) \tag{2.53}$$

$$\mathcal{P}\Phi_{j,k+1}(t) = \mathcal{P}\Phi_{jk}(t+2^{-j}). \tag{2.54}$$

Using A1.2–XIV and A1.2–VII we get for $s, j, k \in \mathbb{Z}$

$$\widehat{\mathcal{P}\Psi_{jk}}(s) = \sqrt{2\pi}\widehat{\Psi_{jk}}(-2\pi s) = \sqrt{2\pi}2^{-j/2}e^{-2\pi isk2^{-j}}\hat{\Psi}(-\frac{2\pi s}{2^j}) \tag{2.55}$$

and

$$\widehat{\mathcal{P}\Phi_{jk}}(s) = \sqrt{2\pi}\widehat{\Phi_{jk}}(-2\pi s) = \sqrt{2\pi}2^{-j/2}e^{-2\pi isk2^{-j}}\hat{\Phi}(-\frac{2\pi s}{2^j}). \tag{2.56}$$

Observe that 2.55 and 2.48 givc that for $j < 0$ and any $k \in \mathbb{Z}$ we have $\widehat{\mathcal{P}\Psi_{jk}}(s) = 0$ for all $s \in \mathbb{Z}$, so

$$\mathcal{P}\Psi_{jk} = 0 \quad \text{for } j < 0 \quad \text{and all } k \in \mathbb{Z}. \tag{2.57}$$

Analogously from 2.56 and Proposition 2.17 we get

$$\mathcal{P}\Phi_{jk} = 2^{j/2} \quad \text{for } j \leq 0 \quad \text{and all } k \in \mathbb{Z}. \tag{2.58}$$

Now for $j = 0, 1, 2, \ldots$ let us denote

$$\tilde{V}_j = \operatorname{span}\left\{\mathcal{P}\Phi_{jk} \;:\; k = 0, 1, \ldots, 2^j - 1\right\}. \tag{2.59}$$

From 2.53 we see that $\tilde{V}_j = \text{span}\{\mathcal{P}\Phi_{jk} \;:\; k \in \mathbb{Z}\}$ and 2.58 means that $\tilde{V}_0$ consists of constant functions.

REMARK 2.9. One usually thinks, and it is very natural thing to do, that $\tilde{V}_j = \mathcal{P}V_j$. One has to be aware that usually there are functions $f \in V_j$ for which $\mathcal{P}f$ is not properly defined.

Proposition 2.21 *With the above definitions and under the assumption that*

$$\mathcal{P}|\Psi| \quad \text{and} \quad \mathcal{P}|\Phi| \quad \text{are bounded functions} \tag{2.60}$$

the following hold

(i) $\tilde{V}_0 \subset \tilde{V}_1 \subset \ldots$

(ii) *for* $j = 0, 1, 2, \ldots$ *the system* $\mathcal{P}\Phi_{jk}$ *with* $k = 0, 1, \ldots, 2^j - 1$ *is an orthonormal basis in* $\tilde{V}_j$

(iii) *for* $j = 0, 1, 2, \ldots$ *the system*

$$1, \mathcal{P}\Psi_{sk} \quad \text{for } s = 0, 1, \ldots, j-1 \quad k = 0, 1, \ldots, 2^{s-1} \tag{2.61}$$

is an orthonormal basis in $\tilde{V}_j$

(iv) $\bigcup_{j=0}^{\infty} \tilde{V}_j$ *is dense in* $L_2[0,1]$, *so the system* $1, \mathcal{P}\Psi_{jk}$ *with* $j = 0, 1, 2, \ldots$ *and* $k = 0, 1, \ldots, 2^{j-1}$ *is a complete orthonormal system in* $L_2[0,1]$.

Proof Since clearly $\int_0^1 \mathcal{P}f(x)\,dx = \int_{-\infty}^{\infty} f(x)\,dx$, condition 2.60 implies that $\Phi, \Psi \in L_1(\mathbb{R})$. Observe that for $s < j$ we have $\Phi_{sk} = \sum_{r\in\mathbb{Z}} \alpha_r \Phi_{jr}$ and $\Psi_{sk} = \sum_{r\in\mathbb{Z}} \beta_r \Phi_{jr}$. From 2.60 we infer that

$$\begin{aligned}
\sum_{r\in\mathbb{Z}} |\alpha_r| &= \sum_{r\in\mathbb{Z}} |\langle \Phi_{sk}, \Phi_{jr} \rangle| \\
&\leq 2^{s/2+j/2} \sum_{r\in\mathbb{Z}} \int_{-\infty}^{\infty} |\Phi(2^s x - k)| \cdot |\Phi(2^j x - r)|\,dx \\
&= 2^{s/2+j/2} \int_{-\infty}^{\infty} |\Phi(2^s x - k)|\, \mathcal{P}|\Phi|(2^j x)\,dx \\
&\leq C 2^{s/2+j/2} \int_{-\infty}^{\infty} |\Phi(2^s x - k)|\,dx < \infty
\end{aligned}$$

and the same argument gives $\sum_{r\in\mathbb{Z}} |\beta_r| < \infty$. This implies that $\mathcal{P}\Phi_{sk} \in \tilde{V}_j$ and $\mathcal{P}\Psi_{sk} \in \tilde{V}_j$. This gives (i). Thus we have to check orthogonality in (ii) and (iii) and (iv). For $j \geq 0$ we have

$$\begin{aligned}\int_0^1 \mathcal{P}\Psi_{jk}(t)\overline{\mathcal{P}\Psi_{j'k'}(t)}\,dt &= \sum_{r,s\in\mathbb{Z}} \int_0^1 \Psi_{jk}(t+s)\overline{\Psi_{j'k'}(t+r)}\,dt \\ &= \sum_{s\in\mathbb{Z}} \int_{-\infty}^{\infty} \Psi_{jk}(t+s)\overline{\Psi_{j'k'}(t)}\,dt \quad (2.62) \\ &= \sum_{s\in\mathbb{Z}} \int_{-\infty}^{\infty} \Psi_{j,k-s2^j}(t)\overline{\Psi_{j'k'}(t)}\,dt.\end{aligned}$$

This shows that $\mathcal{P}\Psi_{jk}$ and $\mathcal{P}\Psi_{j'k'}$ are orthogonal unless $j = j'$ and $k' = k - s2^j$ for some $s \in \mathbb{Z}$. Thus the system 2.61 is orthonormal. Repeating the calculation 2.62 with $j = j'$ and Φ instead of Ψ we infer that (ii) holds. Thus $\dim \tilde{V}_j = 2^j$ so counting dimensions we conclude that the system 2.61 is actually a basis in $\tilde{V}_j$. Note also that all changes of order of summation and integration in calculation 2.62 are easy to justify using 2.60. It remains to show (iv). To this end let us consider P_j, the orthogonal projection from $L_2[0,1)$ onto $\tilde{V}_j$. From (ii) we infer that

$$P_j f = \sum_{k=0}^{2^j-1} \langle f, \mathcal{P}\Phi_{jk}\rangle \mathcal{P}\Phi_{jk}. \quad (2.63)$$

Let us fix an exponential $e^{2\pi irt}$ and let us calculate the r-th Fourier coefficient of $P_j(e^{2\pi irt})$. From 2.63 and 2.56 we get

$$\begin{aligned}P_j(e^{2\pi irt})^\wedge(r) &= \sum_{k=0}^{2^j-1} \langle e^{2\pi irt}, \mathcal{P}\Phi_{jk}\rangle \langle \mathcal{P}\Phi_{jk}, e^{2\pi irt}\rangle \\ &= \sum_{k=0}^{2^j-1} |\langle \mathcal{P}\Phi_{jk}, e^{2\pi irt}\rangle|^2 \quad (2.64) \\ &= \sum_{k=0}^{2^j-1} |\widehat{\mathcal{P}\Phi_{jk}}(r)|^2 \\ &= 2\pi|\hat{\Phi}(-\tfrac{2\pi r}{2^j})|^2.\end{aligned}$$

Using Proposition 2.16 we infer from the above that $P_j(e^{2\pi irt})^\wedge(r) \to 1$ as $j \to \infty$. Since $\|P_j\| = 1$ (because P_j is an orthogonal projection) and $(e^{2\pi is})_{s\in\mathbb{Z}}$ is an orthonormal system in $L_2[0,1)$ we infer that $P_j(e^{2\pi irt})$ tends in $L_2[0,1)$ to $e^{2\pi irt}$ as $j \to \infty$. This implies that for every trigonometric polynomial f $P_j(f) \to f$ in $L_2[0,1)$. Since (by the Weierstrass theorem) trigonometric polynomials are dense in $L_2[0,1)$ we infer that $\bigcup_{j=0}^{\infty} \tilde{V}_j$ is dense in $L_2[0,1)$. □

Sources and comments

The general concept of wavelet emerged in the mid 1980s in the work of Y. Meyer, who constructed wavelets which are called in the next chapter Meyer wavelets; cf. [81]. Almost immediately, different constructions of exponentially decaying spline wavelets were given by P. G. Lemarie [66] and G. Battle [2]. These authors were not aware of Strömberg's paper [107]. Their wavelets were different from Strömberg's but have the same properties. The concept of multiresolution analysis was introduced by S. Mallat [77] and thus the paradigm for constructing wavelets was established. Practically everything we say in Sections 2.1, 2.2 and 2.4 is contained in [77]. Naturally this story was retold many times and I have taken advantage of various improvements and modifications. There are many books and expository papers dealing with the subject of Chapter 2. Our presentation has been greatly influenced by the books [85], [24] and [27] and the expository paper [32]. Everything in this chapter can be found in some or all of those references. The notion of frame defined in Remark 2.3 is used extensively in the theory of wavelets (cf. [24]). For more recent results of a rather abstract nature see [12] and [13].

In our presentation of Theorem 2.13 using Propositions 2.14 and 2.15 we follow Daubechies [24]. Under stronger assumptions these results were known earlier and they follow the folk wisdom of harmonic analysts. In Section 2.5 we follow a well traveled route. Strömberg's wavelets were periodized in [18]. A brief discussion of the general case can be found in [85] or [24]. Also a theory of 'wavelets' on $\mathbb{T}$ independent of the theory on $\mathbb{R}$ (i.e. not involving periodization) can easily be built. For an account see e.g. [93].

There is also a necessary and sufficient condition for a function Ψ to be a wavelet (independent of any multiresolution analysis), formulated in terms of the Fourier transform $\hat{\Psi}$. This can be found, presented in detail, in [51].

Connected with the scaling equation there is also the following problem: given a sequence of numbers $(a_n)_{n\in\mathbb{Z}}$ find (all) functions h on $\mathbb{R}$ such that $h(x/2) = \sum_{n\in\mathbb{Z}} a_n h(x-n)$. This is a special case of a well known general problem with various ramifications. Let us mention that the solution of the equation

$$f(\tfrac{x}{3}) = f(x) + \tfrac{2}{3}f(x-1) + \tfrac{1}{3}f(x-2) + \tfrac{1}{3}f(x-3) + f(x-4)$$

was studied by de Rham [31] as an example of a continuous non-differentiable function. For a reader interested in this problem we suggest [63] or [26] where various aspects of the problem are discussed and some history of the problem and bibliographical references are presented. A more general 'continuous' version of this problem is discussed in [30].

About the exercises. Exercise 2.4 is taken directly from Annex A of [68]. Exercise 2.5 is taken from [24] page 145.

Exercises

2.1 Let L_j denote the space of all functions $f \in L_2(\mathbb{R})$ such that $f|[2^{-j}n, 2^{-j}(n+1)]$ is linear for each $n \in \mathbb{Z}$ (we do *not* assume that they are continuous).

- Show that the subspaces L_j satisfy conditions (i)–(v) of Definition 2.2.
- Show that they do not satisfy condition (vi) of Definition 2.2.

2.2 Find an example of a subspace $V_0 \subset L_2(\mathbb{R})$ satisfying conditions (v) and (vi) of Definition 2.2 and such that the subspaces V_j defined by condition (iv) of Definition 2.2 satisfy neither (i) nor (ii) of the definition.

2.3 Show that there does not exist a function $\Psi \in L_2(\mathbb{R})$ such that $\{\Psi(t-m)\}_{m\in\mathbb{Z}}$ is a Riesz basis in $L_2(\mathbb{R})$.

2.4 Let $N \geq 1$ be given and let $\Phi \in L_2(\mathbb{R})$ be such that

(a) $\left\|x^k \frac{d\Phi}{dx^s}\right\|_\infty < \infty$ for $s = 1, 2, \ldots, N$ and $k \in \mathbb{N}$

(b) $\{\Phi(t-m)\}_{m\in\mathbb{Z}}$ is a Riesz sequence

(c) $\Phi(\frac{x}{2}) = \sum_{k\in\mathbb{Z}} \alpha_k \Phi(x-k)$.

Define $M(x) =: \sum_{k<0} \Phi'(x-k)$. Show that

(a) $\phi'(x) = M(x+1) - M(x)$

(b) $\hat{\Phi}(2\pi + 4k\pi) = 0$ for $k \in \mathbb{Z}$

(c) $m(\xi) =: \frac{1}{2}\sum_{k\in\mathbb{Z}} \alpha_k e^{-ik\xi}$ is a C^∞, 2π-periodic function and $m(\pi) = 0$

(d) M satisfies (a)–(c) above with N replaced by $N-1$.

Show by induction that for every polynomial $p(x)$ of degree $\leq N$ there exists a sequence β_k such that $|\beta_k| \leq C(1+|k|)^N$ and $p(x) = \sum_{k\in\mathbb{Z}} \beta_k \Phi(x-k)$.

2.5 Suppose that $\Phi \in L_2(\mathbb{R}) \cap L_1(\mathbb{R})$ is a scaling function of a multiresolution analysis. Let a_n's be the coefficients appearing in the scaling equation 2.14. Assume additionally that $\sum_{n\in\mathbb{Z}} |a_n| < \infty$. Show that $\sum_{n\in\mathbb{Z}} a_n = 2$ and $\sum_{n\in\mathbb{Z}}(-1)^n a_n = 0$.

2.6 Let S be a Strömberg wavelet. Show that $\{S'(t-m)\}_{m\in\mathbb{Z}}$ is a Riesz sequence.

2.7 Let us consider the multiresolution analysis $V_j = \mathcal{S}(2^j\mathbb{Z})$ discussed in Section 1.2 and in Example 2.1 at the end of Section 2.2.

- Show that there exists an even function $\psi \in V_0$ such that $\{\psi(t-m)\}_{m\in\mathbb{Z}}$ is an orthonormal basis in V_0.
- Show that there exists an even wavelet associated with this multiresolution analysis.
- Show that there does not exists an odd wavelet associated with this multiresolution analysis.

2.8 Show that the function $\mathbf{1} = \mathbf{1}_{[-1/2,1/2]}$ is not a scaling function of a multiresolution analysis.

2.9 Suppose that Φ is a scaling function of a multiresolution analysis such that $\Phi \in L_2(\mathbb{R}) \cap L_1(\mathbb{R})$ and $\mathcal{P}|\Phi| \in L_2[0,1]$. Show that there exists an associated wavelet $\Psi \in L_2(\mathbb{R}) \cap L_1(\mathbb{R})$.

2.10 Let Ψ be a wavelet associated with a multiresolution analysis $\ldots \subset V_{-1} \subset V_0 \subset V_1 \subset \ldots$. Show that there exists a scaling function Φ for this multiresolution analysis, such that Ψ is given by formula 2.47.

2.11 Suppose that Φ is a function such that $\mathcal{P}|\Phi| \in L_\infty[0,1]$ and $\{\Phi(t-m)\}_{m\in\mathbb{Z}}$ is a Riesz sequence. Show that Φ_1 defined by 2.12 also satisfies $\mathcal{P}|\Phi_1| \in L_\infty[0,1]$.

2.12 Let $\Psi \in L_2(\mathbb{R}) \cap L_1(\mathbb{R})$ be a wavelet associated with a multiresolution analysis with a scaling function $\Phi \in L_2(\mathbb{R}) \cap L_1(\mathbb{R})$. Show that for $g(x) = \Psi(x/2)$ we have $\mathcal{P}g = 0$.

2.13 Let Φ_1 be a function in $L_2(\mathbb{R})$ which satisfies (i) and (ii) of Theorem 2.13 and such that

$$|\hat{\Phi}_1(\xi)| \text{ is continuous at } 0 \text{ and } \hat{\Phi}_1(0) \neq 0. \qquad \text{(E2.1)}$$

Show that there exists a function $\Phi \in L_2(\mathbb{R})$ which satisfies (i)–(iii) of Theorem 2.13 and such that $\operatorname{span}\{\Phi_1(t-m)\}_{m\in\mathbb{Z}} = \operatorname{span}\{\Phi(t-m)\}_{m\in\mathbb{Z}}$. Conclude that Theorem 2.13 holds when condition (iii) is replaced by E2.1.

2.14 Improve Proposition 2.15 and show that if P_j is an orthogonal projection onto V_j, then for every $f \in L_2(\mathbb{R})$ $P_j f \to f$ in norm as $j \to \infty$.

2.15 Let $\varphi(x) = \mathbf{1}_{[0,2]}$. Show that $\{\varphi(t-m)\}_{m\in\mathbb{Z}}$ is a frame (cf. Remark 2.3) in the space L_0 defined by 1.5. Let $f(x) = H(\frac{x}{2})$ where H is the Haar wavelet. Show that

$$\left\{2^{j/2} f(2^j t - k)\right\}_{j\in\mathbb{Z},k\in\mathbb{Z}}$$

is a tight frame in $L_2(\mathbb{R})$. Show that the system of vectors $(1,0)$, $(-\frac{1}{2}, \frac{\sqrt{3}}{2})$, $(-\frac{1}{2}, -\frac{\sqrt{3}}{2})$ is a tight frame in $\mathbb{R}^2$.

3
Some important wavelets

In this chapter we will present in detail constructions and properties of some important classes of wavelets. The constructions will follow the general theory established in the previous chapter.

3.1 What to look for in a wavelet?

The answer to the question in the title of this section clearly depends on what we want to use the wavelet for. Our approach taken in Chapter 1 and later in Chapters 8 and 9 is to analyze functions from some function space, very often different from $L_2(\mathbb{R})$, using wavelets. We will base our answer upon the analysis of arguments given later. This however is only a matter of motivation. Our mathematics will in no way rely on things presented in later chapters.

It is clear from our arguments given in Chapters 8 and 9, and has already been mentioned in chapter 1, that good decay of wavelets plays a crucial role in investigating wavelet expansions of a function. It is obviously also crucial in the following question of clear practical importance but not discussed in any detail in this book. Suppose a function f on $\mathbb{R}$ (or on $\mathbb{R}^d$) is given with supp $f \subset [0,1]$ (or some cube Q). How can we recognize it from its wavelet coefficients? Suppose we approximate f by a finite subsum of its wavelet expansion. How will this approximation look outside $[0,1]$? We will use this type of estimate to estimate Σ_B in the proof of the fundamental Proposition 8.8.

The other important property of wavelets used extensively in Chapters 8 and 9 is that $\int \Psi(x)\,dx = 0$ for wavelets and $\int \Phi(x)\,dx = 1$ a for scaling function. If we attempt to analyze smoother functions than we do in Chapter 9 (e.g. using moduli of continuity of higher order) or when we investigate Hardy spaces H_p with $p < 1$ (in Chapters 6 and 8 we talk only about H_1) we encounter the need for more restrictive moment conditions

$$\int_{-\infty}^{\infty} x^k \Psi(x)\,dx = 0 \quad \text{for } k = 0, 1, \ldots, s. \tag{3.1}$$

Clearly, in order to talk about these conditions (even for $s = 0$) we have to assume that $\int_{-\infty}^{\infty} |x|^s |\Psi(x)|\,dx < \infty$, so practically we need to assume some decay of the wavelet Ψ. It is also clear that if we want analyze smooth functions efficiently our wavelets should be smooth.

Thus we are interested in three qualities of wavelets:

- decay
- vanishing moments
- smoothness.

Actually, under appropriate decay assumptions, vanishing moments follow from smoothness. We have:

Proposition 3.1 *Suppose that Ψ is a function on $\mathbb{R}$ such that $\left\{2^{j/2}\Psi(2^j t - k)\right\}_{j\in\mathbb{Z},k\in\mathbb{Z}}$ is orthonormal in $L_2(\mathbb{R})$. Assume that for some $l = 0, 1, 2, \ldots$ we have:*

$$\Psi \text{ is of class } C^l \tag{3.2}$$

$$\text{all derivatives } \Psi^{(s)}(x) \text{ with } s = 0, 1, 2, \ldots, l \text{ are bounded on } \mathbb{R} \tag{3.3}$$

$$|\Psi(x)| \leq \frac{C}{(1 + |x|)^\alpha} \text{ for some } \alpha > l + 1. \tag{3.4}$$

Then $\int_{-\infty}^{\infty} x^s \Psi(x)\,dx = 0$ for $s = 0, 1, \ldots, l$.

Proof Let us choose s to be the smallest integer among $\{0, 1, \ldots, l\}$ such that $\int_{-\infty}^{\infty} x^s \Psi(x)\,dx \neq 0$. If there is no such an s the proposition clearly holds. Since Ψ is not a polynomial $\Psi^{(s)}(x)$ is not identically 0. Let us fix a number $a = k \cdot 2^{-J}$ with $k, J \in \mathbb{Z}$ and $J \geq 0$ such that $\Psi^{(s)}(a) \neq 0$. Using the Taylor formula we write

$$\Psi(x) = \sum_{r=0}^{s} a_r (x - a)^r + R(x) \tag{3.5}$$

where the remainder $R(x)$ satisfies that for each $\varepsilon > 0$ there exists a $\delta > 0$ such that

$$|R(x)| \leq \varepsilon |x - a|^s \text{ for } |a - x| < \delta. \tag{3.6}$$

Note also that

$$|R(x)| \leq C|x-a|^s \text{ for all } x \in \mathbb{R}. \tag{3.7}$$

Since $\Psi^{(s)}(a) \neq 0$ the Taylor formula yields $a_s \neq 0$. For $j > J$ let $k_j = 2^j a = 2^{j-J}k$. It is an integer. From orthogonality we have for $j > J$

$$\int_{-\infty}^{\infty} \Psi(x)\overline{\Psi(2^j x - k_j)}\, dx = 0. \tag{3.8}$$

Using the change of variables $u = x - a$ and substituting 3.5 into 3.8 we get

$$0 = \int_{-\infty}^{\infty} \Big(\sum_{r=0}^{s} a_r u^r + R(u+a)\Big)\overline{\Psi(2^j u)}\, du.$$

Our choice of s gives that for $r < s$ we have $\int_{-\infty}^{\infty} u^r \overline{\Psi(2^j u)}\, du = 0$ so we get

$$-a_s \int_{-\infty}^{\infty} u^s \overline{\Psi(2^j u)}\, du = \int_{-\infty}^{\infty} R(u+a)\overline{\Psi(2^j u)}\, du.$$

Putting $2^j u = x$ in the left hand integral we arrive at

$$\int_{-\infty}^{\infty} x^s \overline{\Psi(x)}\, dx = -\frac{1}{a_s} 2^{j(s+1)} \int_{-\infty}^{\infty} R(a+u)\overline{\Psi(2^j u)}\, du. \tag{3.9}$$

We will show that the right hand side of 3.9 tends to 0 as $j \to \infty$. This will give $\int_{-\infty}^{\infty} x^s \Psi(x)\, dx = 0$, contradicting the choice of s and thus proving the proposition.

From 3.6 and 3.7 we obtain

$$\begin{aligned} &\left|2^{j(s+1)} \int_{-\infty}^{\infty} R(a+u)\overline{\Psi(2^j u)}\, du\right| \\ \leq\ & 2^{j(s+1)} \int_{-\delta}^{\delta} \varepsilon |u|^s \frac{du}{(1+|2^j u|)^\alpha} + 2 \cdot 2^{j(s+1)} \int_{\delta}^{\infty} \frac{C|u|^s\, du}{(1+|2^j u|)^\alpha} \\ \leq\ & \varepsilon \int_{-2^j\delta}^{2^j\delta} \frac{|x|^s\, dx}{(1+|x|)^\alpha} + 2C \int_{2^j\delta}^{\infty} \frac{|x|^s\, dx}{(1+|x|)^\alpha}. \end{aligned}$$

Since $\frac{|x|^s}{(1+|x|)^\alpha} \in L_1(\mathbb{R})$, taking ε sufficiently small and then j sufficiently large we see that the right hand side of 3.9 tends to zero. $\square$

3.2 Meyer's wavelets

In this section we want to present an important class of multiresolution analyses and wavelets introduced by Y. Meyer. Among those wavelets are wavelets belonging to the Schwartz class S. We start with the function $\Theta(\xi)$ defined on $\mathbb{R}$ and satisfying the following conditions:

$$
\begin{aligned}
0 \leq \Theta(\xi) &\leq \frac{1}{\sqrt{2\pi}} && (3.10)\\
\Theta(\xi) &= \Theta(-\xi) && (3.11)\\
\Theta(\xi) &= \frac{1}{\sqrt{2\pi}} \quad \text{for } |\xi| < \frac{2}{3}\pi && (3.12)\\
\Theta(\xi) &= 0 \quad \text{for } |\xi| > \frac{4}{3}\pi && (3.13)\\
\Theta^2(\xi) + \Theta^2(\xi - 2\pi) &= \frac{1}{2\pi} \quad \text{for } 0 \leq \xi \leq 2\pi. && (3.14)
\end{aligned}
$$

Clearly there exists a function $\Phi \in L_2(\mathbb{R})$ such that $\hat{\Phi} = \Theta$. Because Θ is compactly supported Φ is always a C^∞-function, cf. A1.2–IX.

Proposition 3.2 *If Θ satisfies 3.10–3.14 above then the function $\Phi = \check{\Theta}$ is a scaling function of a multiresolution analysis. The corresponding function $m_\Phi(\xi)$ (cf. 2.15–2.17) is a 2π-periodic function which on the interval $[-\pi, \pi)$ equals $\sqrt{2\pi}\Theta(2\xi)$.*

REMARK 3.1. Each such multiresolution analysis is called a Meyer multiresolution analysis.

Proof The easiest way to prove this proposition is to apply Theorem 2.13. Thus we need to check the assumptions of that theorem. Clearly $\Theta = \hat{\Phi}$ is continuous at 0 and $\Theta(0) = \frac{1}{\sqrt{2\pi}} \neq 0$. It follows directly from condition 3.13 that there are at most two non-zero summands in the series $\sum_{l \in \mathbf{Z}} \Theta^2(\xi + 2\pi l)$. Thus condition 3.14 immediately gives that

$$\sum_{l \in \mathbf{Z}} \Theta^2(\xi + 2\pi l) = \frac{1}{2\pi}$$

so from Corollary 2.9 we infer that $\{\Phi(t - m)\}_{m \in \mathbf{Z}}$ is an orthonormal system. Let ψ be a 2π-periodic function which for $\xi \in [-\pi, \pi]$ equals $\sqrt{2\pi}\Theta(2\xi)$. Since supp $\Theta(2\xi) \subset [-\frac{2}{3}\pi, \frac{2}{3}\pi]$ conditions 3.12 and 3.13 imply that $\Theta(2\xi) = \psi(\xi)\Theta(\xi)$. This is condition 2.16 which, as we know, gives the last assumption of Theorem 2.13. This also shows that $m_\Phi = \psi$. □

REMARK 3.2. Theorem 2.13 is not really needed in the above proof. One can easily define spaces V_j directly in terms of the Fourier transform, namely $f \in V_j$ if and only if

$$\hat{f}(\xi) = g(2^{-j}\xi)\Theta(2^{-j}\xi)$$

for some 2π-periodic function g. Then one easily checks conditions (i)–(vi) of Definition 2.2.

To construct a wavelet we apply Theorem 2.20 for $v = 1$. This gives a wavelet Ψ such that

$$\hat{\Psi}(\xi) = e^{i\xi/2} m_\Phi(\xi/2 + \pi)\Theta(\xi/2). \tag{3.15}$$

From Proposition 3.2 we infer that supp $m_\Phi \subset \bigcup_{k\in\mathbb{Z}}[2k\pi - \frac{2}{3}\pi, 2k\pi + \frac{2}{3}\pi]$ so from condition 3.13 we easily get that

$$\text{supp } \hat{\Psi}(\xi) \subset [-\tfrac{8}{3}\pi, -\tfrac{2}{3}\pi] \cup [\tfrac{2}{3}\pi, \tfrac{8}{3}\pi]. \tag{3.16}$$

The following proposition lists formal properties of the wavelet Ψ defined by 3.15.

Proposition 3.3 *With every function $\Theta(\xi)$ satisfying 3.10–3.14 we can associate a wavelet Ψ given by 3.15 which has the following properties*

(i) supp $\hat{\Psi}(\xi) \subset [-\frac{8}{3}\pi, -\frac{2}{3}\pi] \cup [\frac{2}{3}\pi, \frac{8}{3}\pi]$
(ii) Ψ *is a real-valued* C^∞ *function*
(iii) $\Psi(-\frac{1}{2} - x) = \Psi(-\frac{1}{2} + x)$ *for all* $x \in \mathbb{R}$.

REMARK 3.3. Each such wavelet will be called a *Meyer's wavelet.*

Proof Condition (i) is just 3.16. The fact that Ψ is C^∞ follows from (i) and properties of the Fourier transform (cf. A1.2–IX). To see that it is real-valued let us call $\alpha(\xi) = m_\Phi(\xi/2 + \pi)\Theta(\xi/2)$ and observe that

$$\begin{aligned} \alpha(-\xi) &= m_\Phi(-\xi/2 + \pi)\Theta(-\xi/2) = m_\Phi(-\xi/2 + \pi - 2\pi)\Theta(-\xi/2) \\ &= m_\Phi(-\xi/2 - \pi)\Theta(-\xi/2) = m_\Phi(\xi/2 + \pi)\Theta(\xi/2) \\ &= \alpha(\xi). \end{aligned} \tag{3.17}$$

The second equality holds because m_Φ is 2π-periodic, and the last but one because Θ and so also m_Φ is even. From Proposition 3.2 we see that m_Φ is real-valued so also $\alpha(\xi)$ is real-valued. Using 3.15 and 3.17 we have

$$\Psi(x) = \frac{1}{\sqrt{2\pi}} \int_{-\infty}^{\infty} e^{ix\xi} e^{i\xi/2} \alpha(\xi)\, d\xi$$

$$= \frac{1}{\sqrt{2\pi}} \int_{-\infty}^{\infty} \left[\cos(x+\tfrac{1}{2})\xi + i\sin(x+\tfrac{1}{2})\xi\right] \alpha(\xi)\, d\xi$$
$$= \frac{1}{\sqrt{2\pi}} \int_{-\infty}^{\infty} \cos(x+\tfrac{1}{2})\xi\, \alpha(\xi)\, d\xi$$

which is clearly real. This gives (ii). Using 3.15 and 3.17 once more we obtain

$$\begin{aligned}
\Psi(-\tfrac{1}{2}-x) &= \frac{1}{\sqrt{2\pi}} \int_{-\infty}^{\infty} e^{i(-1/2-x)\xi} e^{i\xi/2} \alpha(\xi)\, d\xi \\
&= \frac{1}{\sqrt{2\pi}} \int_{-\infty}^{\infty} e^{-ix\xi} \alpha(\xi)\, d\xi = \frac{1}{\sqrt{2\pi}} \int_{-\infty}^{\infty} e^{ix\xi} \alpha(\xi)\, d\xi \\
&= \frac{1}{\sqrt{2\pi}} \int_{-\infty}^{\infty} e^{i(x-1/2)\xi} e^{i\xi/2} \alpha(\xi)\, d\xi \\
&= \Psi(x-\tfrac{1}{2})
\end{aligned}$$

so (iii) holds. □

The most important special case of the above construction is when Θ, in addition to satisfying 3.10–3.14, is also C^∞. It is well known that such a Θ exists. Some constructions of such functions are given in Exercises 3.2 and 3.3. If Θ is C^∞ then $\hat{\Psi}(\xi)$ is also C^∞ so $\hat{\Psi}$ is in the Schwartz class S. From A1.2–VIII we get that Ψ itself is in S. Thus we have

Theorem 3.4 *There exists a real-valued wavelet Ψ such that*

- *Ψ is in the Schwartz class S, i.e. Ψ is a C^∞ function and for any non-negative integers k and l there exists a constant $C = C(k,l)$ such that*

$$\left|\frac{d^k\Psi}{dt^k}(t)\right| \le C(1+|t|)^{-l}$$

  for all $t \in \mathbb{R}$
- *$\Psi(-\frac{1}{2}+x) = \Psi(-\frac{1}{2}-x)$ for all $x \in \mathbb{R}$*
- *supp $\hat{\Psi} \subset [-\frac{8}{3}\pi, -\frac{2}{3}\pi] \cup [\frac{2}{3}\pi, \frac{8}{3}\pi]$.*

3.3 Spline wavelets

In this section we will describe the construction and properties of multiresolution analyses and wavelets constructed using spline functions

3.3.1 Spline functions.

Let us present first a self-contained account of the basic theory of spline functions. We start with the definition.

DEFINITION 3.5 *Let a be a positive real number and let $n = 0, 1, 2, \ldots$ be an integer. A spline of order n with nodes in $a\mathbb{Z}$ is a function f defined on $\mathbb{R}$ which is of class C^{n-1} and is a polynomial of degree at most n when restricted to each interval $[ja, (j+1)a]$ for $j \in \mathbb{Z}$. The space of all splines of order n with nodes in $a\mathbb{Z}$ will be denoted by $\mathcal{S}^n(a\mathbb{Z})$.*

Some comments about this definition are in order. As usual in this book, by functions of class C^{-1} we mean measurable functions and by functions of class C^0 we mean continuous functions. This means that splines of order 0 are functions which are constant on all intervals of the form $[ja, (j+1)a]$ for $j \in \mathbb{Z}$, and splines of order 1 are continuous piecewise linear functions with nodes in $a\mathbb{Z}$. Thus in this section we will generalize some of the results of Sections 1.1 and 1.2. Clearly we can equally well define splines taking an arbitrary discrete subset of $\mathbb{R}$ as a set of nodes. We can also define splines on intervals. We can also relax smoothness conditions, cf. Exercise 3.12. All this is standard practice in spline theory. Our Definition 3.5 is only a very simple case which however is all that we will need for our construction of wavelets. It is clear from this definition that splines are invariant under appropriate translations and dilations. More precisely we have

$$J_s\left(\mathcal{S}^n(a\mathbb{Z})\right) = \mathcal{S}^n(2^{-s}a\mathbb{Z})$$

and

$$T_{ka}\left(\mathcal{S}^n(a\mathbb{Z})\right) = \mathcal{S}^n(a\mathbb{Z})$$

where the operators J_s and T_h are defined in Definitions 2.3 and 2.4.

We will base our introduction to splines on the concept of B-splines.

DEFINITION 3.6 *For $n = 0, 1, 2, \ldots$ we define functions $N_n(x)$, called B-splines of order n, as follows:*

(i) $N_0 = \mathbf{1}_{[0,1]}$

(ii) *for $n > 0$ we define N_n inductively as $N_{n+1} = N_n * N_0$. It is clear that N_n is a convolution product $N_0 * \ldots * N_0$ with $n+1$ factors.*

The following theorem summarizes elementary properties of B-splines.

Theorem 3.7 *For $n = 0, 1, 2, \dots$ the functions $N_n(x)$ have the following properties:*

$$\begin{aligned} N_n(x) &> 0 \ \textit{for}\ x \in (0, n+1) & (3.18)\\ \operatorname{supp} N_n &= [0, n+1] & (3.19)\\ N_n &\in S^n(\mathbb{Z}) & (3.20)\\ \sum_{k\in\mathbb{Z}} N_n(x-k) &= 1 & (3.21)\\ N_n\left(\tfrac{n+1}{2} - x\right) &= N_n\left(\tfrac{n+1}{2} + x\right) \ \textit{for all}\ x \in \mathbb{R} & (3.22)\\ N'_{n+1}(x) &= N_n(x) - N_n(x-1). & (3.23) \end{aligned}$$

Proof Observe that properties 3.18–3.22 hold for $n = 0$. Thus we will proceed by induction. From Definition 3.6 we can write

$$\begin{aligned} N_{n+1}(x) &= \int_{-\infty}^{\infty} N_n(x-t) N_0(t)\, dt \\ &= \int_0^1 N_n(x-t)\, dt = \int_{x-1}^{x} N_n(u)\, du. \qquad (3.24) \end{aligned}$$

From 3.24 we see immediately that if 3.18 and 3.19 hold for n they also hold for $n+1$. Equation 3.24 also shows that if $N_n \in C^{n-1}$ then $N_{n+1} \in C^n$. To see that $N_{n+1}(x)$ is a polynomial on any interval $[k, k+1]$ take $x \in (k, k+1)$ and use 3.24 to write

$$N_{n+1}(x) = \int_{k-1}^{k} N_n(u)\, du + \int_k^x N_n(u)\, du - \int_{k-1}^{x-1} N_n(u)\, du. \qquad (3.25)$$

Since $N_n(u)$ is a polynomial of degree at most n on the intervals $[k-1, k]$ and $[k, k+1]$ we see that $N_{n+1}(x)$ is a polynomial of degree at most $n+1$ on the interval $(k, k+1)$. This gives 3.20. To get 3.21 we use 3.24 and the inductive hypothesis and obtain

$$\begin{aligned} \sum_{k\in\mathbb{Z}} N_{n+1}(x-k) &= \sum_{k\in\mathbb{Z}} \int_0^1 N_n(x-k-t)\, dt \\ &= \int_0^1 \sum_{k\in\mathbb{Z}} N_n(x-t-k)\, dt = \int_0^1 1\, dt = 1. \end{aligned}$$

Property 3.22 is a general property of convolution. Using changes of variables and the inductive hypothesis we obtain

$$\begin{aligned} N_{n+1}(\frac{n+2}{2}-x) &= \int_{-\infty}^{\infty} N_n(u)N_0(\frac{n+2}{2}-x-u)\,du \\ &= \int_{-\infty}^{\infty} N_n(\frac{n+1}{2}+v)N_0(\frac{1}{2}-x-v)\,dv \\ &= \int_{-\infty}^{\infty} N_n(\frac{n+1}{2}-v)N_0(\frac{1}{2}+x+v)\,dv \\ &= \int_{-\infty}^{\infty} N_n(u)N_0(\frac{n+2}{2}+x-u)\,du \\ &= N_{n+1}(\frac{n+2}{2}+x). \end{aligned}$$

Differentiating 3.25 we obtain 3.23. □

The following easy observation will be used several times in what follows.

Lemma 3.8 *Suppose that* $f \in \mathcal{S}^n(\mathbb{Z})$ *and* $f \mid [-1,0] = 0$. *Then* $f \mid [0,1] = ct^n$ *for some constant* $c \in \mathbb{R}$.

Proof Clearly all left hand derivatives of f at 0 are 0. Since $f \in \mathcal{S}^n(\mathbb{Z})$ we infer that the first $(n-1)$ derivatives of f at 0 exist, so they are equal to 0. This means that $f \mid [0,1]$ is a polynomial of degree at most n whose first $(n-1)$ derivatives are zero at 0. This clearly establishes the claim. □

Theorem 3.9 *If* $f \in \mathcal{S}^n(\mathbb{Z})$ *and* supp $f \subset [0, n+1]$, *then* $f = cN_n$ *for some* $c \in \mathbb{R}$.

Proof We will proceed by induction. The theorem is clearly true for $n = 0$, so let us assume that it is true for $(n-1)$. Translating f by an integer to the left if needed we can assume that $f \mid [0,1]$ is not identically zero. From Lemma 3.8 we see that $f \mid [0,1] = c_1 t^n$ for some $c_1 \neq 0$ and also $N_n \mid [0,1] = c_2 t^n$ for some $c_2 \neq 0$. Thus there exists $\alpha \in \mathbb{R}$ such that $(f - \alpha N_n) \mid [0,1] = 0$. If $f - \alpha N_n = 0$ the proof is finished. Otherwise $f' - \alpha N_n' \in \mathcal{S}^{n-1}(\mathbb{Z})$ and supp $(f' - \alpha N_n') \subset [1, n+1]$. From the inductive hypothesis we get that $f' - \alpha N_n' = c_3 N_{n-1}(x+1)$ for some constant

$c_3 \neq 0$. But this is impossible, since $\int_{-\infty}^{\infty}(f'(x) - \alpha N_n'(x))\,dx = 0$ but it follows from 3.18 and 3.19 that $\int_{-\infty}^{\infty} c_3 N_{n-1}(x+1)\,dx \neq 0$. □

REMARK 3.4. Let us note the following easy consequence of Theorem 3.9: if $f \in \mathcal{S}^n(\mathbb{Z})$ and supp $f \subset I$ for some interval I with $|I| < n+1$, then $f \equiv 0$.

Theorem 3.10 *Suppose that φ is a polynomial of degree at most n. There exists a unique spline $f \in \mathcal{S}^n(\mathbb{Z})$ such that*

$$f = \sum_{k=-n}^{0} a_k N_n(x-k)$$

and $f \mid [0,1] = \varphi \mid [0,1]$. In particular supp $f \subset [-n, n+1]$.

Proof Let $X_n = \text{span}\{N_n(x-k)\}_{k=-n}^{0}$. From 3.19 we easily see that the functions $\{N_n(x-k)\}_{k=-n}^{0}$ are linearly independent, so $\dim X_n = n+1$. Let W_n denote the space of polynomials of degree at most n considered as functions on $[0,1]$. The restriction map $T : X_n \to W_n$ defined as $T(f) = f \mid [0,1]$ is clearly linear. We will show that this map is 1–1. Assume the contrary, that for some sequence $(a_k)_{k=-n}^{0}$, not all of them zero, we have $T(g) = 0$ where $g = \sum_{k=-n}^{0} a_k N_n(x-k)$. This means that

$$\sum_{k=-n}^{0} a_k N_n(x-k) \mid [0,1] = 0. \tag{3.26}$$

But it follows from 3.19 that supp $g \subset [-n, n+1]$, so from 3.26 we obtain supp $g \subset [-n, 0] \cup [1, n+1]$. From the definition of a spline (Definition 3.5) we see that the function g_1 defined as

$$g_1(x) = \begin{cases} g(x) & \text{for } x \leq 0 \\ 0 & \text{for } x > 0 \end{cases}$$

belongs to $\mathcal{S}^n(\mathbb{Z})$. From Remark 3.4 we infer that $g_1 = 0$. Thus supp $g \subset [1, n]$, and the same argument gives that $g = 0$. This shows that T is 1–1, and since $\dim X_n = \dim W_n = n+1$ we infer that T is an isomorphism from X_n onto W_n. This gives the theorem. □

Theorem 3.11 *Suppose that $f \in \mathcal{S}^n(\mathbb{Z})$. Then f can be written in a unique way as*

$$f(x) = \sum_{k \in \mathbb{Z}} a_k N_n(x-k). \tag{3.27}$$

Note that for each $x \in \mathbb{R}$ there are at most $(n+1)$ non-zero summands in the above sum, so there is no problem with convergence.

Proof From Theorem 3.10 we obtain that there exists a spline

$$\varphi = \sum_{k=-n}^{0} \beta_k N_n(x-k)$$

such that $\varphi \mid [0,1] = f \mid [0,1]$, so $f - \varphi \mid [0,1] = 0$. From Lemma 3.8 we see that there are constants c_1 and c_{-1} such that

$$f - \varphi - c_1 N_n(x-1) - c_{-1} N_n(x+1) \mid [-1,2] = 0.$$

Continuing in this way we get a sequence of coefficients $(c_i)_{i \in \mathbb{Z} \setminus 0}$ such that

$$f = \varphi + \sum_{i \in \mathbb{Z} \setminus 0} c_i N_n(x-i) = \sum_{k=-n}^{0} \beta_k N_n(x-k) + \sum_{i \in \mathbb{Z} \setminus 0} c_i N_n(x-i)$$

which is 3.27. If there is no uniqueness of representation, then there exists a sequence of numbers $(a_k)_{k \in \mathbb{Z}}$, not all of them zero, such that $\sum_{k \in \mathbb{Z}} a_k N_n(x-k) = 0$. Translating, we can assume $a_0 \neq 0$. Looking at the supports of $N_n(x-k)$ we infer from 3.19 that

$$\sum_{k \in \mathbb{Z}} a_k N_n(x-k) \mid [0,1] = \sum_{k=-n}^{0} a_k N_n(x-k) \mid [0,1] = 0.$$

From Theorem 3.10 we infer that $\sum_{k=-n}^{0} a_k N_n(x-k) = 0$, so in particular $a_0 = 0$. This contradiction shows that the representation 3.27 is unique. □

3.3.2 Spline wavelets

Now we want to construct wavelets which are splines. We will do this by showing that the spline spaces $S^n(2^{-j}\mathbb{Z})$ form a multiresolution analysis. Let us start with the following proposition.

Proposition 3.12 *For each $n = 0, 1, 2, \ldots$ the system $\{N_n(t-m)\}_{m \in \mathbb{Z}}$ is a Riesz system in $L_2(\mathbb{R})$.*

Proof Let us compute the Fourier transform of N_0. We have

$$\begin{aligned}\hat{N}_0(\xi) &= \frac{1}{\sqrt{2\pi}}\int_0^1 e^{-ix\xi}\,dx \\ &= \frac{1}{\sqrt{2\pi}}\int_0^1 \cos x\xi\,dx - \frac{i}{\sqrt{2\pi}}\int_0^1 \sin x\xi\,dx \\ &= \frac{1}{\sqrt{2\pi}\xi}\int_0^\xi \cos u\,du - \frac{i}{\sqrt{2\pi}\xi}\int_0^\xi \sin u\,du \qquad (3.28) \\ &= \frac{1}{\sqrt{2\pi}\xi}\sin\xi - \frac{i}{\sqrt{2\pi}\xi}(\cos\xi - 1).\end{aligned}$$

Directly from 3.28 one can obtain that for some A and each $n = 1, 2, \ldots$ we have

$$\sum_{l\in\mathbb{Z}} |\hat{N}_0(\xi + 2\pi l)|^{2n} \leq A. \qquad (3.29)$$

The other way to see this is to note that $\{N_0(t - m)\}_{m\in\mathbb{Z}}$ is clearly orthonormal, so Corollary 2.9 gives that $\sum_{l\in\mathbb{Z}} |\hat{N}_0(\xi + 2\pi l)|^2 = \frac{1}{2\pi}$ and 3.29 follows with $A = \frac{1}{2\pi}$. Either from 3.28 or from the fact that $N_0 \in L_1(\mathbb{R})$ we infer that $\hat{N}_0(\xi)$ is continuous (cf. A1.2–II). Note that from 3.28 we clearly see that $|\hat{N}_0(\xi)| > 0$ for $\xi \in (-2\pi, 2\pi)$. This immediately implies that for each $n = 1, 2, \ldots$ there exists a constant $c_n > 0$ such that

$$\sum_{l\in\mathbb{Z}} \left|\hat{N}_0(\xi + 2\pi l)\right|^{2n} \geq c_n. \qquad (3.30)$$

From Definition 3.6 and properties of the Fourier transform and convolution (cf. A1.2–XI) we obtain that $\widehat{N_n}(\xi) = (2\pi)^{n/2}\widehat{N_0}(\xi)^{(n+1)}$, so our Proposition follows from 3.29 and 3.30 using Proposition 2.8. □

By $S_2^n(2^j\mathbb{Z})$ we will denote the spaces $S^n(2^j\mathbb{Z}) \cap L_2(\mathbb{R})$. It follows from Theorem 3.11 and Proposition 3.12 that the $S_2^n(2^j\mathbb{Z})$ are closed subspaces of $L_2(\mathbb{R})$.

Theorem 3.13 *For each $n = 0, 1, 2, \ldots$ spaces $S_2^n(2^{-j}\mathbb{Z})$ with $j \in \mathbb{Z}$ form a multiresolution analysis.*

Proof We have already done all the pieces. Condition (vi) of Definition 2.2 follows from Proposition 3.12 and Proposition 2.11. The remaining conditions are easy to check directly, except perhaps (ii). But this follows easily from Proposition 2.15. □

Thus from the general construction summarized in Theorem 2.20 we know that there are wavelets which are splines of any given order. In the rest of this section we will show that they can be chosen to have exponential decay.

3.3.3 Exponential decay of spline wavelets

Actually the most natural spline wavelets do have exponential decay. We will show that if we start with $N_n(x)$ and construct the scaling function using formula 2.4 and then take the wavelet given by 2.47 we obtain a wavelet with exponential decay. To prove this in this subsection we will use some elementary facts about analytic functions.

Proposition 3.14 *Suppose $g(x)$ is a function on $\mathbb{R}$ satisfying $|g(x)| \leq Ce^{-\gamma|x|}$ for some constants C and $\gamma > 0$. Then there exists a 2π-periodic function $G(z)$ analytic in $|\Im z| < \gamma$ such that for all $\xi \in \mathbb{R}$*

$$G(\xi) = \sum_{l \in \mathbb{Z}} |\hat{g}(\xi + 2\pi l)|^2 .$$

Before we present the proof of this proposition we will prove two lemmas.

Lemma 3.15 *If $g \in L_2(\mathbb{R})$ then*

$$\sum_{l \in \mathbb{Z}} |\hat{g}(\xi - 2\pi l)|^2 = \frac{1}{2\pi} \sum_{k \in \mathbb{Z}} \left(\int_{-\infty}^{\infty} g(x-k)\overline{g(x)}\, dx \right) e^{ik\xi}.$$

Proof Let us compute the Fourier coefficients of the left hand side. For each $k \in \mathbb{Z}$ we have from A1.2–III and A1.2–IV

$$\begin{aligned} &\int_0^{2\pi} \left(\sum_{l \in \mathbb{Z}} |\hat{g}(\xi - 2\pi l)|^2 \right) e^{-ik\xi}\, d\xi \\ &\quad = \int_{-\infty}^{\infty} |\hat{g}(\xi)|^2 e^{-ik\xi}\, d\xi = \int_{-\infty}^{\infty} \hat{g}(\xi) e^{-ik\xi} \overline{\hat{g}(\xi)}\, d\xi \\ &\quad = \int_{-\infty}^{\infty} \widehat{(T_k g)}(\xi)\overline{\hat{g}(\xi)}\, d\xi = \int_{-\infty}^{\infty} g(x-k)\overline{g(x)}\, dx. \end{aligned}$$

Thus the equality follows. □

REMARK 3.5. This Lemma clearly provides another proof of Corollary 2.9.

Lemma 3.16 *Suppose that $|g(x)| \le Ce^{-\gamma|x|}$ for some constants C and $\gamma > 0$. Then*

$$\left|\int_{-\infty}^{\infty} g(x-k)\overline{g(x)}\,dx\right| \le C_\beta e^{-\beta k} \quad \textit{for each } 0 < \beta < \gamma \tag{3.31}$$

and conversely, if a sequence $(a_k)_{k\in\mathbb{Z}}$ satisfies $|a_k| \le Ce^{-\alpha|k|}$ then also

$$\left|\sum_{k\in\mathbb{Z}} a_k g(x-k)\right| \le C_\beta e^{-\beta|x|} \quad \textit{for each } \beta < \min(\alpha,\gamma). \tag{3.32}$$

Proof Assume $k \ge 0$. The argument for $k \le 0$ is exactly the same or we can use a change of variables. Since

$$|x| + |x-k| = \begin{cases} k & \text{if } 0 \le x \le k \\ 2x - k & \text{if } k \le x \\ k - 2x & \text{if } x < 0 \end{cases}$$

we obtain

$$\begin{aligned} &\left|\int_{-\infty}^{\infty} g(x-k)\overline{g(x)}\,dx\right| \\ \le\ & C^2 \int_{-\infty}^{\infty} e^{-\gamma|x|}\cdot e^{-\gamma|x-k|}\,dx = C^2\int_{-\infty}^{\infty} e^{-\gamma(|x|+|x-k|)}\,dx \\ =\ & C^2\left(\int_0^k e^{-\gamma k}\,dx + \int_{-\infty}^0 e^{-\gamma(k-2x)}\,dx + \int_k^{\infty} e^{-\gamma(2x-k)}\,dx\right) \\ \le\ & C_1 k e^{-\gamma k}. \end{aligned}$$

This gives 3.31. To show the other statement we observe that

$$\begin{aligned} \left|\sum_{k\in\mathbb{Z}} a_k g(x-k)\right| &\le C^2 \sum_{k\in\mathbb{Z}} e^{-\alpha|k|} e^{\gamma|x-k|} \\ &= \sum_{k\in\mathbb{Z}} e^{-\alpha|k|-\gamma|x-k|}. \end{aligned}$$

And this, by a similar argument, implies the claim. □

Proof of Proposition 3.14 From Lemma 3.15 and 3.31 we see that

$$\sum_{l\in\mathbb{Z}} |\hat{g}(\xi + 2\pi l)|^2 = \sum_{n\in\mathbb{Z}} a_n e^{in\xi}$$

where $|a_n| \leq C_\beta e^{-\beta|n|}$ for every $\beta < \gamma$. We define

$$G(z) = \sum_{n\in\mathbb{Z}} a_n e^{inz}. \tag{3.33}$$

Because $|a_n e^{inz}| \leq C_\beta e^{-\beta|n|} \cdot e^{-n|\Im z|}$ we easily see that the series in 3.33 converges almost uniformly in the strip $|\Im z| < \gamma$, so it defines an analytic function. Clearly $G(z)$ is 2π-periodic. □

Proposition 3.17 *There exists a function $\Phi \in S_2^n(\mathbb{Z})$ such that the system $\{\Phi(t-m)\}_{m\in\mathbb{Z}}$ is an orthonormal basis in the space $S_2^n(\mathbb{Z})$ and $|\Phi(x)| \leq Ce^{-\alpha|x|}$ for some C and $\alpha > 0$.*

Proof We define Φ by the condition

$$\hat{\Phi}(\xi) = \left(\sum_{l\in\mathbb{Z}} |\hat{N}_n(\xi - 2\pi l)|^2\right)^{-1/2} \hat{N}_n(\xi). \tag{3.34}$$

From Proposition 3.12 and Remark 2.4 we see that $\{\Phi(t-m)\}_{m\in\mathbb{Z}}$ is orthonormal. Since $N_n(x)$ has compact support, Proposition 3.14 shows that there exists a 2π-periodic function $G(z)$ analytic in $\mathbb{C}$, such that for $\xi \in \mathbb{R}$

$$G(\xi) = \sum_{l\in\mathbb{Z}} |\hat{N}_n(\xi - 2\pi l)|^2.$$

Since $\{N_n(t-m)\}_{m\in\mathbb{Z}}$ is a Riesz system (cf. Proposition 3.12) we infer from Proposition 2.10 that $G(\xi) > \delta$ for all $\xi \in \mathbb{R}$ and some $\delta > 0$. Because $G(z)$ is 2π-periodic, this implies that there exists an $\alpha > 0$ such that the analytic branch of $G(z)^{-1/2}$ exists in $|\Im z| < \alpha$. From 3.27 and Proposition 2.10 we know that

$$\Phi(x) = \sum_{k\in\mathbb{Z}} a_k N_n(x-k)$$

where

$$a_k = \int_{-\pi}^{\pi} G^{-1/2}(\xi) e^{-ik\xi}\, d\xi.$$

For any fixed β with $|\beta| < \alpha$ we apply the Cauchy integral theorem to the function $G^{-1/2}(z)e^{-ikz}$ in the rectangle with vertices

$$-\pi, \pi, \pi + i\beta, -\pi + i\beta.$$

Since this function is 2π-periodic (because G was), the integrals along the vertical sides cancel each other and we obtain

$$a_k = \int_{-\pi}^{\pi} G^{-1/2}(\xi + i\beta)e^{-ik(\xi+i\beta)}\,d\xi$$

so we have

$$|a_k| \le 2\pi e^{\beta k} \cdot \sup_{\xi} |G^{-1/2}(\xi + i\beta)| \le Ce^{\beta k}.$$

Since this holds for β both positive and negative we obtain $|a_k| \le C_\beta e^{-\beta|k|}$ for any $0 < \beta < \alpha$. So from Lemma 3.16 we obtain that $\Phi(x)$ decays exponentially. □

If we define a wavelet by the formula 2.47 we obtain

Theorem 3.18 *For each* $n = 0, 1, 2, \ldots$ *one can construct a wavelet* $\Psi \in S^n(\frac{1}{2}\mathbb{Z})$ *such that* $|\Psi(x)| \le Ce^{-\alpha|x|}$ *for some* $\alpha > 0$.

REMARK 3.6. Our construction does not give directly any estimate for α in Theorem 3.18. The value of α depends on the possibility of defining the analytical square root of G in the strip $|\Im z| < \alpha$. Note however that Lemma 3.15 and the fact that basic splines have compact support show that $G(z)$ is a polynomial in e^{ikz}. It is also not difficult to calculate $G(z)$ explicitly, at least for splines of low order.

REMARK 3.7. It follows from Proposition 3.1 that there is no wavelet $\Psi(x)$ with exponential decay such that $\Psi \in C^\infty$ and has all derivatives bounded. Here is a sketch of an argument. We note that the inequality $|\Psi(x)| \le Ce^{-\gamma|x|}$ implies that the integral $\frac{1}{\sqrt{2\pi}}\int_{-\infty}^{\infty} \Psi(x)e^{-ixz}\,dx$ makes sense for all $z \in \mathbb{C}$ with $|\Im z| < \gamma$ and defines an analytic function $F(z)$ which clearly on $\mathbb{R}$ coincides with $\hat{\Psi}(\xi)$, so is not identically 0. On the other hand Proposition 3.1 implies that $\hat{\Psi}^{(s)}(0) = 0$ for $s = 0, 1, \ldots$, so also $F^{(s)}(0) = 0$ for $s = 0, 1, \ldots$, which is possible only when $F \equiv 0$. This shows that both Meyer's wavelets and spline wavelets are almost at the boundary of the possible if we want *both* decay and smoothness.

3.3.4 Exponential decay of spline wavelets – another approach

In this subsection we will present a different argument for exponential decay of spline wavelets. As we present it, this approach uses almost no analytic functions and avoids almost entirely the use of the Fourier transform. It relies on the theory of operators on Hilbert space and on the theory of matrices. Naturally in this subsection we will use more of the theory of operators on Hilbert spaces than is explained in Appendix

A.1. Nonetheless all that we are going to use is well known and can be found in many textbooks, e.g. [98], [119].

In our discussion we will identify and denote by the same symbol an operator A on the Hilbert space $\ell_2(\mathbb{Z})$ and its matrix $A = (a_{jk})_{j,k\in\mathbb{Z}}$. Naturally, this identification is given by the formula

$$A\left((\alpha_k)_{k\in\mathbb{Z}}\right) = \left(\sum_{k\in\mathbb{Z}} a_{jk}\alpha_k\right)_{j\in\mathbb{Z}}. \tag{3.35}$$

For a fixed integer $n = 0, 1, 2, \ldots$ let us consider the Gram matrix of the system $\{N_n(t-m)\}_{m\in\mathbb{Z}}$, that is the matrix $A = (a_{jk})_{jk\in\mathbb{Z}}$ where

$$a_{jk} = \langle N_n(x-k), N_n(x-j)\rangle = \int_{-\infty}^{\infty} N_n(x-k)N_n(x-j)\,dx. \tag{3.36}$$

Proposition 3.19 *For each $n = 0, 1, 2, \ldots$ the matrix A defined by 3.36 defines a positive, self-adjoint invertible operator on $\ell_2(\mathbb{Z})$.*

Proof Let us consider an operator $S : \ell_2(\mathbb{Z}) \to S_2^n(\mathbb{Z})$ defined by $S(e_k) = N_n(x-k)$ where e_k is the unit vector in $\ell_2(\mathbb{Z})$, i.e. e_k is a sequence of zeros except that in the k-th place there is 1 (cf. A1.1–IV). Since $\{N_n(t-m)\}_{m\in\mathbb{Z}}$ is a Riesz basis in $S_2^n(\mathbb{Z})$ (see Proposition 3.12) we know that S is an isomorphism. Since

$$\langle S^*S(e_k), e_j\rangle = \langle S(e_k), S(e_j)\rangle = \langle N_n(x-k), N_n(x-j)\rangle = a_{jk}$$

we see that $A = S^*S$. This shows that A is a positive, self-adjoint isomorphism. □

This proposition and the spectral theorem allow us to form functions of an operator A. We will be interested in A^{-1} and $A^{-1/2}$. The reason for this is made clear by the following proposition.

Proposition 3.20 *Let $A = (a_{jk})_{j,k\in\mathbb{Z}}$ be given by 3.36 and suppose that A^{-1} is given by the matrix $(b_{jk})_{j,k\in\mathbb{Z}}$ and that $A^{-1/2}$ is given by the matrix $(c_{jk})_{j,k\in\mathbb{Z}}$. Let*

$$\psi_j = SA^{-1}(e_j) = \sum_{k\in\mathbb{Z}} b_{jk}N_n(x-k) \in S_2^n(\mathbb{Z}) \tag{3.37}$$

and let

$$\phi_j = SA^{-1/2}(e_j) = \sum_{k\in\mathbb{Z}} c_{jk}N_n(x-k) \in S_2^n(\mathbb{Z}). \tag{3.38}$$

Then $(\psi_j)_{j\in\mathbb{Z}}$ are biorthogonal functionals (for definition see Lemma 2.7) to $\{N_n(t-m)\}_{m\in\mathbb{Z}}$ and $(\phi_j)_{j\in\mathbb{Z}}$ is an orthonormal basis in $S_2^n(\mathbb{Z})$.

Proof Using the definitions and the above Proposition 3.19 we obtain for each $j,k\in\mathbb{Z}$

$$\begin{aligned}\langle\psi_j, N_n(x-k)\rangle &= \langle SA^{-1}(e_j), S(e_k)\rangle = \langle A^{-1}(e_j), S^*S(e_k)\rangle \\ &= \langle A^{-1}(e_j), A(e_k)\rangle = \langle e_j, e_k\rangle\end{aligned}$$

and

$$\begin{aligned}\langle\phi_j,\phi_k\rangle &= \left\langle SA^{-1/2}(e_j), SA^{-1/2}(e_k)\right\rangle \\ &= \left\langle S^*SA^{-1/2}(e_j), A^{-1/2}(e_k)\right\rangle \\ &= \left\langle A^{1/2}(e_j), A^{-1/2}(e_k)\right\rangle = \langle e_j, e_k\rangle\end{aligned}$$

so indeed $(\psi_j)_{j\in\mathbb{Z}}$ is a system of biorthogonal functionals to the system $\{N_n(t-m)\}_{m\in\mathbb{Z}}$ and $(\phi_j)_{j\in\mathbb{Z}}$ is an orthonormal system. It is an orthonormal basis in $S_2^n(\mathbb{Z})$ because $SA^{-1/2}$ is an isomorphism. □

In order to estimate the decay of the functions $(\psi_j)_{j\in\mathbb{Z}}$ and $(\phi_j)_{j\in\mathbb{Z}}$ we must study the decay of the elements of the matrices $(b_{jk})_{j,k\in\mathbb{Z}}$ and $(c_{jk})_{j,k\in\mathbb{Z}}$. To do this we will need the following easy approximation lemma.

Lemma 3.21 *Let $f(t)$ be either t^{-1} or $t^{-1/2}$. Given $0<a<b<\infty$ there exist constants C and q, $0<q<1$, such that for each natural number k there exists an algebraic polynomial p_k of degree at most k such that*

$$\sup_{t\in[a,b]}|f(t)-p_k(t)|\le Cq^k.$$

Proof The only property of the function $f(t)$ which we are going to use in this proof is that it can be expanded into a power series

$$f(t)=\sum_{s=0}^{\infty}a_s\left(t-\frac{a+b}{2}\right)^s \tag{3.39}$$

convergent for all t such that $\left|t-\frac{a+b}{2}\right|<\frac{a+b}{2}$. The elementary theory of power series shows that both t^{-1} and $t^{-1/2}$ have this property. Let η be any real number between $\frac{b-a}{2}$ and $\frac{b+a}{2}$. From our assumption about the convergence of the series 3.39 and from elementary properties of power series it follows that $\sup_s |a_s|\eta^s < \infty$. We put $q = \frac{b-a}{2\eta} < 1$. Given k we take $p_k(t) = \sum_{s=0}^{k} a_s(t - \frac{a+b}{2})^s$ and we have

$$\begin{aligned}
\sup_{t\in[a,b]} |f(t) - p_k(t)| &= \sup_{t\in[a,b]} \left| \sum_{s=k+1}^{\infty} a_s(t - \frac{a+b}{2})^s \right| \\
&\leq \sup_{t\in[a,b]} \sum_{s=k+1}^{\infty} |a_s| \left|t - \frac{a+b}{2}\right|^s \\
&\leq \sum_{s=k+1}^{\infty} |a_s| \left(\frac{b-a}{2}\right)^s \\
&\leq \sum_{s=k+1}^{\infty} |a_s|\eta^s q^s \leq Cq^k.
\end{aligned}$$

□

DEFINITION 3.22 *A matrix $A = (a_{jk})_{j,k\in\mathbb{Z}}$ is called banded if there exists a constant C such that $a_{jk} = 0$ whenever $|j-k| > C$.*

Theorem 3.23 *Suppose that A is a banded matrix which induces an invertible, positive, self-adjoint operator A on $\ell_2(\mathbb{Z})$. Let the operator A^{-1} be given by the matrix $(b_{jk})_{j,k\in\mathbb{Z}}$ and let the operator $A^{-1/2}$ be given by the matrix $(c_{jk})_{j,k\in\mathbb{Z}}$. There exist constants C and $\alpha > 0$ such that $|b_{jk}| \leq Ce^{-\alpha|j-k|}$ and $|c_{jk}| \leq Ce^{-\alpha|j-k|}$.*

Proof We will give the proof for $A^{-1/2}$. The proof for A^{-1} is exactly the same. Since A is a positive, invertible operator its spectrum $\sigma(A)$ is contained in some interval $[a,b]$ with $0 < a < b$. Let C be the constant appearing in Definition 3.22. Given $(j,s) \in \mathbb{Z}\times\mathbb{Z}$ let us fix the largest integer k such that $k \cdot C < |j-s|$. Let p_k be the polynomial given by Lemma 3.21 with $f(t) = t^{-1/2}$. Since p_k is a polynomial of degree at most k, we easily see that the matrix $p_k(A)$ has the entry 0 at the place (j,s). Thus from the spectral theorem and Lemma 3.21 we obtain

$$|c_{js}| \leq \left\|A^{-1/2} - p_k(A)\right\| \leq \sup_{t\in[a,b]} |t^{-1/2} - p_k(t)| \leq C_1 q^k. \tag{3.40}$$

This readily implies the theorem. □

DEFINITION 3.24 *A matrix* $A = (a_{jk})_{j,k\in\mathbb{Z}}$ *is called a Toeplitz matrix if for each* $r, j, k \in \mathbb{Z}$ *we have* $a_{j-r,k-r} = a_{jk}$.

Lemma 3.25 *A product of Toeplitz matrices is a Toeplitz matrix.*

Proof This is a straightforward and well known exercise in matrix multiplication. □

Now we have all the pieces needed to construct an alternative proof of Proposition 3.17.

Proof of Proposition 3.17 Let us take the matrix A given by 3.36 and form the orthonormal basis $(\phi_j)_{j\in\mathbb{Z}}$ in $\mathcal{S}_2^n(\mathbb{Z})$ given by 3.38. This is possible since Proposition 3.19 says that A is invertible and positive. Since $A^{-1/2}$ is the limit of a sequence of polynomials in A (cf. 3.40) and A is clearly a Toeplitz matrix, we infer from Lemma 3.25 that $A^{-1/2}$ is a Toeplitz matrix. This gives immediately that $\phi_j(x) = \phi_0(x-j)$ for every $j \in \mathbb{Z}$. The matrix A is clearly banded (cf. 3.19), so Theorem 3.23 and Lemma 3.16 show that there exist constants C and $\alpha > 0$ such that $|\phi_0(x)| \leq Ce^{-\alpha|x|}$. □

3.4 Unimodular wavelets

In this section we want to discuss a series of examples of wavelets $\Psi(x)$ such that $\sqrt{2\pi}\hat{\Psi}(\xi)$ is the characteristic function of a set. I do not think that these wavelets are of great practical importance, but they are interesting and provide examples of wavelets *not* associated with any multiresolution analysis.

To define our series of examples we start (naturally enough) with definitions of some sets. For $r = 1, 2, 3, \ldots$ let

$$I^r = \left[\frac{2^r}{2^{r+1}-1}\pi, \pi\right] \tag{3.41}$$

and

$$J^r = \left[2^r\pi, 2^r\pi + \frac{2^r}{2^{r+1}-1}\pi\right] = 2^r\left[\pi, \frac{2^{r+1}}{2^{r+1}-1}\pi\right]. \tag{3.42}$$

Next we define

$$K_r^+ = I^r \cup J^r \tag{3.43}$$

and

$$K_r = K_r^+ \cup -K_r^+. \tag{3.44}$$

We define the function $\Psi^r(x)$ by the condition

$$\hat{\Psi}^r(\xi) = \frac{1}{\sqrt{2\pi}} \mathbf{1}_{K_r}. \tag{3.45}$$

Since $\hat{\Psi}^r(\xi) \in L_2(\mathbb{R})$ and is compactly supported we infer that $\Psi^r \in L_2(\mathbb{R}) \cap C^\infty(\mathbb{R}) \cap C_0(\mathbb{R})$. Since $\hat{\Psi}^r(\xi)$ is not continuous we see that $\Psi^r \notin L_1(\mathbb{R})$. One easily checks (cf. calculation 3.28) that for any finite interval I we have $|\check{\mathbf{1}}_I(x)| \leq \frac{C}{1+|x|}$, so we can conclude that

$$|\Psi^r(x)| \leq \frac{C_r}{1+|x|}.$$

The following theorem is one of the main reasons for introducing the functions $\Psi^r(x)$.

Theorem 3.26 *The function $\Psi^r(x)$ for $r = 1, 2, 3, \ldots$ is a wavelet. This wavelet is associated with a multiresolution analysis only for $r = 1$.*

Naturally the proof of this theorem depends on properties of the sets K_r so we will isolate them in a lemma.

Lemma 3.27 *Let us fix $r = 1, 2, 3, \ldots$. The sets $2^j K_r$, $j \in \mathbb{Z}$ do not overlap and $\bigcup_{j\in\mathbb{Z}} 2^j K_r = \mathbb{R}$ up to a set of measure zero.*

Proof Consider the family of intervals

$$I_j =: 2^j I^r, \quad J_s =: 2^s J^r, \quad j, s \in \mathbb{Z}. \tag{3.46}$$

From 3.41 and 3.42 we see that

$$I_j \cup J_{j-r} = 2^j \left[\frac{2^r}{2^{r+1}-1}\pi, \frac{2^{r+1}}{2^{r+1}-1}\pi \right] =: L_j.$$

It is easy to see that $\bigcup_{j\in\mathbb{Z}} L_j = \frac{2^r}{2^{r+1}-1} \bigcup_{j\in\mathbb{Z}} 2^j[\pi, 2\pi] = (0, \infty)$ so

$$\begin{aligned} (0,\infty) &= \bigcup_{j\in\mathbb{Z}} I_j \cup J_{j-r} = \bigcup_{j\in\mathbb{Z}} I_j \cup J_j \\ &= \bigcup_{j\in\mathbb{Z}} 2^j (I^r \cup J^r) = \bigcup_{j\in\mathbb{Z}} 2^j K_r^+. \end{aligned}$$

This implies that $\bigcup_{j\in\mathbb{Z}} -2^j K_r^+ = (-\infty, 0)$, so $\bigcup_{j\in\mathbb{Z}} 2^j K_r = \mathbb{R} \setminus \{0\}$.

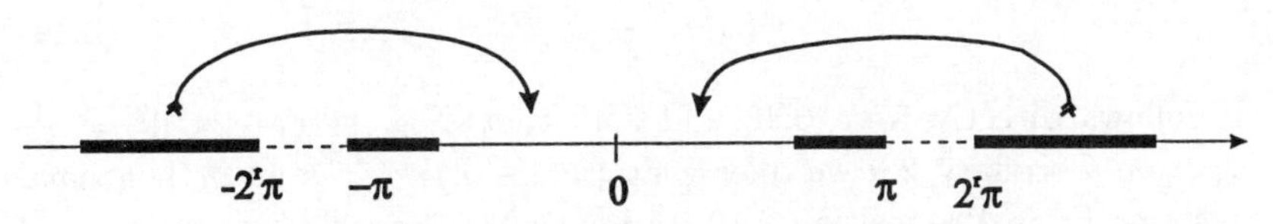

Fig. 3.1. The sets I_r and J_r.

Note also that L_j's do not overlap, so the family 3.46 does not overlap and this implies that $2^j K_r$, $j \in \mathbb{Z}$ do not overlap. □

Proof of Theorem 3.26 To show that $\Psi^r(x)$ is a wavelet we have to show that the system $\left\{2^{j/2}\Psi^r(2^j t - k)\right\}_{j\in\mathbb{Z},k\in\mathbb{Z}}$ is an orthonormal basis in $L_2(\mathbb{R})$. From Plancherel's theorem A1.2–IV we infer that this is the same as checking that the system $\left\{2^{j/2}\hat{\Psi}^r(2^j t - k)\right\}_{j\in\mathbb{Z},k\in\mathbb{Z}}$ is an orthonormal basis in $L_2(\mathbb{R})$. From A1.2–VII we see that

$$\widehat{\Psi}^r_{jk}(\xi) = 2^{-j/2} e^{ik2^{-j}\xi}\hat{\Psi}^r(2^{-j}\xi) = \frac{1}{\sqrt{2\pi}} 2^{-j/2} e^{ik2^{-j}\xi}\mathbf{1}_{2^j K_r}. \tag{3.47}$$

From Lemma 3.27 and 3.47 we see that it suffices to make sure that for each $j \in \mathbb{Z}$ the functions $\widehat{\Psi}^r_{jk}(\xi)$ with $k \in \mathbb{Z}$ form an orthonormal basis in $L_2(2^j K_r)$. Using an appropriate dilation (or a change of variables) we see that it suffices to check it for $j = 0$, i.e. we have to make sure that $\{\frac{1}{\sqrt{2\pi}} e^{ik\xi}\mathbf{1}_{K_r}\}_{k\in\mathbb{Z}}$ is an orthonormal basis in $L_2(K_r)$. Note that the set K_r is a disjoint union of four intervals. Shifting J_r by $-2^r\pi$ and $-J_r$ by $2^r\pi$ we can transform K_r onto $[-\pi, \pi]$ (see Figure 3.1). This transformation induces an isometry between $L_2(K_r)$ and $L_2[-\pi, \pi]$ which maps $e^{ik\xi}\mathbf{1}_{K_r}$ onto $e^{ik\xi} \mid [-\pi, \pi]$. This implies in particular that $\{\frac{1}{\sqrt{2\pi}} e^{ik\xi}\mathbf{1}_{K_r}\}_{k\in\mathbb{Z}}$ is an orthonormal basis in $L_2(K_r)$. Thus $\Psi^r(x)$ is a wavelet for $r = 1, 2, 3, \ldots$.

Now let us check that Ψ^1 is associated with a multiresolution analysis. The only candidate for such a multiresolution analysis is given by

$$V_r = \operatorname{span}\left\{\Psi^1_{jk}\right\}_{j<r,k\in\mathbb{Z}}.$$

These spaces clearly satisfy conditions (i)–(v) of Definition 2.2, so we have to exhibit a scaling function. It follows from our previous arguments that $f \in V_0$ if and only if

$$\operatorname{supp} \hat{f} \subset \bigcup_{j<0} 2^j K_1 = [-\tfrac{4}{3}\pi, -\pi] \cup [-\tfrac{2}{3}\pi, \tfrac{2}{3}\pi] \cup [\pi, \tfrac{4}{3}\pi] =: L. \tag{3.48}$$

This indicates that a natural candidate for a scaling function is

$$\Phi(x) = \frac{1}{\sqrt{2\pi}} \check{\mathbf{1}}_L(x). \tag{3.49}$$

It follows directly from 3.48 and 3.49 that $\sum_{l\in\mathbb{Z}} |\hat{\Phi}(\xi + 2\pi l)|^2 = \frac{1}{2\pi}$ so from Corollary 2.9 we infer that $\{\Phi(t-m)\}_{m\in\mathbb{Z}}$ is an orthonormal system. From Proposition 2.10 we see that $f \in \operatorname{span}\{\Phi(t-m)\}_{m\in\mathbb{Z}}$ if and only if

$$\hat{f}(\xi) = m(\xi) \cdot \mathbf{1}_L(\xi) \tag{3.50}$$

for some 2π-periodic function $m(\xi)$. Thus $m \mid [-\frac{4}{3}\pi, \pi]$ is the same as $m \mid [\frac{2}{3}\pi, \pi]$ and $m \mid [\pi, \frac{4}{3}\pi]$ is the same as $m \mid [-\pi, -\frac{2}{3}\pi]$. This implies that every $g \in L_2(L)$ is of the form 3.50. This shows that $\Phi(x)$ is a scaling function and so Ψ^1 is indeed associated with a multiresolution analysis.

Now suppose that for some $r = 2, 3, \ldots$ the wavelet $\Psi^r(x)$ is associated with a multiresolution analysis with a scaling function $\Phi(x)$. Then $\Psi^r(x/2)$ and $\Psi^r(x/4)$ belong to span $\{\Phi(t-m)\}_{m\in\mathbb{Z}}$. From Proposition 2.10(ii) using A1.2–VII we obtain

$$\hat{\Psi}^r(2\xi) = m_1(\xi)\hat{\Phi}(\xi) \tag{3.51}$$

and

$$\hat{\Psi}^r(4\xi) = m_2(\xi)\hat{\Phi}(\xi) \tag{3.52}$$

for some 2π-periodic functions $m_1(\xi)$ and $m_2(\xi)$. Since $\hat{\Psi}^r(2\xi) = 1$ for $\xi \in \frac{1}{2}K_r$ we must have $m_1(\xi) \neq 0$ on

$$\frac{1}{2}K_r \supset [2^{r-1}\pi, 2^{r-1}\pi + \frac{2^{r-1}}{2^{r+1}-1}\pi].$$

Because $m_1(\xi)$ is 2π-periodic and $r \geq 2$ we infer that $m_1(\xi) \neq 0$ on

$$[0, \frac{2^{r-1}}{2^{r+1}-1}\pi] \supset [0, \pi/4].$$

Using this we can solve 3.51 and 3.52 and obtain

$$\hat{\Psi}^r(4\xi) = \frac{m_2(\xi)}{m_1(\xi)}\hat{\Psi}^r(2\xi) \tag{3.53}$$

for $\xi \in [0, \pi/4]$. But 3.53 *cannot be true.* It follows readily from 3.41–3.45 that for ξ in the interval

$$\left[\frac{2^{r-2}}{2^{r+1}-1}\pi, \tfrac{1}{4}\pi\right] \subset [0, \pi/4]$$

$\hat{\Psi}^r(4\xi) \neq 0$ while $\hat{\Psi}^r(2\xi) = 0$. This shows that for $r = 2, 3, \ldots$ the wavelet Ψ^r is not associated with any multiresolution analysis. □

REMARK 3.8. Formulas 3.41–3.45 make perfect sense for $r = 0$ also. The wavelet we obtain in this case is one of Meyer's wavelets discussed in Proposition 3.3. This particular wavelet is the subject of Exercise 3.4.

Sources and comments

Philosophically Proposition 3.1 is well known. In the context of wavelets it can be found in Meyer [85], Battle [3] or Daubechies [24], page 153. Our proof is a version of the last two proofs. The Meyer's wavelets presented in Section 3.2 were constructed by Y. Meyer [81], and are the starting point of the whole theory. Their construction and properties are presented in detail in [83], [85]. The spline wavelets are also primary examples discussed in [85] and in [24]. They are also discussed in great detail in [15]. The introduction to the spline functions given in Section 3.3.1 is of course well known to specialists. I have tried to make the presentation here as easy and accessible as possible. The basic book about spline functions is Schumaker [100]. As already remarked, spline wavelets were first constructed by Strömberg [107] and next by Lemarie [66] and Battle [2]. All those constructions gave exponential decay. Strömberg's argument for exponential decay was an extension of arguments presented in Section 1.2. Fourier transform methods to show exponential decay like those presented in Section 3.3.3 were already used in [66], [2] and [83] and are also presented in detail in [24], where such wavelets are called Battle–Lemarie wavelets.

The method presented in Section 3.3.4 showing exponential decay by using inverses of banded matrices appeared in the work of Ciesielski and Domsta [19] on orthogonal spline systems on the interval. $[0,1]$. The fundamental Theorem 3.23 was proved by S. Demko [28] but the proof we present is due to S. Demko, W. F. Moss and Ph. Smith [29]. This technique was also used by P. G. Lemarie [67] to construct wavelet bases on Lie groups. The Fourier transform techniques are not available in this context.

The results of Section 3.4 are taken from the paper [51], where a detailed study of wavelets Ψ such that $\sqrt{2\pi}\hat{\Psi}(\xi) = \mathbf{1}_K(\xi)$ is presented. It should be pointed out that the wavelet Ψ^2 discussed in this section was discovered earlier by J. L. Journe and presented in Mallat [77] as an example of a wavelet not associated with a multiresolution analysis.

Let us also remark that wavelets Ψ with $\hat{\Psi}(\xi)$ compactly supported were studied by A. Bonami, F. Soria and G. Weiss in [7]. This covers both Meyer's wavelets and unimodular wavelets. The wavelets $\Psi^r(x)$ with $r = 2, 3, \ldots$ are not associated with any multiresolution analysis. On the other hand, all the other wavelets we construct are associated with a multiresolution analysis and are constructed using one. Actually good wavelets are always associated with a multiresolution analysis. This is made precise by the following theorem proved by P. G. Lemarie-Rieusset in [69] and by P. Auscher in [1].

Theorem 3.28 *If for some $\eta > 1$ a wavelet $\Psi(x)$ satisfies*

$$\int_{-\infty}^{\infty} |\Psi(x)|^2 (1 + |x|)^\eta \, dx < \infty$$

and

$$\int_{-\infty}^{\infty} |\hat{\Psi}(\xi)|^2 (1 + |\xi|)^\eta \, d\xi < \infty$$

then the wavelet $\Psi(x)$ is associated with a multiresolution analysis.

The following necessary and sufficient condition for a wavelet to be associated with a multiresolution analysis was given by G. Gripenberg [46].

Theorem 3.29 *A wavelet Ψ in $L_2(\mathbb{R})$ is associated with a multiresolution analysis if and only if*

$$\sum_{p=1}^{\infty} \sum_{k \in \mathbb{Z}} \left| \hat{\Psi}(2^p(\xi + 2\pi k)) \right|^2 > 0 \quad a.e.$$

In this case it is actually true that

$$\sum_{p=1}^{\infty} \sum_{k \in \mathbb{Z}} \left| \hat{\Psi}(2^p(\xi + 2\pi k)) \right|^2 = \frac{1}{2\pi} \quad a.e.$$

About the exercises. Exercises 3.6 and 3.7 can be found in [28] or [29]. Exercise 3.15 is taken from [51]. Exercise 3.14 is largely taken from [107].

Exercises

3.1 Give in detail the direct proof of Proposition 3.2 on page 49 as indicated in Remark 3.2 after the proof this proposition.

3.2 Construct a C^∞ function Θ satisfying 3.10–3.14 by the following procedure:

(a) Show that

$$f(x) = \begin{cases} e^{-1/x^2} & \text{if } x \geq 0 \\ 0 & \text{if } x \leq 0 \end{cases}$$

is a C^∞ function.

(b) Let $f_1(x) = f(x) \cdot f(1-x)$ where f is defined above. Show that the function

$$g(x) = \left(\int_{-\infty}^{\infty} f_1(t)\, dt \right)^{-1} \int_{-\infty}^{x} f_1(t)\, dt$$

satisfies

1. $0 \leq g(x) \leq 1$
2. $g(x) = 0$ for $x \leq 0$ and $g(x) = 1$ for $x \geq 1$
3. $g(x) + g(1-x) = 1$ for all $x \in \mathbb{R}$
4. g is a C^∞ function.

(c) Suppose that the function $g(x)$ satisfies 1–4 above. Show that the function

$$\Theta(\xi) = \frac{1}{\sqrt{2\pi}} \cos\left(\frac{\pi}{2} g\left(\frac{3}{2\pi}|\xi| - 1\right) \right)$$

is a C^∞ function satisfying 3.10–3.14.

3.3 Another construction of a C^∞ function $\Theta(\xi)$ satisfying 3.10–3.14.

(a) Using the previous exercise construct a C^∞ function f such that $\frac{1}{2} \leq f(x) \leq 1$ and $f(x) = 1$ for $x \leq 0$ and $f(x) = \frac{1}{2}$ for $x \geq \frac{1}{2}$.

(b) Show that Θ defined by the following conditions is a C^∞ function satisfying 3.10–3.14.

$$\Theta(x) = \begin{cases} \frac{1}{\sqrt{2\pi}} & \text{for } 0 \leq x \leq \frac{2}{3}\pi \\ \sqrt{\frac{1}{2\pi} f(\frac{3}{\pi}x - 2)} & \text{for } \frac{2}{3}\pi \leq x \leq \pi \\ \Theta(2\pi - x) & \text{for } \pi \leq x \leq \frac{4}{3}\pi \\ 0 & \text{for } x \geq \frac{4}{3}\pi \\ \Theta(-x) & \text{for } x \leq 0. \end{cases}$$

3.4 Take $\Theta(x) = \mathbf{1}_{[-\pi,\pi]}$. Check that it satisfies conditions 3.10–3.14. Compute explicitly the scaling function and the Meyer wavelet Ψ corresponding to this Θ. Show that $\Psi \in L_p(\mathbb{R})$ for all $1 < p \leq \infty$ but $\Psi \notin L_1(\mathbb{R})$.

3.5 Let Ψ be any Meyer wavelet from the Schwartz class S. Show without using Proposition 3.1 that for each $k = 0, 1, 2, \dots$ we have $\int_{-\infty}^{\infty} x^k \Psi(x)\, dx = 0$.

3.6 Let A be an invertible (not necessarily self-adjoint) operator on $\ell_2(\mathbb{Z})$ with a banded matrix. Show that $A^{-1} = (b_{jk})_{j,k\in\mathbb{Z}}$ satisfies $|b_{jk}| \leq Ce^{-\alpha|j-k|}$ for some constants C and $\alpha > 0$.

3.7 Prove Theorem 3.23 on page 64, replacing the assumption that the matrix A is banded by the assumption that it has exponential decay.

3.8 Let $a_0 < a_1 < \dots < a_n < a_{n+1}$ be real numbers. Define $\mathbf{1}_i = \mathbf{1}_{[a_i, a_{i+1}]}$ for $i = 0, 1, 2, \dots n$.

(a) Show that there exists a function $g = \sum_{i=0}^{n} a_i \mathbf{1}_i$ such that $\int_{-\infty}^{\infty} g(t) t^s\, dt = 0$ for $s = 0, 1, 2, \dots, n-1$.

(b) Define $B_1(x) = \int_{-\infty}^{x} g(t)\, dt$ and inductively $B_{s+1}(x) = \int_{-\infty}^{x} B_s(t)\, dt$. Show that $B_n(x)$ satisfies

- $B_n(x)$ has $(n-1)$ continuous derivatives
- $B_n \mid [a_i, a_{i+1}]$ is a polynomial of degree n for each $i = 0, 1, 2, \dots, n$.
- $\operatorname{supp} B_n(x) = [a_0, a_{n+1}]$
- $B_n(x) \geq 0$.

3.9 Let $(\psi_j)_{j\in\mathbb{Z}}$ be given by 3.37.

(a) Show that $\psi_j(x) = \psi_0(x - j)$.

(b) Show that the orthogonal projection P onto $\mathcal{S}_2^n(\mathbb{Z})$ can be expressed as $Pf(x) = \sum_{j\in\mathbb{Z}} \langle f, \psi_j \rangle N_n(x - j)$.

(c) Show that

$$\sum_{j\in\mathbb{Z}} N_n(x-j)\overline{\psi_j(t)} \equiv \sum_{j\in\mathbb{Z}} \phi_j(x)\overline{\phi_j(t)}$$

where the ϕ_j's are defined by 3.38.

(d) Write $Pf(x)$ as $\int_{-\infty}^{\infty} K(x, y) f(y)\, dy$ and show that

$$|K(x, y)| \leq Ce^{-\alpha|x-y|}$$

for some $\alpha > 0$.

3.10 For $g(x) = N_1(x)$ (B-spline of order 1) compute explicitly $G(z)$ given in Proposition 3.14. Show that $G(z)^{-1/2}$ exists in each strip $|\Im z| < \gamma$ with $\gamma < \ln(2+\sqrt{3})$ and this estimate for γ is best possible. Trace this through the construction and arguments leading to Theorem 3.18 and estimate the decay of a piecewise linear wavelet. Compare this with the decay of the Strömberg wavelet. For $g(x) = N_2(x)$ (the B-spline of order 2) compute explicitly $G(z)$ given in Proposition 3.14 and estimate in what strip $G(z)^{-1/2}$ exists.

3.11 Show without using the Fourier transform that $\{N_n(t-m)\}_{m\in\mathbb{Z}}$ is a Riesz sequence in $L_2(\mathbb{R})$.

3.12 Let V_0 be the following space: $f \in V_0$ if and only if $f \in L_2(\mathbb{R}) \cap C^1(\mathbb{R})$ and $f'(k) = 0$ for all $k \in \mathbb{Z}$ and for each $k \in \mathbb{Z}$, $f \mid [k, k+1]$ is a polynomial of degree at most 3. Show that there exists a multiresolution analysis generated by this V_0.

3.13 Show that when we apply 2.12 to $\Phi = N_1$ we obtain a scaling function in $S_2^1(\mathbb{Z})$ different from the one obtained in Exercise 1.8. Show also that applying 2.47 to the scaling function in $S_2^1(\mathbb{Z})$ obtained from N_1 using 2.12 we get a wavelet different from the Strömberg wavelet.

3.14 Show that there does not exist a function $g \in S_2^n(\mathbb{Z})$ with compact support and such that $\{g(t-m)\}_{m\in\mathbb{Z}}$ is an orthonormal sequence. Let $\phi \in S_2^n(\mathbb{Z})$ be such that $\|\phi\|_2 = 1$, $\phi \mid (1,\infty) = 0$ and ϕ is orthogonal to all $f \in S_2^n(\mathbb{Z})$ such that $f \mid (0,\infty) = 0$. Show that such a ϕ exists and is unique up to a unimodular constant. Show that $\{\phi(t-m)\}_{m\in\mathbb{Z}}$ is an orthonormal basis in $S_2^n(\mathbb{Z})$. Show that ϕ has exponential decay.

3.15 Let $0 < a < 2\pi$ and let $K(a) =: [2a-4\pi, a-2\pi] \cup [a, 2a]$. Show that $\Psi_a =: \mathcal{F}^{-1}(\frac{1}{\sqrt{2\pi}}\mathbf{1}_{K(a)})$ is a wavelet associated with a multiresolution analysis.

3.16 Fill in the details of the argument skeched in Remark 3.7 on page 61.

3.17 Using Theorem 3.29 show that the wavelets Ψ^r (cf. 3.45) for $r = 2, 3, \ldots$ are not associated with any multiresolution analysis.

3.18 Let $\alpha(\xi)$ be a real-valued, measurable function. Show that for each $r = 1, 2, 3, \ldots$ the function Ψ defined by

$$\hat{\Psi}(\xi) = \frac{1}{\sqrt{2\pi}} e^{i\alpha(\xi)} \mathbf{1}_{K_r},$$

where the set K_r is defined in 3.44, is a wavelet.

3.19 Formula for the natural piecewise linear wavelet in terms of the Fourier transform.

(a) Show that $\hat{\mathbf{1}}_{[-1/2,1/2]}(\xi) = 2(\sqrt{2\pi}\xi)^{-1}\sin\xi/2$ and that $\sum_{k\in\mathbb{Z}}(\xi+2k\pi)^{-2} = (4\sin^2\xi/2)^{-1}$.

(b) Let $\Lambda_0(t)$ be the simple tent defined in 1.20. Show that

$$\hat{\Lambda}_0(\xi) = \frac{1}{\sqrt{2\pi}}\frac{4\sin^2\xi/2}{\xi^2}.$$

(c) Show that

$$\sum_{k\in\mathbb{Z}}(\xi+2k\pi)^{-4} = (16\sin^4\xi/2)^{-1}(1-\tfrac{2}{3}\sin^2\xi/2).$$

(d) Show that the function $\Phi(x)$ such that

$$\hat{\Phi}(\xi) = \frac{1}{\sqrt{2\pi}}\frac{4\sin^2\xi/2}{\xi^2}\frac{1}{\sqrt{1-\frac{2}{3}\sin^2\xi/2}}$$

is a scaling function of multiresolution analysis $S_2^1(2^{-j}\mathbb{Z})$, $j\in\mathbb{Z}$. Note that this is a translation of the function Φ given by 3.34 with $n=1$

(e) Show that the corresponding function $m_\Phi(\xi)$ (see 2.16) is given by

$$m_\Phi(\xi) = \cos^2\xi/2\sqrt{\frac{1-\frac{2}{3}\sin^2\xi/2}{1-\frac{2}{3}\sin^2\xi}}.$$

(f) Show that this gives the Fourier transform of the natural piecewise linear wavelet $\Psi(x)$ as

$$\hat{\Psi}(\xi) = \frac{e^{-i\xi/2}}{\sqrt{2\pi}}\frac{16(\sin^2\xi/2)\left(1-\frac{2}{3}\cos^2\xi/2\right)^{1/2}}{\xi^2(1-\frac{2}{3}\sin^2\xi/2)^{1/2}(1-\frac{2}{3}\sin^2\xi/4)^{1/2}}.$$

This is the same wavelet that was discussed in Subsection 3.3.2 for $n=1$

4
Compactly supported wavelets

In this chapter we will discuss wavelets which have compact support. It is a surprising fact that such wavelets, other than Haar wavelet, exist and moreover can be chosen arbitrarily smooth.

4.1 General constructions

In this section we will present a general construction of a compactly supported wavelet. The question of the smoothness of these wavelets will be discussed in the next section. We will follow our usual approach and construct an appropriate multiresolution analysis. From formula 2.47 we see that it suffices to construct a multiresolution analysis with a compactly supported scaling function Φ. This is done in the following theorem.

Theorem 4.1 *Suppose that* $m(\xi) = \sum_{k=T}^{S} a_k e^{-ik\xi}$ *is a trigonometric polynomial such that*

$$|m(\xi)|^2 + |m(\xi+\pi)|^2 = 1 \text{ for all } \xi \in \mathbb{R} \tag{4.1}$$

$$m(0) = 1 \tag{4.2}$$

$$m(\xi) \neq 0 \text{ for } \xi \in [-\tfrac{\pi}{2}, \tfrac{\pi}{2}]. \tag{4.3}$$

Then the infinite product

$$\Theta(\xi) = \prod_{j=1}^{\infty} m(2^{-j}\xi) \tag{4.4}$$

converges almost uniformly. The function $\Theta(\xi)$ *is thus continuous. Moreover it is in* $L_2(\mathbb{R})$*. The function* Φ *given by* $\hat{\Phi} = \frac{1}{\sqrt{2\pi}}\Theta$ *has* *the support contained in* $[T, S]$ *and is a scaling function of a multiresolution analysis. In particular it has orthonormal translates. The formula*

$$\Psi(x) = 2\sum_{k=T}^{S} \overline{a_k}(-1)^k \Phi(2x+k+1) \tag{4.5}$$

gives a compactly supported wavelet with supp $\Psi \subset [\frac{T-S-1}{2}, \frac{S-T-1}{2}]$.

REMARK 4.1. The formula for $m(\xi)$ with '$-$' in the exponents is used to conform with the notation appearing in the scaling equation 2.17

REMARK 4.2. We know from Remark 4.6 below that $S-T$ is odd. This implies that supp Ψ is an interval with integer endpoints of the same length as supp Φ.

REMARK 4.3. 1. Hidden in this theorem is another approach to the construction of a scaling function of multiresolution analysis, already alluded to in Section 2.1. We start from the scaling equation 2.14. Suppose that we are given a sequence of numbers $(a_k)_{k\in\mathbb{Z}}$ and we want to find the function $\Phi(x)$ such that

$$\Phi(x) = \sum_{k\in\mathbb{Z}} a_k \Phi(2x-k).$$

We know that this is equivalent to

$$\hat{\Phi}(\xi) = m(\xi/2)\hat{\Phi}(\xi/2) \tag{4.6}$$

where $m(\xi) = \sum_{k\in\mathbb{Z}} a_k e^{-ik\xi}$. Substituting 4.6 into itself (and assuming that $\hat{\Phi}$ is continuous at 0 and $\hat{\Phi}(0) = 1$), we obtain $\hat{\Phi}(\xi) = \prod_{j=1}^{\infty} m(2^{-j}\xi)$. This shows that 4.4 is a very natural thing to analyze.

2. Observe that we have to assume 4.1 and 4.2 because if Φ is a scaling function then 4.6 will hold, so taking $\xi = 0$ we get 4.2. From Lemma 2.12 we see that 4.1 must also hold.

3. It is clear from the scaling equation 4.6 that different polynomials $m(\xi)$ give different scaling functions. Note however that if we take $m_1(\xi) = e^{ir\xi}m(\xi)$ then we get

$$\begin{aligned} \Theta_1(\xi) &=: \prod_{j=1}^{\infty} m_1(2^{-j}\xi) \\ &= \prod_{j=1}^{\infty} e^{ir\xi 2^{-j}}\Theta(\xi) = e^{ir\xi}\Theta(\xi) \end{aligned}$$

which gives $\check{\Theta}_1(x) = \check{\Theta}(x+r)$. This shows that the scaling function obtained with $m_1(\xi)$ is a translate of the scaling function obtained with $m(\xi)$. Thus essentially we get the same scaling function and the same wavelet given by 4.5. Note also that it follows from Lemma 4.4 on page 81 that unless $m_1(\xi) = e^{ir\xi}m(\xi)$, we get different multiresolution analyses.

Lemma 4.2 *If $m(\xi)$ is a trigonometric polynomial and 4.1 and 4.2 hold, then the product 4.4 converges almost uniformly and*

$$\int_{-\infty}^{\infty} \Big|\prod_{j=1}^{\infty} m(2^{-j}\xi)\Big|^2 \, d\xi \leq 2\pi. \tag{4.7}$$

In particular $\Theta(\xi)$ is continuous and $\Theta(0) = 1$. If we assume also 4.3 then for each $k \in \mathbb{Z}$

$$\int_{-\infty}^{\infty} \Big|\prod_{j=1}^{\infty} m(2^{-j}\xi)\Big|^2 e^{-2\pi i k\xi} \, d\xi = \begin{cases} 2\pi & \textit{if } k = 0 \\ 0 & \textit{otherwise.} \end{cases} \tag{4.8}$$

Proof Since $m(\xi)$ is a trigonometric polynomial, it satisfies the Lipschitz condition, i.e. there exists a constant C such that $|m(\xi) - 1| \leq C|\xi|$, so in particular for each $\xi \in \mathbb{R}$ we have $|m(2^{-j}\xi) - 1| \leq C2^{-j}|\xi|$. This shows that the product 4.4 converges almost uniformly on $\mathbb{R}$ and thus Θ is continuous. Now let us introduce some notation to be used in the proof of this Lemma:

$$\begin{aligned} \Pi_N(\xi) &=: \prod_{j=1}^{N} m(2^{-j}\xi) \\ g_N(\xi) &=: \Pi_N(\xi)\mathbf{1}_{[-2^N\pi, 2^N\pi]}(\xi) = \Pi_N(\xi)\mathbf{1}_{[-\pi/2,\pi/2]}(2^{-N-1}\xi) \\ I_N^k &=: \int_{-2^N\pi}^{2^N\pi} |\Pi_N(\xi)|^2 e^{-2\pi i k\xi} \, d\xi = \int_{-\infty}^{\infty} |g_N(\xi)|^2 e^{-2\pi i k\xi} \, d\xi. \end{aligned}$$

Observe that $\Pi_N(\xi)$ is a $2^N \cdot 2\pi$-periodic function for each $N = 1, 2, \ldots$. Using this and 4.1 we get

$$\begin{aligned} I_N^k &= \int_0^{2^{N+1}\pi} |\Pi_N(\xi)|^2 e^{-2\pi i k\xi} \, d\xi \\ &= \int_0^{2^N\pi} |\Pi_N(\xi)|^2 e^{-2\pi i k\xi} \, d\xi + \int_{2^N\pi}^{2^{N+1}\pi} |\Pi_N(\xi)|^2 e^{-2\pi i k\xi} \, d\xi \\ &= \int_0^{2^N\pi} |\Pi_{N-1}(\xi)|^2 e^{-2\pi i k\xi} |m(2^{-N}\xi)|^2 \, d\xi \\ &\quad + \int_0^{2^N\pi} |\Pi_{N-1}(\xi)|^2 e^{-2\pi i k\xi} |m(2^{-N}\xi + \pi)|^2 \, d\xi \\ &= \int_0^{2^N\pi} |\Pi_{N-1}(\xi)|^2 e^{-2\pi i k\xi} (|m(2^{-N}\xi)|^2 + |m(2^{-N}\xi + \pi)|^2) \, d\xi \\ &= I_{N-1}^k. \end{aligned}$$

Continuing in this way we get

$$\begin{aligned} I_N^k &= I_1^k = \int_{-2\pi}^{2\pi} |m(\xi/2)|^2 e^{-2\pi i k\xi}\, d\xi \\ &= 2\int_{-\pi}^{0} (|m(\xi)|^2 + |m(\xi+\pi)|^2) e^{-4\pi i k\xi}\, d\xi \qquad (4.9) \\ &= \begin{cases} 2\pi & \text{if } k = 0 \\ 0 & \text{otherwise.} \end{cases} \end{aligned}$$

Since 4.1 clearly implies $|m(\xi)| \leq 1$ for all $\xi \in \mathbb{R}$ we get for each natural number N

$$\int_{-2^N\pi}^{2^N\pi} \left| \prod_{j=1}^{\infty} m(2^{-j}\xi) \right|^2 d\xi \leq I_N^0 \leq 2\pi$$

so 4.7 holds. To obtain 4.8 note that because $m(\xi)$ is continuous 4.3 implies that there exists a $c > 0$ such that $|m(\xi)| > c$ for $\xi \in [-\pi/2, \pi/2]$. Since $\prod_{j=1}^{\infty} m(2^{-j}\xi)$ converges uniformly on $[-\pi, \pi]$ there exists an integer M such that $\left|\prod_{j=M}^{\infty} m(2^{-j}\xi)\right| > \frac{1}{2}$ for $\xi \in [-\pi, \pi]$. This implies that for $\xi \in [-\pi, \pi]$

$$|\Theta(\xi)| = \prod_{j=1}^{M-1} |m(2^{-j}\xi)| \cdot \prod_{j=M}^{\infty} |m(2^{-j}\xi)| \geq c^{M-1}\tfrac{1}{2} = c' > 0.$$

Writing $\Theta(\xi) = \Pi_N(\xi) \cdot \Theta(2^{-N}\xi)$ we see that for $\xi \in [-2^N\pi, 2^N\pi]$ we have $|\Pi_N(\xi)| \leq \frac{1}{c'}|\Theta(\xi)|$, which implies

$$|g_N(\xi)| \leq \frac{1}{c'}|\Theta(\xi)| \quad \text{for} \quad \xi \in \mathbb{R}. \qquad (4.10)$$

It is clear that $g_N(\xi) \to \Theta(\xi)$ pointwise, so using the Lebesgue dominated convergence theorem (which is valid because of 4.7 and 4.10) we get

$$\begin{aligned} \int_{-\infty}^{\infty} |\Theta(\xi)|^2 e^{-2\pi i k\xi}\, d\xi &= \int_{-\infty}^{\infty} \lim_{N\to\infty} |g_N(\xi)|^2 e^{-2\pi i k\xi}\, d\xi \\ &= \lim_{N\to\infty} \int_{-\infty}^{\infty} |g_N(\xi)|^2 e^{-2\pi i k\xi}\, d\xi \\ &= \lim_{N\to\infty} I_N^k. \end{aligned}$$

Comparing this with 4.9 we get 4.8. □

Lemma 4.3 *Let* $m(\xi) = \sum_{k=T}^{S} a_k e^{-ik\xi}$ *be a trigonometric polynomial which satisfies 4.1 and 4.2 and let* $\Theta(\xi) = \prod_{j=1}^{\infty} m(2^{-j}\xi)$. *Then* supp $\check{\Theta} \subset [T, S]$. *Moreover if all the coefficients a_k are real then $\check{\Theta}$ is also real.*

Proof In this proof we will be using the Fourier transform of measures (although in the most simple case). For a measure μ of bounded variation on $\mathbb{R}$ we have (cf. A1.2–XI)

$$\mathcal{F}_1(\mu)(\xi) = \sqrt{2\pi}\hat{\mu}(\xi) = \int_{-\infty}^{\infty} e^{-ix\xi}\, d\mu(x).$$

As we know, this normalization gives

$$\mathcal{F}_1(\mu * \nu) = \mathcal{F}_1(\mu) \cdot \mathcal{F}_1(\nu). \tag{4.11}$$

Let us define measures $\mu_j = \sum_{k=T}^{S} a_k \delta(k2^{-j})$ where $\delta(a)$ is the Dirac delta measure concentrated at the point a. A straightforward calculation shows that $\mathcal{F}_1(\mu_j) = m(2^{-j}\xi)$. Thus 4.11 gives that for each natural number N we get

$$\mathcal{F}(\mu_1 * \mu_2 * \ldots * \mu_N) = \prod_{j=1}^{N} m(2^{-j}\xi).$$

On the other hand a direct calculation using the linearity of the convolution gives

$$\begin{aligned} &\mu_1 * \mu_2 * \ldots * \mu_N \\ &\quad = \sum_{k_1,\ldots,k_N \,:\, T \le k_i \le S} A_{k_1,k_2,\ldots,k_N} \delta(2^{-1}k_1) * \ldots * \delta(2^{-N}k_N) \qquad (4.12) \\ &\quad = \sum_{k_1,\ldots,k_N \,:\, T \le k_i \le S} A_{k_1,k_2,\ldots,k_N} \delta(2^{-1}k_1 + 2^{-2}k_2 + \ldots + 2^{-N}k_N). \end{aligned}$$

This implies in particular that

$$\operatorname{supp} \mu_1 * \mu_2 * \ldots * \mu_N \subset [T, S] - 2^{-N-1}.$$

Now let us take f, a C^∞ function with compact support such that supp $f \cap [T, S] = \emptyset$. Since $\hat{f} \in S$, from Plancherel's theorem A1.2–IV, Lemma 4.2 and the above observations we have

$$\begin{aligned} \int_{-\infty}^{\infty} \check{\Theta}(t)\overline{f(t)}\, dt &= \int_{-\infty}^{\infty} \Theta(\xi)\overline{\hat{f}(\xi)}\, d\xi \\ &= \lim_{N\to\infty} \int_{-\infty}^{\infty} \prod_{j=1}^{N} m(2^{-j}\xi)\overline{\hat{f}(\xi)}\, d\xi \qquad (4.13) \end{aligned}$$

$$= \lim_{N\to\infty}\int_{-\infty}^{\infty}\mathcal{F}_1(\mu_1 * \mu_2 * \ldots * \mu_N)(\xi)\overline{\hat{f}(\xi)}\,d\xi$$

$$= \lim_{N\to\infty}\frac{1}{\sqrt{2\pi}}\int_{-\infty}^{\infty}\overline{f(x)}d(\mu_1 * \mu_2 * \ldots * \mu_N)(x)$$

$$= 0.$$

Since f was arbitrary we infer that supp $\check{\Theta} \subset [T, S]$. If we assume that all a_k's are real then from the calculation leading to 4.12 we see that the measure $\mu_1 * \mu_2 * \ldots \mu_N$ is real. If we take f a real-valued C^∞ function with compact support, then calculation 4.13 shows that $\int_{-\infty}^{\infty}\check{\Theta}(t)f(t)\,dt$ is real. This implies that $\check{\Theta}(t)$ is real. □

REMARK 4.4. For a reader familiar with some functional analysis it should be clear that $\check{\Theta}$ is the weak* limit of the measures $\mu_1 * \mu_2 * \ldots * \mu_N$.

Proof of Theorem 4.1 We will use Theorem 2.13 on page 28. We know from Lemma 4.2 that $\hat{\Phi}(\xi)$ is continuous and $\hat{\Phi}(0) \neq 0$. It is clear from 4.4 and the definition of Φ that $\hat{\Phi}(\xi) = m(\xi/2)\cdot\hat{\Phi}(\xi/2)$ so conditions (ii) and (iii) of Theorem 2.13 hold. Lemma 4.3 shows that supp $\Phi \subset [T, S]$. This means that the only thing left to show is that $\{\Phi(t-m)\}_{m\in\mathbb{Z}}$ is an orthonormal system. But writing the left hand side of 4.8 as $\int_0^{2\pi}\sum_{l\in\mathbb{Z}}|\Theta(\xi+2\pi l)|^2 e^{-2\pi k\xi}\,d\xi$ we infer from properties of Fourier coefficients (cf. A1.2–XII) that $\sum_{l\in\mathbb{Z}}|\Theta(\xi+2\pi l)|^2 = 1$. From Corollary 2.9 we infer that $\{\Phi(t-m)\}_{m\in\mathbb{Z}}$ is orthonormal. Formula 4.5 is just formula 2.47 adapted to the present situation. The evaluation of supp Ψ follows directly from 4.5 and the evaluation of supp Φ. □

If we look at Theorem 4.1 it is clear that there is no problem with finding polynomials $m(\xi)$ satisfying 4.2 and 4.3. It may not be clear that we can find any polynomials satisfying 4.1. We will produce examples of polynomials satisfying 4.1–4.3 in the next section and in the Exercises. Let us note here one easy example $m(\xi) = \frac{1}{2}(1 + e^{i\xi})$, which however is not very exciting since it gives our old friend the Haar wavelet (see Exercise 4.1). We want to conclude this section with some general remarks about compactly supported wavelets (assuming they do exist). First we will show that for a given multiresolution analysis there is at most one (up to translation) scaling function and wavelet with compact support. This immediately follows from the following lemma.

Lemma 4.4 *If Φ_1 and Φ_2 are compactly supported functions such that $\{\Phi_1(t-m)\}_{m\in\mathbb{Z}}$ and $\{\Phi_2(t-m)\}_{m\in\mathbb{Z}}$ are orthonormal bases in the same subspace of $L_2(\mathbb{R})$, then $\Phi_1(x) = \Phi_2(x-k)$ for some $k \in \mathbb{Z}$.*

Proof Since $\{\Phi_2(t-m)\}_{m\in\mathbb{Z}}$ is orthonormal and both Φ_1 and Φ_2 have compact supports we can write

$$\Phi_1(x) = \sum_{j\in A} a_j \Phi_2(x-j)$$

for some finite subset $A \subset \mathbb{Z}$. Taking the Fourier transform we get

$$\hat{\Phi}_1(\xi) = m(\xi)\hat{\Phi}_2(\xi)$$

for some trigonometric polynomial $m(\xi)$. Using orthonormality and Corollary 2.9 we get

$$\begin{aligned} \frac{1}{2\pi} &= \sum_{l\in\mathbb{Z}} |\hat{\Phi}_1(\xi + 2\pi l)|^2 \\ &= |m(\xi)|^2 \sum_{l\in\mathbb{Z}} |\hat{\Phi}_2(\xi + 2\pi l)|^2 = \frac{1}{2\pi}|m(\xi)|^2. \end{aligned}$$

Thus $|m(\xi)|^2 = m(\xi)\cdot\overline{m(\xi)} = 1$. From this we easily get that $m(\xi) = e^{ik\xi}$ for some $k \in \mathbb{Z}$. This implies $\Phi_1(x) = \Phi_2(x-k)$. □

REMARK 4.5. Note that it follows from Exercise 3.13 that the above lemma is false even for functions with exponential decay. This is connected with the fact that we can have nontrivial rational function $f(z)$ analytic in $|z| < 1 + \delta$ (for some $\delta > 0$) such that for $|z| = 1$ we have $|f(z)| = 1$. A simple example of such a function is $f(z) = (z-a)/(1-\bar{a}z)$ with $|a| < 1$.

The following theorem contains some information about the support of a compactly supported scaling function. To formulate this theorem we will use the following notation: $\operatorname{Supp} f$ will denote the smallest closed *interval* containing the support of f.

Theorem 4.5 *Let $\Phi(x)$ be a scaling function of a multiresolution analysis. If Φ has compact support, then* $\operatorname{Supp}\Phi = [B, B+C]$ *for some $B, C \in \mathbb{Z}$ and C odd.*

Proof Let us write $\operatorname{Supp}\Phi = [a, b]$. Let us look at the scaling equation $\Phi(x/2) = \sum_{n\in\mathbb{Z}} a_n \Phi(x-n)$. Since $\{\Phi(t-m)\}_{m\in\mathbb{Z}}$ is orthonormal and

Φ is compactly supported, this is a finite sum, so

$$\Phi(x/2) = \sum_{n=B}^{E} a_n \Phi(x-n) \tag{4.14}$$

with $B, E \in \mathbb{Z}$ and $a_B \neq 0$ and $a_E \neq 0$. Since $\operatorname{Supp}\Phi(x-n) = [a+n, b+n]$ we infer from 4.14 that $\operatorname{Supp}\Phi(x/2) \subset [a+B, b+E]$ and also that

$$\begin{aligned} \Phi(x/2) \mid [a+B, a+B+1] &= a_B\Phi(x-B) \mid [a+B, a+B+1] \\ \Phi(x/2) \mid [b+E-1, b+E] &= a_E\Phi(x-E) \mid [b+E-1, b+E]. \end{aligned}$$

This implies that

$$\operatorname{Supp}\Phi(x/2) = [a+B, b+E]. \tag{4.15}$$

On the other hand, we see directly that $\operatorname{Supp}\Phi(x/2) = [2a, 2b]$. Comparing this with 4.15 we get $a = B$ and $b = E$. In particular $\operatorname{Supp}\Phi = [B, E]$ with $B, E \in \mathbb{Z}$. Taking an appropriate translation we can assume $B = 0$. We still need to show that E (in general $E - B$) is odd. After this translation equation 4.14 becomes

$$\Phi(x/2) = \sum_{n=0}^{E} a_n \Phi(x-n) \tag{4.16}$$

with $a_0 \neq 0$ and $a_E \neq 0$. This gives that for any integer l

$$\Phi\left(\frac{x-2l}{2}\right) = \sum_{n=0}^{E} a_n \Phi(x-2l-n) = \sum_{n=2l}^{E+2l} a_{n-2l}\Phi(x-n). \tag{4.17}$$

An easy change of variables yields that for $l \neq 0$ the function $\Phi(x/2)$ is orthogonal to $\Phi(\frac{x-2l}{2})$. Since $\{\Phi(t-m)\}_{m\in\mathbb{Z}}$ is orthonormal, this together with 4.16 and 4.17 gives

$$\sum_n a_n \bar{a}_{n-2l} = 0. \tag{4.18}$$

Now suppose that E is even and take $2l = E$ in 4.18. With this choice of l equation 4.18 becomes $a_E \cdot a_0 = 0$, which is impossible. □

REMARK 4.6. Note that the above argument also shows that a polynomial $m(\xi)$ satisfying 4.1–4.3 must have the form $m(\xi) = \sum_{k=B}^{B+C} a_k e^{-ik\xi}$ with C odd. This is a reformulation of 4.16 just as 2.16 and 2.17 are reformulations of 2.13. Clearly the Supp of the corresponding scaling function equals $[B, B+C]$.

4.2 Smooth wavelets

If we look at Theorem 4.1 it is clear that it is very easy to find polynomials $m(\xi)$ satisfying 4.2 and 4.3. It may not be clear at first glance that we can find polynomials satisfying 4.1. This is possible. Otherwise we would not bother with compactly supported wavelets. A very useful tool to produce trigonometric polynomials satisfying 4.1 is the following classical lemma.

Lemma 4.6 (Riesz) *If $g(\xi) = \sum_{k=-T}^{T} \gamma_k e^{ik\xi}$ is a non-negative trigonometric polynomial with the γ_k's real, then there exists a polynomial $m(\xi) = \sum_{k=0}^{T} a_k e^{ik\xi}$ with all a_k's real and such that $|m(\xi)|^2 = g(\xi)$ for all $\xi \in \mathbb{R}$.*

Proof Since $g(\xi)$ is real and the γ_k's are also real we infer that

$$\gamma_k = \gamma_{-k} = \bar{\gamma}_k \quad \text{for all } k \in \mathbb{Z}. \tag{4.19}$$

To fix our notation let us assume also that $\gamma_T \neq 0$. We consider a rational function on $\mathbb{C}$,

$$G(z) = \sum_{k=-T}^{T} \gamma_k z^k = z^{-T}(\gamma_{-T} + \gamma_{-T+1} z + \ldots + \gamma_T z^{2T}). \tag{4.20}$$

Using the fundamental theorem of algebra we can factor $G(z)$ as

$$G(z) = z^{-T} \prod_{j=1}^{2T} (z - c_j).$$

Since we have $\gamma_T \neq 0$ we infer from 4.19 that $\gamma_{-T} \neq 0$, so zero is not a root of the polynomial $\gamma_{-T} + \gamma_{-T+1} z + \ldots + \gamma_T z^{2T}$. Thus we see that $c_j \neq 0$ for $j = 1, 2, \ldots, 2T$. Note that 4.19 and 4.20 imply that for any $z \in \mathbb{C}$, $z \neq 0$

$$G(\bar{z}) = \overline{G(z)} \tag{4.21}$$

and

$$G(\bar{z}^{-1}) = \overline{G(z)}. \tag{4.22}$$

This implies that if z_0 is a zero of the function $G(z)$, i.e. if $z_0 = c_j$ for some j, then $\bar{z}_0$, ${z_0}^{-1}$ and $\bar{z}_0^{-1}$ are also zeros of $G(z)$. Note also that if $|z_0| = 1$ and $G(z_0) = 0$, then z_0 is a zero of $G(z)$ of even multiplicity. To see this write $z_0 = e^{i\xi_0}$ and note that $g(\xi_0) = 0$. Since $g(\xi) \geq 0$ this implies that $g(\xi)$ has a local minimum at ξ_0. By a well known calculus criterion for local minima, this implies that the polynomial $g(\xi)$ has a zero of even multiplicity at ξ_0. Since $g(\xi) = G(e^{i\xi})$ we see that $G(z)$ has zero of even multiplicity at $z_0 = e^{i\xi_0}$. Thus we can write

$$\prod_{j=1}^{2T}(z - c_j) = \prod_s g_s(z) \tag{4.23}$$

where either

$$g_s(z) = (z-c)(z-\bar{c})(z-c^{-1})(z-\bar{c}^{-1}) \text{ for some } c \in \mathbb{C}\setminus\mathbb{R} \tag{4.24}$$

or

$$g_s(z) = (z-c)(z-c^{-1}) \quad \text{for some } c \in \mathbb{R}. \tag{4.25}$$

When $|z| = 1$ and $c \in \mathbb{C}\setminus\{0\}$ we can write

$$\begin{aligned} \left|(z-c)(z-\frac{1}{\bar{c}})\right| &= \left|(z-c)\frac{z\bar{c}-1}{\bar{c}}\right| \\ &= \frac{1}{|c|}|(z-c)(\bar{c}-\bar{z})| \\ &= \frac{1}{|c|}|z-c|^2. \end{aligned} \tag{4.26}$$

From 4.26 we see that if $g_s(z)$ is given by 4.24 then for $|z| = 1$ we can write

$$|g_s(z)| = \left|\frac{1}{|c|}(z-c)(z-\bar{c})\right|^2 \tag{4.27}$$

and the polynomial $(z-c)(z-\bar{c})$ has real coefficients. From 4.26 we see that if $g_s(z)$ is given by 4.25 then for $|z| = 1$ we can write

$$|g_s(z)| = \left|\frac{1}{\sqrt{|c|}}(z-c)\right|^2. \tag{4.28}$$

From 4.24–4.28 we see that for each g_s which appears in 4.23 there exists a polynomial $p_s(z)$ (either linear or quadratic) with real coefficients, such that for $|z| = 1$ we have $|g_s(z)| = |p_s(z)|^2$. Since

$$g(\xi) = G(e^{i\xi}) = \prod_s |g_s(e^{i\xi})|$$

we see that we can take $m(\xi) = \prod_s p_s(e^{i\xi})$. □

REMARK 4.7. 1. Observe that in general the polynomial $m(\xi)$ given by Lemma 4.6 is not unique. Indeed the polynomials p_s are not uniquely determined by the polynomials g_s. This happens if $|c| \neq 1$ because in formulas 4.24 and 4.25 we can choose c to be either one of the roots with absolute value greater than 1 or one of the roots with absolute value less then 1. This choice results in different polynomials $m(\xi)$.

2. Comparing Lemma 4.6 with Theorem 4.1 we see that in order to construct a compactly supported wavelet we need to construct a positive trigonometric polynomial $m(\xi)$ satisfying 4.2 and 4.3 and

$$m(\xi) + m(\xi + \pi) = 1. \tag{4.29}$$

Observe that condition 4.29 is very easy to check on coefficients. It is easy and well known (we used it in the above proof) that $m(\xi) = \sum_{k=-N}^{N} a_k e^{ik\xi}$ is a real trigonometric polynomial if and only if $a_k = \bar{a}_{-k}$ for $k = 1, 2, \ldots, N$. If 4.29 holds then

$$1 = \sum_{k=-N}^{N} a_k e^{ik\xi} + \sum_{k=-N}^{N} a_k (-1)^k e^{ik\xi}$$

which is clearly equivalent to $a_0 = \frac{1}{2}$ and $a_k = 0$ for all other even k's.

Now we will state and start the proof of the main result of this chapter.

Theorem 4.7 *There exists a constant C such that for each $r = 1, 2, \ldots$ there exists a multiresolution analysis in $L_2(\mathbb{R})$ with scaling function $\Phi(x)$ and an associated wavelet $\Psi(x)$ such that*

(i) *$\Phi(x)$ and $\Psi(x)$ are C^r functions*

(ii) *$\Phi(x)$ and $\Psi(x)$ are compactly supported and both* $\operatorname{supp}\Phi$ *and* $\operatorname{supp}\Psi$ *are contained in $[-Cr, Cr]$.*

REMARK 4.8. It follows directly from Remark 3.7 that there does not exist a C^∞ wavelet with compact support.

The rest of this section will be devoted to the proof of this theorem. In order to give the reader the general idea we first present the strategy.

Strategy of the proof. We will give an explicit definition of a polynomial $f_k(\xi) = \sum_{n=-N}^{N} a_n e^{in\xi} \geq 0$ such that f_k satisfies 4.29, 4.2 and 4.3. From Lemma 4.6 we will get a polynomial $m(\xi) = \sum_{n=0}^{N} b_n e^{in\xi}$ satisfying 4.1–4.3. We will make sure that $N \leq Ck$ so Theorem 4.1 will give condition (ii) of Theorem 4.7 for the scaling function. Condition (i) for the scaling function will follow from the estimate $|\prod_{j=1}^{\infty} m(2^j\xi)| \leq C(1 + |\xi|)^{-k-1}$. This will also show (cf. Theorem 4.1) that the wavelet satisfies (i) and (ii). **End of strategy**

Now let us start the execution of the above strategy. For $k = 1, 2, 3, \ldots$ let us define trigonometric polynomials

$$g_k(\xi) = 1 - c_k \int_0^{\xi} (\sin t)^{2k+1}\, dt \tag{4.30}$$

where $c_k = \left(\int_0^\pi (\sin t)^{2k+1}\, dt\right)^{-1}$.

It is clear that $g_k(\xi)$ is a trigonometric polynomial of degree $2k+1$. Note also that

$$g_k(\xi) = c_k \int_\xi^\pi (\sin t)^{2k+1}\, dt = c_k \int_\xi^\pi (1-\cos^2 t)^k \sin t\, dt. \tag{4.31}$$

Making the substitution $u = \cos t$ we get

$$g_k(\xi) = p_k(\cos \xi) \tag{4.32}$$

where $p_k(x)$ is an algebraic polynomial of degree $2k+1$ defined as

$$p_k(x) = c_k \int_{-1}^x (1-u^2)^k\, du. \tag{4.33}$$

REMARK 4.9. In what follows we will not be very careful with constants, but we will show that there exists an integer L such that we can take f_k described in the above 'Strategy of the proof' to be g_{Lk}.

In the following lemma we will collect for future use some properties of the polynomials g_k.

Lemma 4.8 *The polynomials $g_k(\xi)$ defined by 4.30 have the following properties*

$$0 \le g_k(\xi) \le 1 \quad \text{and} \quad g_k(\xi) = g_k(-\xi) \tag{4.34}$$

$$g_k(\xi) \neq 0 \text{ for } \xi \in (-\pi, \pi) \tag{4.35}$$

$$g_k(0) = 1 \tag{4.36}$$

$$1 = g_k(\xi) + g_k(\xi+\pi) \text{ for all } \xi \in \mathbb{R} \tag{4.37}$$

$$c_k \le 3\sqrt{k}. \tag{4.38}$$

Also we can factor

$$g_k(\xi) = \left(\frac{1+\cos\xi}{2}\right)^{k+1} \varphi_k(\xi) \tag{4.39}$$

where $\varphi_k(\xi)$ is a trigonometric polynomial.

Proof Properties 4.34–4.36 for every $k = 1, 2, 3, \ldots$ immediately follow from the definition of g_k and elementary properties of the sin function. Also for each $k = 1, 2, 3, \ldots$ we have

$$g_k(\xi) + g_k(\xi+\pi) = 2 - c_k \int_0^\xi (\sin t)^{2k+1}\, dt - c_k \int_0^{\xi+\pi} (\sin t)^{2k+1}\, dt$$

$$= 1 - c_k \int_0^\xi (\sin t)^{2k+1}\, dt - c_k \int_\pi^{\xi+\pi} (\sin t)^{2k+1}\, dt$$
$$= 1$$

where the last equality follows from the relation $\sin t = -\sin(t+\pi)$. This gives 4.37. We have

$$\begin{aligned}
\int_0^\pi (\sin t)^{2k+1}\, dt &\geq \int_{\frac{\pi}{2}-\frac{1}{\sqrt{2k+1}}}^{\frac{\pi}{2}+\frac{1}{\sqrt{2k+1}}} (\sin t)^{2k+1}\, dt \\
&\geq \frac{2}{\sqrt{2k+1}} \left[\sin(\frac{\pi}{2} + \frac{1}{\sqrt{2k+1}}) \right]^{2k+1} \\
&= \frac{2}{\sqrt{2k+1}} (\cos \frac{1}{\sqrt{2k+1}})^{2k+1} \\
&\geq \frac{2}{\sqrt{2k+1}} (1 - \frac{1}{2k+1})^{2k+1} \\
&\geq \frac{2}{e\sqrt{2k+1}} \geq \frac{1}{3\sqrt{k}}
\end{aligned}$$

so 4.38 follows. It follows directly from 4.33 that $p_k(x)$ has zero of order $k+1$ at $x=-1$ so we can factor $p_k(x) = (x+1)^{k+1}\tilde{p}_k(x)$ with $\tilde{p}_k(x)$ an algebraic polynomial. Thus 4.32 gives 4.39. □

Lemma 4.9 *Denote $m = \frac{k}{2}$ and rewrite 4.39 as*

$$g_k(\xi) = \left(\frac{1+\cos\xi}{2} \right)^m M_k(\xi). \tag{4.40}$$

Then there exists an integer N and a constant $\alpha < 1$ such that for $k \geq N$

$$\sup_{\xi\in\mathbb{R}} |M_k(\xi)| < 2^{\alpha k}. \tag{4.41}$$

Proof From 4.40 we see that

$$M_k(\xi) = 2^m g_k(\xi)(1+\cos\xi)^{-m}$$

so using 4.32 we infer that

$$\sup_{\xi\in\mathbb{R}} |M_k(\xi)| = 2^m \sup_{-1\leq x\leq 1} p_k(x)(1+x)^{-m}. \tag{4.42}$$

Using 4.33 we get for $-1 \leq x \leq 1$

$$\begin{aligned}(1+x)^{-m}p_k(x) &= c_k \int_{-1}^{x} \frac{(1-u^2)^k}{(1+x)^m}\, du \\ &= c_k \int_{-1}^{x} \left(\frac{1+u}{1+x}\right)^m (1+u)^m(1-u)^k\, du \\ &\leq c_k \int_{-1}^{x} (1+u)^m(1-u)^k\, du \qquad (4.43) \\ &= c_k \int_{-1}^{x} \left[\sqrt{1+u}\,(1-u)\right]^k du\end{aligned}$$

A standard calculation shows that on the interval $[-1,1]$ the function $\sqrt{1+u}(1-u)$ has an absolute maximum equal to $\frac{4}{3}\sqrt{\frac{2}{3}}$. This and 4.43 imply that

$$\max_{-1\leq x\leq 1} (1+x)^{-m}p_k(x) \leq 2c_k \left(\frac{4}{3}\sqrt{\frac{2}{3}}\right)^k. \qquad (4.44)$$

Since $\frac{4}{3}\sqrt{\frac{2}{3}} < \sqrt{2}$ the lemma follows easily from 4.38,4.44 and 4.42. □

REMARK 4.10. Looking more carefully at these calculations we can conclude that $N = 12$ suffices. It also follows that for large k we can take as α any number $> \log_2 \frac{8}{3\sqrt{3}} \simeq 0.299$.

Proposition 4.10 *Let $g_k(\xi)$, $k = 1, 2, \ldots$ be trigonometric polynomials defined by 4.30 and let us define $G_k(\xi) = \prod_{j=1}^{\infty} g_k(2^{-j}\xi)$. Then for $|\xi| > 1$ and $k \geq N$ we have*

$$|G_k(\xi)| \leq C_k|\xi|^{(\alpha-1)k} \qquad (4.45)$$

where N and α are given in Lemma 4.9.

Proof Using 4.40 we can write

$$G_k(\xi) = \left(\prod_{j=1}^{\infty} \frac{1+\cos 2^{-j}\xi}{2}\right)^m \prod_{j=1}^{\infty} M_k(2^{-j}\xi) \qquad (4.46)$$

which makes sense provided both products converge. The first product can be computed explicitly; note first that

$$\prod_{j=1}^{\infty} \cos 2^{-j}x = \frac{\sin x}{x} \qquad (4.47)$$

because using the elementary formula $\sin 2x = 2\cos x \sin x$ we can write

$$\prod_{j=1}^{m} \cos 2^{-j}x = \prod_{j=1}^{m} \frac{\sin 2^{-j+1}x}{2\sin 2^{-j}x} = \frac{\sin x}{2^m \sin 2^{-m}x}$$

and because $\lim_{m\to\infty} 2^m \sin 2^{-m}x = x$ we get 4.47. Because $\frac{1+\cos x}{2} = \cos^2 \frac{x}{2}$ we get from 4.47

$$\prod_{j=1}^{\infty} \frac{1+\cos 2^{-j}\xi}{2} = \frac{4\sin^2 \xi/2}{\xi^2}. \tag{4.48}$$

Now let us look at the second product. It follows directly from 4.40 and 4.39 that $M_k(\xi)$ is a continuous function which satisfies

$$|M_k(\xi) - 1| \leq C|\xi|.$$

This implies (exactly as in the proof of Lemma 4.2) that the product $\prod_{j=1}^{\infty} M_k(2^{-j}\xi)$ converges almost uniformly to a continuous function. In particular

$$\sup_{|\xi|\leq 1} \Big|\prod_{j=1}^{\infty} M_k(2^{-j}\xi)\Big| \leq C_k. \tag{4.49}$$

For $|\xi| > 1$ let us fix an integer r such that $2^{r-1} \leq |\xi| < 2^r$. From 4.49 and 4.41 we obtain

$$\begin{aligned} \Big|\prod_{j=1}^{\infty} M_k(2^{-j}\xi)\Big| &= \prod_{j=1}^{r} |M_k(2^{-j}\xi)| \cdot \prod_{j=1}^{\infty} |M_k(2^{-j}2^{-r}\xi)| \\ &\leq 2^{\alpha k r} C_k \leq 2C_k|\xi|^{\alpha k}. \end{aligned} \tag{4.50}$$

Putting together 4.46, 4.48 and 4.50 we get 4.45. □

Proof of Theorem 4.7. Let us start with the polynomials $g_k(\xi)$ defined in 4.30. Since $g_k(\xi) \geq 0$ we use Lemma 4.6 to obtain a polynomial $m_k(\xi)$ of degree $2k+1$ such that $|m_k(\xi)|^2 = g_k(\xi)$. It follows from 4.37, 4.36 and 4.35 that the polynomial $m_k(\xi)$ satisfies 4.1–4.3. We apply Theorem 4.1 to get a scaling function Φ_k and a wavelet Ψ_k supported on the interval $[-2k-1, 2k+1]$. From 4.4 and Proposition 4.10 we infer that

$$\left|\hat{\Phi}_k(\xi)\right| \leq C_k|\xi|^{(\alpha-1)k/2}$$

for $|\xi| \geq 1$. From properties of the Fourier transform (cf. A1.2–IV) we infer that $\Phi_k \in C^r$ with $r < \frac{1-\alpha}{2}k - 1$. From 4.5 it is clear that also $\Psi_k \in C^r$. This readily gives the theorem. □

REMARK 4.11. It follows from Lemma 4.6 and Remark 4.6 that actually Supp Φ_k and Supp Ψ_k have length $2k+1$. Thus the argument above and Remark 4.10 show that in order to have a compactly supported wavelet of class C^r the above construction requires $k > \frac{2(r+1)}{1-\alpha}$ so we get Supp Ψ_k of length $> \frac{4(r+1)}{1-\alpha} - 1$. For large r we get $k \geq 14(r+1)$. There is no reason to believe that these numbers are close to the optimum.

REMARK 4.12. The proof of Theorem 4.7 uses *only* estimates for $g_k(\xi)$. The actual scaling function and wavelet depend on the polynomial $m(\xi)$ obtained from $g_k(\xi)$ using Lemma 4.6. This polynomial, as we know from Remark 4.7, is not unique. Thus for each k our construction actually applies to the whole family of wavelets.

4.3 Bare hands construction

Our aim in this section is to present an 'elementary' construction of a continuous, compactly supported wavelet. The construction is elementary in the sense that it does not use any tools developed so far. It also allows extremely easy calculation of an approximation to the resulting scaling function and wavelet. On the other hand the numerical values from which the construction starts are lifted directly from the general theory. Otherwise we would have to invoke illumination to get them. Actually we are constructing one of the scaling functions associated with $g_1(\xi)$ from the previous section (cf. Exercise 4.6). This construction also makes evident the self-similar nature of the graph of a compactly supported scaling function. This is present in our general construction in formula 4.12. The construction proceeds through a series of elementary lemmas with relatively easy and straightforward proofs. Thus we will be more brief in this section than usual in this book. Let us start with the following notation: $D_j = \{k2^{-j} \ : \ k \in \mathbb{Z}\}$ and

$$D = \bigcup_{j\in\mathbb{Z}} D_j = \bigcup_{j=0}^{\infty} D_j. \tag{4.51}$$

Observe that $D \subset \mathbb{R}$ is dense. It is also a ring, i.e. sums, differences and products of elements of D are also in D. In the construction we will use two explicit numbers

$$a =: \frac{1+\sqrt{3}}{4} \tag{4.52}$$

and

$$b =: \frac{1-\sqrt{3}}{4} \tag{4.53}$$

Clearly

$$\tfrac{1}{2} < a < 1 \quad \text{and} \quad -\tfrac{1}{4} < b < 0. \tag{4.54}$$

Proposition 4.11 *There exists a unique function $\Phi : D \to \mathbb{R}$ satisfying the following three conditions:*

$$\Phi(x) = a\Phi(2x) + (1-b)\Phi(2x-1) + (1-a)\Phi(2x-2) + b\Phi(2x-3) \tag{4.55}$$

$$\sum_{k\in\mathbb{Z}} \Phi(k) = 1 \tag{4.56}$$

$$\Phi(d) = 0 \quad \textit{if } d < 0 \quad \textit{or} \quad d > 3. \tag{4.57}$$

Proof We first try to calculate the values of our function on the integers $\mathbb{Z} = D_0$. Condition 4.57 says that we need to find values of Φ at $k = 0, 1, 2, 3$. Thus condition 4.55 leads to the following system of equations:

$$\begin{bmatrix} \Phi(0) \\ \Phi(1) \\ \Phi(2) \\ \Phi(3) \end{bmatrix} = \begin{bmatrix} a & 0 & 0 & 0 \\ (1-a) & (1-b) & a & 0 \\ 0 & b & (1-a) & (1-b) \\ 0 & 0 & 0 & b \end{bmatrix} \begin{bmatrix} \Phi(0) \\ \Phi(1) \\ \Phi(2) \\ \Phi(3) \end{bmatrix} \tag{4.58}$$

If A denotes the 4×4 matrix on the right hand side of 4.58 then

$$\begin{aligned} \det(A - \lambda I) &= (a-\lambda)(b-\lambda)\det \begin{bmatrix} 1-b-\lambda & a \\ b & 1-a-\lambda \end{bmatrix} \\ &= (a-\lambda)(b-\lambda)(1-\lambda)(1-\lambda-a-b). \end{aligned}$$

If we compare this with 4.52 and 4.53 we see that $\lambda = 1$ is an eigenvalue of A with multiplicity 1, so there exists a unique (up to scalar multiplication) non-zero solution of 4.58. Such a unique solution satisfying 4.56 is given by

$$\begin{aligned} \Phi(0) &= 0 & \Phi(1) &= \tfrac{1+\sqrt{3}}{2} \\ \Phi(2) &= \tfrac{1-\sqrt{3}}{2} & \Phi(3) &= 0 \end{aligned} \tag{4.59}$$

Observe that once we have determined Φ on the integers (i.e. on D_0) we can use 4.55 to find values of Φ on D_1. Observe that in this process we also obtain 4.57. Continuing in this way we obtain values on D_2 and we easily check that they satisfy 4.57. We continue in this way and by induction we get $\Phi : D \to \mathbb{R}$. Since $\Phi \mid D_0$ is unique, it is unique in general. □

Lemma 4.12 *For every $x \in D$ we have*

$$\sum_{k\in\mathbb{Z}} \Phi(x-k) = 1 \tag{4.60}$$

and

$$\sum_{k\in\mathbb{Z}} \left(\frac{3-\sqrt{3}}{2} + k\right) \Phi(x-k) = x. \tag{4.61}$$

Proof In order to simplify the notation in this proof we will introduce the number c defined by

$$c =: \frac{3-\sqrt{3}}{2}. \tag{4.62}$$

The proof is by induction on j. Let us start with $j = 0$. For $x \in D_0 = \mathbb{Z}$ 4.60 is already proven (it is 4.56). When $x \in D_0$ we use 4.57 and 4.59 and see that the left hand side of 4.61 becomes

$$(c+x-1)\,\frac{1+\sqrt{3}}{2} + (c+x-2)\,\frac{1-\sqrt{3}}{2}.$$

After a routine calculation using 4.62 this gives x, so 4.61 holds for $x \in D_0$. Now assume that 4.60 holds for $x \in D_j$ and take $x_0 \in D_{j+1}$. Then $2x_0 - k \in D_j$ for every $k \in \mathbb{Z}$, so substituting 4.55 into 4.60 we easily get 4.60 for x_0. Thus 4.60 holds for all $x \in D$ by induction. The proof of 4.61 also proceeds by induction. Assume that it holds for all $x \in D_j$ and take $x_0 \in D_{j+1} \setminus D_j$. Write $x_0 = N + \alpha$ with $0 < \alpha < 1$ and $N \in \mathbb{Z}$. Then using 4.57 equation 4.61 becomes

$$(c+N)\,\Phi(\alpha) + (c+N-1)\,\Phi(\alpha+1) + (c+N-2)\,\Phi(\alpha+2).$$

Using 4.60 which we have already proved, this can be written as

$$c\Phi(\alpha) + (c-1)\,\Phi(\alpha+1) + (c-2)\,\Phi(\alpha+2) + N.$$

Into this we substitute 4.55 and using 4.57 we get

$$\begin{aligned} N \;&+\; \Phi(2\alpha-1)\,[(1-b)c + b\,(c-1)] \\ &+\; \Phi(2\alpha)\,[ac + (1-a)\,(c-1)] \\ &+\; \Phi(2\alpha+1)\,[(1-b)\,(c-1) + b\,(c-2)] \\ &+\; \Phi(2\alpha+2)\,[a\,(c-1) + (1-a)\,(c-2)]\,. \end{aligned}$$

Into this we substitute 4.52, 4.53 and 4.62 to obtain

$$\begin{aligned} N \;&+\; \tfrac{1}{2}\,[(c+1)\,\Phi(2\alpha-1) + c\Phi(2\alpha) \\ &+\; (c-1)\,\Phi(2\alpha+1) + (c-2)\,\Phi(2\alpha+2)]\,. \end{aligned}$$

Since $2\alpha + k \in D_j$ for all $k \in \mathbb{Z}$, we apply the inductive hypothesis to obtain

$$N + \tfrac{1}{2}2\alpha = N + \alpha = x_0.$$

This shows that 4.61 holds. □

Lemma 4.13 *If $x \in D$ and $0 \le x \le 1$, then*

$$2\Phi(x) + \Phi(x+1) = x + \frac{1+\sqrt{3}}{2} \tag{4.63}$$

$$2\Phi(x+2) + \Phi(x+1) = -x + \frac{3-\sqrt{3}}{2} \tag{4.64}$$

$$\Phi(x) - \Phi(x+2) = x + \frac{-1+\sqrt{3}}{2}. \tag{4.65}$$

Proof For $x \in D$, $0 \le x \le 1$ relations 4.60 and 4.61 give us two equations connecting $\Phi(x)$, $\Phi(x+1)$ and $\Phi(x+2)$. Eliminating $\Phi(x+2)$ from those equations we get 4.63, eliminating $\Phi(x)$ we get 4.64, and eliminating $\Phi(x+1)$ we get 4.65. □

Lemma 4.14 *For $0 \le x \le 1$ and $x \in D$ the following relations hold:*

$$\Phi\left(\frac{0+x}{2}\right) = a\Phi(x)$$

$$\Phi\left(\frac{1+x}{2}\right) = b\Phi(x) + ax + \frac{2+\sqrt{3}}{4}$$

$$\Phi\left(\frac{2+x}{2}\right) = a\Phi(1+x) + bx + \frac{\sqrt{3}}{4}$$

$$\Phi\left(\frac{3+x}{2}\right) = b\Phi(1+x) - ax + \frac{1}{4}$$

$$\Phi\left(\frac{4+x}{2}\right) = a\Phi(2+x) - bx + \frac{3-2\sqrt{3}}{4}$$

$$\Phi\left(\frac{5+x}{2}\right) = b\Phi(2+x)$$

Proof The proof of each of the six equations is basically the same and consists of the following three steps:

- Using 4.55 and 4.57 we write the left hand side in terms of $\Phi(x)$, $\Phi(x+1)$ and $\Phi(x+2)$.

- Using the appropriate two of equations 4.63–4.65 we eliminate two of those values and retain only the desired one. An appropriate linear function appears.
- Using 4.52 and 4.53 we explicitly calculate the numbers appearing.

□

This lemma will be the basis of the proof that the function Φ extends to a continuous function on $\mathbb{R}$. To do this formally we introduce a non-linear operator K acting on functions on $\mathbb{R}$. It is defined as follows: when $x \in [0,1]$ then we define $K(f)$ by the following set of conditions:

$$\begin{aligned}
K(f)\left(\frac{0+x}{2}\right) &= af(x) \\
K(f)\left(\frac{1+x}{2}\right) &= bf(x) + ax + \frac{2+\sqrt{3}}{4} \\
K(f)\left(\frac{2+x}{2}\right) &= af(1+x) + bx + \frac{\sqrt{3}}{4} \\
K(f)\left(\frac{3+x}{2}\right) &= bf(1+x) - ax + \frac{1}{4} \\
K(f)\left(\frac{4+x}{2}\right) &= af(2+x) - bx + \frac{3-2\sqrt{3}}{4} \\
K(f)\left(\frac{5+x}{2}\right) &= bf(2+x) \\
K(f)(y) &= 0 \quad \text{for } y \notin [0,3].
\end{aligned}$$

This definition can produce two values for $K(f)$ at the points 0, $\frac{1}{2}$, 1, $1\frac{1}{2}$, 2, $2\frac{1}{2}$, 3. Let us denote by Φ_j, $j = 0,1,2,\ldots$, the continuous, piecewise linear function on $\mathbb{R}$ which on D_j equals Φ. It follows from Lemma 4.14 and Proposition 4.11 that for all $j = 0,1,2,\ldots$ the function $K(\Phi_j)$ is well defined at each point and actually $K(\Phi_j) = \Phi_{j+1}$. Let $x \in [0,3]$ and $j > 0$. From the definition of $K(f)$ we immediately see that

$$\Phi_{j+1}(x) - \Phi_j(x) = K(\Phi_j)(x) - K(\Phi_{j-1})(x) = \eta(\Phi_j(y) - \Phi_{j-1}(y)) \quad (4.66)$$

where $\eta = a$ or $\eta = b$ and $y \in \mathbb{R}$ is a point which depends on x (and for given x can be easily deduced from the definition of K). Since $K(f)(x) = 0$ for $x \notin [0,3]$ and $\max(|a|,|b|) = a$, from 4.66 we get

$$\|\Phi_{j+1} - \Phi_j\|_\infty \le a\|\Phi_j - \Phi_{j-1}\|_\infty$$

so inductively we get

$$\|\Phi_{j+1} - \Phi_j\|_\infty \le a^j \|\Phi_1 - \Phi_0\|_\infty.$$

Since $\|\Phi_1 - \Phi_0\| < \infty$, this shows that the sequence $(\Phi_j)_{j=0}^\infty$ converges uniformly to a continuous function which we will call Φ. Thus we have:

Theorem 4.15 *The function Φ defined on the set $D \subset \mathbb{R}$ by Proposition 4.11 extends to a continuous function on $\mathbb{R}$, which we will also denote by Φ.*

Clearly all formulas which we have proved for function $\Phi(x)$ with $x \in D$ extend by continuity to $x \in \mathbb{R}$.

Theorem 4.16 *If $\Phi(x)$ is the function on $\mathbb{R}$ given by Theorem 4.15 then*

$$\int_{-\infty}^{\infty} \Phi(x)\, dx = 1 \tag{4.67}$$

and

$$\int_{-\infty}^{\infty} \Phi(x) \cdot \Phi(x-k)\, dx = \begin{cases} 1 & \text{if } k = 0 \\ 0 & \text{if } k \neq 0. \end{cases} \tag{4.68}$$

Proof We know that 4.60 holds for each $x \in \mathbb{R}$. Also, since supp $\Phi \subset [0,3]$, for each $x \in \mathbb{R}$ there are at most three non-zero terms in the sum $\sum_{k \in \mathbb{Z}} \Phi(x-k)$. Given a positive integer K let $F_K(x) = \sum_{k=-K}^{K} \Phi(x-k)$. From the above remarks and 4.56 we infer that $|F_K(x)| \le C$ for some constant C and

$$F_K(x) = \begin{cases} 1 & \text{if } |x| \le K-3 \\ 0 & \text{if } |x| \ge K+3. \end{cases}$$

Thus for every integer K

$$2(K-3) - 12C \le \int_{-\infty}^{\infty} F_K(x)\, dx \le 2(K-3) + 12C. \tag{4.69}$$

From the definition of F_K we also infer that

$$\int_{-\infty}^{\infty} F_K(x)\, dx = (2K+1) \int_{-\infty}^{\infty} \Phi(x)\, dx. \tag{4.70}$$

Since 4.69 and 4.70 hold for every positive integer K, letting K tend to infinity we obtain

$$\int_{-\infty}^{\infty} \Phi(x)\, dx = 1.$$

In order to prove 4.68 let us denote

$$L_k = \int_{-\infty}^{\infty} \Phi(x) \cdot \Phi(x-k)\, dx. \tag{4.71}$$

Since supp $\Phi \subset [0,3]$ we see that

$$L_k = 0 \quad \text{for } |k| \geq 3. \tag{4.72}$$

It is also clear that

$$L_k = L_{-k} \tag{4.73}$$

so we need to investigate only L_0, L_1 and L_2. An easy change of variables gives that for any $k, r, s \in \mathbb{Z}$ we have

$$\int_{-\infty}^{\infty} \Phi(2x-r) \cdot \Phi(2x-2k-s)\, dx = \tfrac{1}{2} L_{2k+s-r} \tag{4.74}$$

Substituting 4.55 into 4.71 for $k = 0, 1, 2$ and using 4.72–4.74 we obtain the following three equations:

$$\begin{aligned}
\big(a(1-a)+b(1-b)\big)L_0 &= (1-ab)L_1 + \big(a(1-a)+b(1-b)\big)L_2 \\
2L_1 &= \big(a(1-a)+b(1-b)\big)L_0 + (1-b)L_1 \\
&\quad +\big((1-b)^2+(1-a)^2+b^2\big)L_2 \\
2L_2 &= abL_1 + \big(a(1-a)+b(1-b)\big)L_2.
\end{aligned}$$

A direct calculation using 4.52 and 4.53 gives

$$a(1-a)+b(1-b) = 0 \tag{4.75}$$

so the above system of equations becomes

$$\begin{aligned}
0 &= (1-ab)L_1 \\
2L_1 &= (1-b)L_1 + ((1-b)^2+(1-a)^2+b^2)L_2 \\
2L_2 &= abL_1.
\end{aligned}$$

So we obtain $L_1 = L_2 = 0$ and no condition on L_0. This implies that $L_k = 0$ for all $k \neq 0$. To compute L_0 let us use 4.67, 4.56 and 4.68 and write

$$\begin{aligned}
1 &= \int_{-\infty}^{\infty} \Phi(x)\, dx = \int_{-\infty}^{\infty} \Phi(x) \cdot \sum_{k\in\mathbb{Z}} \Phi(x-k)\, dx \\
&= \sum_{k\in\mathbb{Z}} L_k = L_0.
\end{aligned}$$

□

If we now recall the material explained in Section 2.3 we see that $\Phi(x)$ is a scaling function of a multiresolution analysis. As we know from Theorem 4.1, the corresponding wavelet can be given by the formula

$$\Psi(x) = -b\Phi(2x)+(1-a)\Phi(2x-1)-(1-b)\Phi(2x-2)+a\Phi(2x-3). \quad (4.76)$$

However, we can proceed in our elementary spirit and show:

Theorem 4.17 *The function $\Psi(x)$ defined by the formula 4.76 satisfies the following conditions:*

$$\operatorname{supp} \Psi(x) \subset [0,3] \quad (4.77)$$

$$\int_{-\infty}^{\infty} \Psi(x) \cdot \Psi(x-k)\, dx = \begin{cases} 0 & \text{if } k \neq 0 \\ 1 & \text{if } k = 0 \end{cases} \quad (4.78)$$

$$\int_{-\infty}^{\infty} \Psi(x-k) \cdot \Phi(x)\, dx = 0 \quad \text{for all } k \in \mathbb{Z}. \quad (4.79)$$

Thus $\left\{2^{j/2}\Psi(2^j t - k)\right\}_{j\in\mathbb{Z},k\in\mathbb{Z}}$ *is an orthonormal system in* $L_2(\mathbb{R})$.

Proof Property 4.77 immediately follows from 4.76 and the fact that $\operatorname{supp} \Phi \subset [0,3]$. To obtain 4.78 we substitute 4.76 into the left hand side of 4.78 and calculate using 4.74, 4.68, 4.75 and the values of a and b. To obtain 4.79 we proceed analogously, but we substitute both 4.76 and 4.55 into the left hand side of 4.79. The fact that $\left\{2^{j/2}\Psi(2^j t - k)\right\}_{j\in\mathbb{Z},k\in\mathbb{Z}}$ is orthonormal follows directly from 4.78 and 4.79. □

Sources and comments

The first compactly supported wavelets different from the Haar wavelet, in particular smooth compactly supported wavelets, were constructed by I. Daubechies in [25]. Almost immediately the whole theory developed and the construction was simplified in various ways. Our argument given in Sections 4.1 and 4.2 is an adaptation of arguments presented in [25], [85], [27] and [24]. A very detailed presentation of the theory of compactly supported wavelets is given in Chapters 6–8 of [24]. The interested reader should consult that book and references given there.

It should be pointed out that for each $N = 2, 3, \ldots$ there are many trigonometric polynomials of degree N satisfying 4.1 and 4.2 and giving wavelets. Various choices give compactly supported wavelets with different properties; examples are discussed extensively in [24]. A full parametrization of such polynomials is given by R. O. Wells [112]. In Section 4.2, following [27] and [85] we concentrate on one concrete sequence of polynomials which is easy to write down and to manipulate.

The elementary construction presented in Section 4.3 is taken from Pollen [94]. As is clear from our presentation and Exercises 4.4, 4.5 and 4.7 the detailed study of the regularity of a compactly supported wavelet is a delicate problem. For some recent results on this subject see e.g. [110]. Cohen's condition in Exercise 4.8 was introduced by A. Cohen in [20]. It is also necessary, i.e. if $m(\xi)$ satisfies 4.1 and 4.2 and if Φ given by $\hat{\Phi}(\xi) = \frac{1}{\sqrt{2\pi}}\Theta(\xi)$, where $\Theta(\xi)$ is given by 4.4, is a scaling function then $m(\xi)$ satisfies Cohen's condition. Other equivalent conditions were given by W. Lawton in [64]. A nice presentation of various such equivalent conditions is given (with simple proofs) in [48].

About the exercises. As already mentioned, Exercise 4.8 is part of a result of Cohen [20]. The results of Exercise 4.9 can be found in [71] and [70]. Exercise 4.10 can be found in [94], while Exercise 4.12 is a result of W. Lawton [65] which can also be found in [24] Proposition 6.2.3.

Exercises

4.1 Let $m(\xi) = \frac{1}{2}(1 + e^{i\xi})$.

- Show that $m(\xi)$ satisfies 4.1–4.3.
- Show that $\prod_{j=1}^{\infty} m(2^{-j}\xi) = e^{i\xi/2}\frac{\sin \xi/2}{\xi/2}$.
- Conclude that $m(\xi)$ gives $\mathbf{1}_{[-1,0]}$ as a compactly supported scaling function, so it gives a Haar wavelet as a compactly supported wavelet.
- Take $m_1(\xi) = m(3\xi)$. Show that it satisfies 4.1 and 4.2 but not 4.3. Show that $\prod_{j=1}^{\infty} m_1(2^{-j}\xi)$ is the Fourier transform of $\mathbf{1}_{[-3,0]}$. This shows that some condition like 4.3 is needed to ensure the orthogonality of translates.

4.2 Suppose that $m_1(\xi)$ and $m_2(\xi)$ are two polynomials satisfying 4.1–4.3 which via Theorem 4.1 yield the same multiresolution analysis. Show that $m_1(\xi) = e^{is\xi}m_2(\xi)$ for some $s \in \mathbb{Z}$.

4.3 Let Φ with a compact support be a scaling function of a multiresolution analysis.

- Show that if $\mathrm{supp}\Phi \subset [0,1]$ then $\Phi = \mathbf{1}_{[0,1]}$.
- Show that if $\Phi(a+x) = \Phi(a-x)$ for some $a \in \mathbb{R}$ and all $x \in \mathbb{R}$ then $\Phi(x) = \mathbf{1}_{[k,k+1]}$ for some $k \in \mathbb{Z}$.

4.4 Consider g_1 defined by 4.30.

- Factor it explicitly as in 4.39.
- Show that $\max_{\xi\in\mathbf{R}} |\varphi_1(\xi)| = 3$.
- Use the factorization 4.39 and the above to show that for $|\xi| > 1$ one has $\prod_{j=1}^{\infty} g_1(2^{-j}\xi) \leq C|\xi|^{-4+\ln_2 3}$.
- Note that any corresponding scaling function and wavelet are supported on an interval of length 3. Show that they are continuous and satisfy Hölder's condition of any order less than $1 - \ln_2 \sqrt{3}$.
- Find all trigonometric polynomials $m(\xi)$ satisfying 4.1–4.3 such that $|m(\xi)|^2 = g_1(\xi)$.

4.5 Arguing as in the previous exercise show that any wavelet corresponding to $g_2(\xi)$ satisfies Hölder's condition of any order less than $2 - \ln_2 \sqrt{10}$. Note that $1 - \ln_2 \sqrt{3} < 2 - \ln_2 \sqrt{10}$.

4.6 Let $\Phi(x)$ denote the scaling function constructed in Section 4.3. Find the polynomial $m_0(\xi)$ such that

$$\hat{\Phi}(\xi) = \frac{1}{\sqrt{2\pi}} \prod_{j=1}^{\infty} m_0(2^{-j}\xi).$$

Show that $|m_0(\xi)|^2 = g_1(\xi)$ where g_1 is defined in 4.30.

4.7 Suppose that $\Phi(x)$ is a continuous scaling function such that $\mathrm{Supp}\Phi = [0, 3]$, so that

$$\Phi(x) = c_0\Phi(2x) + c_1\Phi(2x-1) + c_2\Phi(2x-2) + c_3\Phi(2x-3).$$

- Observe that $c_0 \neq 0$ and $c_3 \neq 0$.
- Show that $\Phi(1) \neq 0$ and $\Phi(2) \neq 0$.
- Note that $\Phi(2^{-j}) = c_0^j\Phi(1)$ so Φ does not satisfy Hölder's condition at 0 with any exponent $> -\ln_2 |c_0|$.
- Note that c_0, c_1, c_2, c_3 satisfy the following equations

$$\begin{aligned} c_0 + c_1 + c_2 + c_3 &= 2 \\ c_0^2 + c_1^2 + c_2^2 + c_3^2 &= 2 \\ \frac{c_0}{1 - c_1} &= \frac{1 - c_2}{c_3}. \end{aligned}$$

- Suppose that Φ has continuous derivatives at integer points and that $\Phi'(1) \neq 0$. Show that this implies $|c_0| < \frac{1}{2}$ and $|c_3| < \frac{1}{2}$ and also

$$\frac{2c_0}{1 - 2c_1} = \frac{1 - 2c_2}{2c_3}.$$

- Show that the above conditions on c_0, c_1, c_2, c_3 are contradictory, so Φ can not be continuously differentiable.

4.8 Prove Theorem 4.1 when condition 4.3 is replaced by the following condition called Cohen's condition: there exists a compact set $K \subset \mathbb{R}$ such that

- $|K| = 2\pi$
- for each $t \in [-\pi, \pi]$ there exists $l \in \mathbb{Z}$ such that $t + 2\pi l \in K$
- K contains a neighborhood of 0
- $\inf_{j=1,2,\ldots} \inf_{t \in K} |m(2^{-j}\xi)| > 0$.

Using this show (by constructing an appropriate K) that we can replace $\pi/2$ in 4.3 by $\pi/3$. Observe (use Exercise 4.1) that the constant $\pi/3$ is smallest possible.

4.9 Suppose we have a multiresolution analysis with compactly supported scaling function Φ with Supp $\Phi = [0, N]$.

- Suppose $f \in V_0$ and $f \mid [0, 2N+1] = 0$. Show that $f\mathbf{1}_{[0,\infty)} \in V_0$.
- Suppose $f \in V_0$ and $f \mid [a, b] = 0$ with $a < b$. Show that $f\mathbf{1}_{[b,\infty)} \in V_0$.
- Conclude that supp $\Phi = [0, N]$ and that for the associated wavelet Ψ we have Supp Ψ = supp Ψ.
- Show that $V_0 \mid [-1, 1]$ is a linear space of dimension $N + 1$.

4.10 Let Ψ be the wavelet given by 4.76. Show that

$$\int_{-\infty}^{\infty} x\Psi(x)\, dx = 0.$$

4.11 Show that the function $f(x) =: \max(0, x)$ cannot be written as $f(x) = \sum_{l \in \mathbb{Z}} a_l \Phi(x - l)$ where Φ is a compactly supported scaling function of a multiresolution analysis.

4.12 Let $m(\xi)$ be a trigonometric polynomial satisfying 4.1 and 4.2, let $\Theta(\xi)$ be given by 4.4, let $\Phi(x)$ be given by $\hat{\Phi} = \frac{1}{\sqrt{2\pi}}\Theta$ and let $\Psi(x)$ be given by 4.5. Show that

$$\left\{2^{j/2}\Psi(2^j t - k)\right\}_{j \in \mathbb{Z}, k \in \mathbb{Z}}$$

is a tight frame in $L_2(\mathbb{R})$. For the definition of tight frame see Remark 2.3.

5
Multivariable wavelets

Our aim in this Chapter is to obtain multivariable generalizations of one-variable wavelets. This can be done in many different ways. The most natural way to pass from one variable to several is to use tensors, i.e. functions of the form $f(x_1, \ldots, x_d) = f_1(x_1) \cdot \ldots \cdot f_d(x_d)$. This idea we can employ at two different levels: for wavelets and for scaling functions. We will present this in Section 5.1. In Section 5.2 we will present a genuinely multivariate theory of multiresolution analyses on $\mathbb{R}^d$, together with some examples. Actually we will present our theory in such generality that even for $d = 1$ we will get a more general theory than presented so far. The fundamental difference between the above three aproaches is the way we generalize the one-dimensional dyadic dilations $J_s f(x) = f(2^s x)$. Tensoring at the level of wavelets corresponds to dilations

$$J_{s_1, s_2, \ldots, s_d}(f)(x_1, \ldots, x_d) = f(2^{s_1} x_1, \ldots, 2^{s_d} x_d). \tag{5.1}$$

Tensoring at the level of the scaling function corresponds to dilations

$$J_s f(x_1, \ldots, x_d) = f(2^s x_1, \ldots, 2^s x_d). \tag{5.2}$$

Our more general approach uses dilations of the form

$$J_A f(x_1, \ldots, x_d) = f(A(x_1, \ldots, x_d)) \tag{5.3}$$

where A is a suitable linear transformation of $\mathbb{R}^d$. The last two approaches force us to use instead of one wavelet a finite 'wavelet set'. Our translations will always be the same as before: for $h \in \mathbb{R}^d$ we define

$$T_h f(x) =: f(x - h). \tag{5.4}$$

To generate wavelets we will use $h \in \mathbb{Z}^d$.

In Section 5.2 we will show how to construct wavelet sets from multiresolution analysis in our most general framework. In our last Section 5.3 we will construct many examples of multiresolution analyses and in particular we will give the construction of smooth, fast decaying wavelets on $\mathbb{R}^d$.

It should be stated at the outset place that the multivariable theory is much less developed and much more complicated than the one-variable theory presented in Chapters 1–4. Nevertheless in many places in this chapter we will use arguments which are similar to arguments used in earlier chapters. In such cases we will be more brief than usual. This should not prevent more mathematically experienced readers from starting their reading from this chapter if they wish to do so. The reader who is familiar with earlier chapters should have no problems at all.

5.1 Tensor products

The most natural way to pass from the one-variable situation to a multivariable situation is to form tensor products. This is a very general concept which we will use in the most simple context only – really we will only use the notation.

Given d functions of one variable $f^j(x)$ for $j = 1, \dots, d$ we will form the function of d variables $f^1 \otimes f^2 \otimes \dots \otimes f^d = \bigotimes_{j=1}^d f^j$ defined as

$$\bigotimes_{j=1}^{d} f^j(x_1, \dots, x_d) = \prod_{j=1}^{d} f^j(x_j).$$

If we have d closed subspaces $X_j \subset L_2(\mathbb{R})$ for $j = 1, 2, \dots, d$ we can form a closed subspace of $L_2(\mathbb{R}^d)$ denoted by $\bigotimes_{j=1}^d X_j$ or by $X_1 \otimes X_2 \otimes \dots \otimes X_d$ and defined as the closed linear span in $L_2(\mathbb{R}^d)$ of all functions of the form $f^1(x_1) \cdot \dots \cdot f^d(x_d)$ where $f^j \in X_j$ for all $j = 1, 2, \dots, d$.

It is easy to check that if the systems $\left(f_s^j\right)_{s \in A_j}$ are orthonormal bases in subspaces $X_j \subset L_2(\mathbb{R})$ for $j = 1, 2, \dots, d$ then the system

$$\left(\bigotimes_{j=1}^{d} f_{s_j}^j\right)_{(s_1, \dots, s_d) \in A_1 \times \dots \times A_d}$$

is an orthonormal basis in $\bigotimes_{j=1}^d X_j$. It is also easy and well known that $\bigotimes_{j=1}^d L_2(\mathbb{R}) = L_2(\mathbb{R}^d)$.

If our aim is to obtain an orthonormal basis in $L_2(\mathbb{R}^d)$, then the most natural approach is to take d orthonormal bases $\left(\varphi_n^j(x)\right)_{n \in A_j}$ in $L_2(\mathbb{R})$

for $j = 1, 2, \ldots, d$ and to form an orthonormal system $\bigotimes_{j=1}^{d} \varphi_{n_j}^{j}$ indexed by the set $A = A_1 \times \ldots \times A_d$. It follows from what we have said above that this system is an orthonormal basis in $L_2(\mathbb{R}^d)$. When we apply this procedure to wavelet bases we get the following:

Proposition 5.1 *Let $(\Psi_j)_{j=1}^{d}$ be wavelets on $\mathbb{R}$ and let*

$$\Psi(x_1, \ldots, x_d) = \prod_{j=1}^{d} \Psi_j(x_j).$$

Then the system

$$2^{\frac{j_1+\ldots+j_d}{2}} \Psi(2^{j_1} x_1 - k_1, \ldots, 2^{j_d} x_d - k_d) \tag{5.5}$$

for all $j_1, \ldots, j_d$ and $k_1, \ldots, k_d$ in $\mathbb{Z}$ forms an orthonormal basis in $L_2(\mathbb{R}^d)$.

The basis given by 5.5 looks very appealing: it is naturally generated by one function Ψ. It is actually quite useful in many instances. Its drawback, which is quite serious in many situations, is that the integers $j_1, \ldots, j_d$ are totally independent, so the decay of elements of the basis in different directions can be markedly different. The easiest way to see what I mean is to take $d = 2$ and $\Psi_1 = \Psi_2 = H$, the Haar wavelet defined in Definition 1.1 on page 1. In this special case the supports of the functions given by 5.5 are all dyadic rectangles, so we have squares like $[0, 2^n] \times [0, 2^n]$ but also rectangles arbitrarily narrow in one direction like $[0, 1] \times [0, 2^n]$. This may cause problems. The way to avoid this is not to tensor wavelets but multiresolution analyses. As an introductory example of this approach let us try to construct a two-dimensional Haar wavelet.

EXAMPLE 5.1. When we try to build a two-dimensional Haar wavelet basis the natural choice is to use squares in the plane $\mathbb{R}^2$. When we want to divide a square into equal squares we need to divide it into at least *four* squares. To be more precise let V_0 be the space of all functions in $L_2(\mathbb{R}^2)$ which are constant on each square $(n, n+1) \times (k, k+1)$. When we divide each square into four equal squares we obtain the space V_1 of all functions in $L_2(\mathbb{R}^2)$ which are constant on all squares $(\frac{n}{2}, \frac{n+1}{2}) \times (\frac{k}{2}, \frac{k+1}{2})$. Thus to complement V_0 to V_1 we need three functions on each square $(n, n+1) \times (k, k+1)$. These three orthogonal functions can be given as

$$\begin{aligned} \Psi_1(x, y) &= H(x) \cdot \mathbf{1}_{[0,1]}(y) \\ \Psi_2(x, y) &= \mathbf{1}_{[0,1]}(x) \cdot H(y) \end{aligned}$$

$$\Psi_3(x,y) \quad = \quad H(x)\cdot H(y)$$

where $H(t)$ is the Haar wavelet defined by 1.1. Clearly these three functions are in V_1. The functions $\Psi_j(x-k,y-l)$ for all $k,l\in\mathbb{Z}$ and $j=1,2,3$ form an orthonormal system. This system spans the orthogonal complement of V_0 in V_1. To see this let us observe that two such functions $\Psi_i(x-k_1,y-l_1)$ and $\Psi_j(x-k_2,y-l_2)$ have disjoint supports unless $k_1=k_2$ and $l_1=l_2$. But in this case they are orthogonal whenever $i\neq j$. Since every function from V_0 is constant on the support of each function from our system and $\int_{\mathbb{R}^2}\Psi_i(x,y)\,dx\,dy=0$ for $i=1,2,3$, we see that all functions from our system are orthogonal to V_0. The fact that our system spans the whole complement of V_0 in V_1 can be checked on each square separately, where it is obvious. •

We will try to follow the procedure indicated in the above example in the general setting, assuming for simplicity of notation that $d=2$.

Suppose that on $\mathbb{R}$ we are given two multiresolution analyses, say $\ldots\subset V_{-1}^i\subset V_0^i\subset V_1^i\subset\ldots$ with scaling functions $\Phi_i(x)$ and corresponding wavelets $\Psi_i(x)$ where $i=1,2$. Let us define subspaces $F_j\subset L_2(\mathbb{R}^2)$ as

$$F_j=V_j^1\otimes V_j^2. \tag{5.6}$$

The sequence of subspaces $(F_j)_{j\in\mathbb{Z}}$ has the following properties:

$$\ldots\subset F_{-1}\subset F_0\subset F_1\subset\ldots \tag{5.7}$$

$$\overline{\bigcup_{j\in\mathbb{Z}}F_j}=L_2(\mathbb{R}^2) \tag{5.8}$$

$$\bigcap_{j\in\mathbb{Z}}F_j=\{0\} \tag{5.9}$$

$$f(x,y)\in F_j\Longleftrightarrow f(2^{-j}x,2^{-j}y)\in F_0 \tag{5.10}$$

$$f(x,y)\in F_0\Longleftrightarrow f(x-k,y-l)\in F_0 \text{ for all } k,l\in\mathbb{Z} \tag{5.11}$$

the system $\{\Phi_1(x-k)\Phi_2(y-l)\}_{k,l\in\mathbb{Z}}$ is an orthonormal basis in F_0. (5.12)

If we write $V_1^i=V_0^i\oplus W_0^i$ for $i=1,2$ then we infer that

$$\begin{aligned} F_1 &= V_1^1\otimes V_1^2=(V_0^1\oplus W_0^1)\otimes(V_0^2\oplus W_0^2)\\ &= (V_0^1\otimes V_0^2)\oplus(V_0^1\otimes W_0^2)\oplus(W_0^1\otimes V_0^2)\oplus(W_0^1\otimes W_0^2)\\ &= F_0\oplus(V_0^1\otimes W_0^2)\oplus(W_0^1\otimes V_0^2)\oplus(W_0^1\otimes W_0^2). \end{aligned}$$

We also infer that

$$\begin{aligned}
\{\Psi_1(x-k)\Psi_2(y-l)\}_{k,l\in\mathbb{Z}} &\quad \text{is an orthonormal basis in} \quad W_0^1\otimes W_0^2\\
\{\Psi_1(x-k)\Phi_2(y-l)\}_{k,l\in\mathbb{Z}} &\quad \text{is an orthonormal basis in} \quad W_0^1\otimes V_0^2\\
\{\Phi_1(x-k)\Psi_2(y-l)\}_{k,l\in\mathbb{Z}} &\quad \text{is an orthonormal basis in} \quad V_0^1\otimes W_0^2.
\end{aligned}$$

Using properties 5.7–5.12 we get three functions $f_1 = \Psi_1\otimes\Psi_2$, $f_2 = \Psi_1\otimes\Phi_2$ and $f_3 = \Phi_1\otimes\Psi_2$ such that the system $\{f_i(2^jx-k, 2^jy-l)\}$ with $j,k,l\in\mathbb{Z}$ and $i=1,2,3$ is an orthonormal basis in $L_2(\mathbb{R}^2)$.

The above considerations indicate that if we use dilations of the form 5.2 in the multivariable situation, we should not require one wavelet but several.

To conclude this section we will state without detailed proof what the above procedure gives for d variables.

Proposition 5.2 *Suppose we have d multiresolution analyses in $L_2(\mathbb{R})$ with scaling functions $\Phi^{0,j}(x)$ and associated wavelets $\Phi^{1,j}(x)$ for $j=1,2,\ldots,d$. Let $E=\{0,1\}^d\backslash(0,\ldots,0)$. For $e=(e_1,\ldots,e_d)\in E$ let $\Psi^e=\bigotimes_{j=1}^d\Phi^{e_j,j}$. Then the system*

$$\left\{2^{\frac{dj}{2}}\Psi^e(2^jx-\gamma)\right\}_{e\in E,\, j\in\mathbb{Z},\,\gamma\in\mathbb{Z}^d}$$

is an orthonormal basis in $L_2(\mathbb{R}^d)$. Here, as usual we use the notation $2^jx-\gamma=(2^jx_1-\gamma_1,\ldots,2^jx_d-\gamma_d)$.

REMARK 5.1. If we apply this procedure to a fixed multiresolution analysis with a C^k compactly supported scaling function and a corresponding compactly supported C^k wavelet we obtain a wavelet set $(\Psi^s)_{s=1}^{2^d-1}$ consisting of C^k functions such that $\operatorname{supp}\Psi^s\subset[-C,C]^d$ for some constant C. It follows from Theorem 4.7 that C depends linearly on k.

5.1.1 Multidimensional notation

Naturally the notation in $\mathbb{R}^d$ tends to be more cumbersome than in $\mathbb{R}$. We will try to make it as user friendly as possible by employing the following conventions:

- Generally we will use one letter, e.g. t, x, ξ, to denote a point in $\mathbb{R}^d$. Only when coordinates are really essential will we write $x=(x_1,\ldots,x_d)$.
- The same will apply to elements of $\mathbb{Z}^d$, which will be generically denoted by γ.

- Addition and subtraction of points from $\mathbb{R}^d$ will be understood coordinatewise, i.e. $x + y$ denotes $(x_1 + y_1, \ldots, x_d + y_d)$. For a number c and $x \in \mathbb{R}^d$ by cx we mean $(cx_1, \ldots, cx_d)$.
- For a set $A \subset \mathbb{R}^d$ and a point $x \in \mathbb{R}^d$, by $x + A$ we mean the set $\{a + x : a \in A\}$. Analogously for a set $A \subset \mathbb{R}^d$ and a real number c, by cA we mean the set $\{ca : a \in A\}$.
- A $d \times d$ matrix A acts on $\mathbb{R}^d$ naturally.
- For a point $x \in \mathbb{R}^d$, by $|x|$ we mean a euclidean norm of x, i.e. $|x| =: \sqrt{\sum_{j=1}^d x_j^2}$.
- For a $d \times d$ matrix A, by $\|A\|$ we mean the norm of this matrix treated as an operator on $(\mathbb{R}^d, |\,.\,|)$, i.e.

$$\|A\| = \sup\{|Ax| : x \in R^d, \ |x| = 1\}.$$

- For a function $f(x)$ defined on some subset $U \subset \mathbb{R}^d$ (in particular on the whole of $\mathbb{R}^d$) $\int_U f(x)\,dx$ means the natural d-dimensional Lebesgue integral of f, so $dx = dx_1\,dx_2 \ldots dx_d$.
- For a subset $A \subset \mathbb{R}^d$, by $|A|$ we will mean its Lebesgue measure. This should not come into any conflict with the notation $|x|$ to denote the euclidean norm of a point x.
- Suppose $G \subset \mathbb{R}^d$ is an additive subgroup, i.e. if $a, b \in G$ then $a+b$ and $a-b$ are also in G. A function $f(x)$ defined on $\mathbb{R}^d$ is called G-periodic if for each $a \in G$ we have $f(x + a) = f(x)$ for almost all $x \in \mathbb{R}^d$. We will use this notation most often when $G = \mathbb{Z}^d$ or $G = 2\pi\mathbb{Z}^d$.

5.2 Multiresolution analyses

It would be quite natural (and is often done) to define a multiresolution analysis in $L_2(\mathbb{R}^d)$ as a sequence of subspaces satisfying 5.7–5.12, with obvious modifications required by the passage from $\mathbb{R}^2$ to $\mathbb{R}^d$. We will however adopt a more general definition. I see at least three reasons to do so:

- we will have many more interesting examples
- we will avoid trivial repetitions from previous chapters
- we will get a theory which for $d = 1$ will allow us to treat dilations of the form e.g. $x \mapsto 3x$.

Underlying our considerations in one variable were two types of maps of $\mathbb{R}$ – dilations defined in Definition 2.4 and discrete translations defined in Definition 2.3. Those two classes of maps are related. In order to present a coherent generalization we need to preserve the general outlines of this structure.

Our translations will be given by elements of $\mathbb{Z}^d$. Our basic dilation will be $x \mapsto Ax$ with A a fixed linear map $A : \mathbb{R}^d \to \mathbb{R}^d$ such that

$$A(\mathbb{Z}^d) \subset \mathbb{Z}^d \tag{5.13}$$

and

$$\text{all (complex) eigenvalues of } A \text{ have absolute values greater than 1.} \tag{5.14}$$

Condition 5.13 ensures the proper coordination between translations and dilations, while 5.14 means that the dilations are in fact expansions. Condition 5.13 clearly implies that the matrix A has integer entries.

Every invertible linear map $A : \mathbb{R}^d \longrightarrow \mathbb{R}^d$ induces a unitary operator on $L_2(\mathbb{R}^d)$ by the formula

$$U_A f(x) = |\det A|^{1/2} f(Ax). \tag{5.15}$$

Now we are ready to define the multivariable generalization of our basic concept.

DEFINITION 5.3 *A wavelet set associated with a dilation matrix A is a* finite *set of functions $\Psi^r(x) \in L_2(\mathbb{R}^d)$, $r = 1, 2, \ldots, s$ such that the system*

$$\left\{ |\det A|^{j/2} \Psi^r(A^j x - \gamma) \right\}$$

with $r = 1, 2, \ldots, s$, $j \in \mathbb{Z}$ and $\gamma \in \mathbb{Z}^d$ is an orthonormal basis in $L_2(\mathbb{R}^d)$.

REMARK 5.2. This is clearly a generalization of the notion of wavelet, which is simply a wavelet set consisting of one element. In the previous section we saw the need for this generalization. We will use the word *wavelet* to denote any element of some wavelet set.

By analogy with the one-dimensional case we will use the following notation: for a function F (Φ, Ψ etc.) on $\mathbb{R}^d$, by $F_{j\gamma}$ we will mean

$$F_{j\gamma}(x) =: |\det A|^{j/2} F(A^j x - \gamma)$$

where naturally $j \in \mathbb{Z}$ and $\gamma \in \mathbb{Z}^d$. The dilation matrix A has to be understood from the context.

A multidimensional multiresolution analysis is defined as follows.

DEFINITION 5.4 *A multiresolution analysis associated with a dilation matrix A is a sequence of closed subspaces $(V_j)_{j\in\mathbb{Z}}$ of $L_2(\mathbb{R}^d)$ satisfying*

(i) $\ldots \subset V_{-1} \subset V_0 \subset V_1 \subset \ldots$

(ii) $\bigcup_{j\in\mathbb{Z}} V_j$ *is dense in* $L_2(\mathbb{R}^d)$

(iii) $\bigcap_{j\in\mathbb{Z}} V_j = \{0\}$

(iv) $f \in V_j \iff f(Ax) \in V_{j+1}$, *i.e.* $V_j = U_A^j V_0$

(v) $f \in V_0 \iff f(x-\gamma) \in V_0$ *for all* $\gamma \in \mathbb{Z}^d$

(vi) *there exists a function $\Phi \in V_0$ called a scaling function, such that the system $\{\Phi(t-\gamma)\}_{\gamma\in\mathbb{Z}^d}$ is an orthonormal basis in V_0.*

REMARK 5.3. The reader may wonder if it is possible to generalize the notion of translations. One can replace $\mathbb{Z}^d$ everywhere above by any discrete lattice $\Gamma \subset \mathbb{R}^d$, i.e. $\Gamma = S(\mathbb{Z}^d)$ for some invertible matrix S, and define a multiresolution analysis associated with A and Γ. This is not a real generalization, however, because the spaces $\tilde{V}_j = U_{S^{-1}} V_j$ would form a multiresolution analysis in the sense of Definition 5.4, associated with the dilation matrix $A_1 = S^{-1}AS$.

We will produce many examples of multiresolution analyses later in this chapter. For the moment let us indicate only few simple examples.

1. Let us take the dilation on $\mathbb{R}$ given by $Ax = 3x$. Starting from this dilation we can produce spline multiresolution analyses. We simply start with the same V_0 and the same scaling function as for the dilation $x \mapsto 2x$ and produce the other V_j's using the dilation A.

2. We can also observe that if $(V_j)_{j\in\mathbb{Z}}$ is a multiresolution analysis associated with the matrix A then $(V_{2j})_{j\in\mathbb{Z}}$ is a multiresolution analysis associated with the matrix A^2.

3. On $\mathbb{R}^2$ let us consider the dilation $A = 2Id$. Let us fix a triangulation of the plane $\mathbb{R}^2$ given by the family of all triangles with vertices $\{(k,l),(k+1,l),(k,l+1)\}$ or $\{(k,l),(k-1,l),(k,l-1)\}$ for $k,l \in \mathbb{Z}$, see Figure 5.1. The space V_0 is defined as the space of all continuous, $L_2(\mathbb{R}^2)$ functions which are affine on each of the above triangles. The Riesz basis in V_0 is given by translations of the 'hexagonal pyramid' that is the function f from V_0 given by the conditions $f(0,0) = 1$ and $f(k,l) = 0$ for all other $k,l \in \mathbb{Z}$.

Now we want to present the general theorem about the existence of L_2-wavelets associated with a multiresolution analysis. Our general strategy will be parallel to the one-dimensional construction given in Chapter 2.

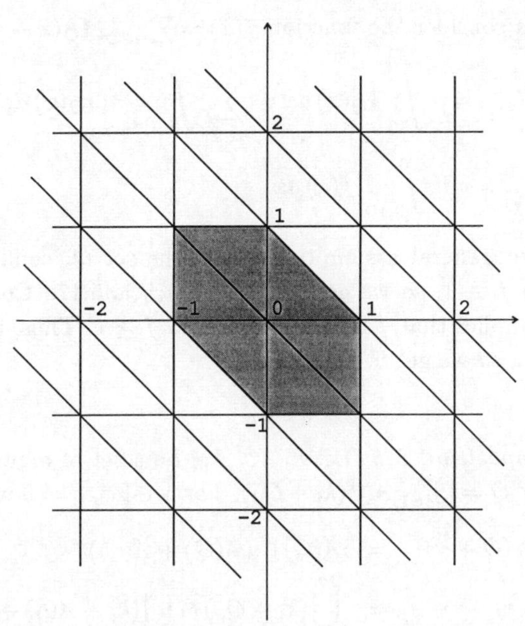

Fig. 5.1. The triangulation of the plane giving the 'hexagonal pyramid' and its support

An important role will be played in our consideration by the number $q =: |\det A|$. Since $A(\mathbb{Z}^d) \subset \mathbb{Z}^d$ we see that A is an integer matrix so q is an integer and $q > 1$. This last inequality follows from our assumption 5.14 about the eigenvalues of A. We will treat $\mathbb{Z}^d$ as an additive group. Then $A(\mathbb{Z}^d)$ is a normal subgroup, so we can form the cosets of $A(\mathbb{Z}^d)$ in $\mathbb{Z}^d$. They naturally form a group. A subset of $\mathbb{Z}^d$ consisting of exactly one element of each coset of $A(\mathbb{Z}^d)$ in $\mathbb{Z}^d$ will be called a *set of digits.*

Proposition 5.5 *The number of different cosets of $A(\mathbb{Z}^d)$ in $\mathbb{Z}^d$ equals $q = |\det A|$.*

Surprisingly enough the proof of this algebraic fact is entircly analytic. It is based on the following lemma which will be also used later.

Lemma 5.6 *Suppose that $Q \subset \mathbb{R}^d$ is a measurable subset such that $\bigcup_{\gamma \in \mathbb{Z}^d}(Q+\gamma) = \mathbb{R}^d$. Then the following conditions are equivalent*

$$|Q \cap (Q+\gamma)| = 0 \quad \text{for every } \gamma \in \mathbb{Z}^d, \ \gamma \neq 0 \tag{5.16}$$

$$|Q| = 1. \tag{5.17}$$

Proof Let us consider the function $f(x) = \sum_{\gamma \in \mathbb{Z}^d} \mathbf{1}_Q(x-\gamma)$. We have

$$\begin{aligned} |Q| &= \int_{\mathbb{R}^d} \mathbf{1}_Q(x)\,dx = \sum_{\gamma \in \mathbb{Z}^d} \int_{[0,1]^d+\gamma} \mathbf{1}_Q(x)\,dx \\ &= \int_{[0,1]^d} f(x)\,dx. \end{aligned}$$

In view of our general assumption about the set Q, condition 5.16 is equivalent to $f \equiv 1$, so we get $|Q| = 1$ which is 5.17. Conversely our assumption implies that f is $\mathbb{Z}^d$ periodic and $f \geq 1$. Thus, if $|Q| = 1$ we get $f = 1$ a.e., so we get 5.16. □

Proof of Proposition 5.5 Let $k_1, \ldots, k_m$ be a set of digits. Let $Q_0 = [0,1]^d$ and let $Q = \bigcup_{i=1}^m A^{-1}(k_i + Q_0)$. For $\gamma \in \mathbb{Z}^d$, $\gamma \neq 0$ we have

$$\begin{aligned} A(Q \cap (Q+\gamma)) &= A(Q) \cap (A(Q) + A(\gamma)) \\ &= \bigcup_{i=1}^m (k_i + Q_0) \cap \bigcup_{i=1}^m (k_i + A(\gamma) + Q_0). \end{aligned}$$

Since translations of Q_0 are disjoint and $k_i \neq k_j + A(\gamma)$ for all i, j, (remember $\gamma \neq 0$) we infer that $A(Q \cap (Q+\gamma)) = \emptyset$ so also $Q \cap (Q+\gamma) = \emptyset$. Since also

$$\begin{aligned} \bigcup_{\gamma \in \mathbb{Z}^d} (Q+\gamma) &= A^{-1}\Big(\bigcup_{\gamma \in \mathbb{Z}^d} \bigcup_{i=1}^m (k_i + A(\gamma) + Q_0) \Big) \\ &= A^{-1}\Big(\bigcup_{\gamma \in \mathbb{Z}^d} (Q_0 + \gamma) \Big) = A^{-1}(\mathbb{R}^d) = \mathbb{R}^d \end{aligned}$$

we infer from Lemma 5.6 that $|Q| = 1$. But Q is the union of m disjoint sets $A^{-1}(k_i + Q_0)$, each having measure $1/q$. This gives $m = q$ □

Before we proceed we need to state an analog of Propositions 2.8–2.11.

Proposition 5.7 *Let F be a function in $L_2(\mathbb{R}^d)$. Then*

(i) $\{F(t-\gamma)\}_{\gamma \in \mathbb{Z}^d}$ *is a Riesz sequence if and only if*

$$0 < c \leq \sum_{l \in \mathbb{Z}^d} |\hat{F}(\xi + 2\pi l)|^2 \leq C < \infty \tag{5.18}$$

for some constants c and C and almost all $\xi \in \mathbb{R}^d$.

(ii) $\{F(t-\gamma)\}_{\gamma\in\mathbb{Z}^d}$ *is an orthonormal sequence if and only if*

$$\sum_{l\in\mathbb{Z}^d} |\hat{F}(\xi+2\pi l)|^2 = (2\pi)^{-d} \quad a.e. \tag{5.19}$$

(iii) *If* $\{F(t-\gamma)\}_{\gamma\in\mathbb{Z}^d}$ *is a Riesz basis in a subspace* $X \subset L_2(\mathbb{R}^d)$ *then there exists a function* F_1 *such that* $\{F_1(t-\gamma)\}_{\gamma\in\mathbb{Z}^d}$ *is an orthonormal basis in* X.

Proof The proof that follows is basically the same as given in Chapter 2. Let us start with the following calculation, which follows from basic properties of the Fourier transform (cf. A1.2–IV and A1.2–III) and the fact that $\sum_{k\in\mathbb{Z}^d} a_k e^{-i\langle k,x\rangle}$ is always a $2\pi\mathbb{Z}^d$-periodic function:

$$\int_{\mathbb{R}^d} \Big|\sum_{k\in\mathbb{Z}^d} a_k F(x-k)\Big|^2 dx = \int_{\mathbb{R}^d} \Big|\sum_{k\in\mathbb{Z}^d} a_k e^{-i\langle\xi,k\rangle}\hat{F}(\xi)\Big|^2 d\xi$$

$$= \int_{[0,2\pi]^d} \Big|\sum_{k\in\mathbb{Z}^d} a_k e^{-i\langle\xi,k\rangle}\Big|^2 \sum_{l\in\mathbb{Z}^d} |\hat{F}(\xi+2\pi l)|^2 \, d\xi.$$

From the above calculation (i) and (ii) routinely follow, while to get (iii) we define

$$\hat{F}_1(\xi) = \Big(\sum_{l\in\mathbb{Z}^d} |\hat{F}(\xi+2\pi l)|^2\Big)^{-1/2} \hat{F}(\xi).$$

□

Lemma 5.8 *Let* $(V_j)_{j\in\mathbb{Z}}$ *be a multiresolution analysis with a scaling function* $\Phi(x)$. *The function* f *belongs to* V_1 *if and only if* $\hat{f}(A^*\xi) = m_f(\xi)\hat{\Phi}(\xi)$ *where* $m_f(\xi)$ *is a* $2\pi\mathbb{Z}^d$*-periodic function and we have*

$$\int_{[0,2\pi]^d} |m_f(\xi)|^2 \, d\xi = \frac{(2\pi)^d}{q} \int_{\mathbb{R}^d} |f(x)|^2 \, dx. \tag{5.20}$$

REMARK 5.4. This is a direct analog of 2.35.

Proof From Definition 5.4 we get that $f \in V_1$ if and only if $f(A^{-1}x) \in V_0$. This means that

$$f(A^{-1}x) = \sum_{\gamma\in\mathbb{Z}^d} a_\gamma \Phi(x-\gamma). \tag{5.21}$$

Taking the Fourier transform and using A1.2–VII we obtain

$$q\hat{f}(A^*\xi) = \sum_{\gamma\in\mathbb{Z}^d} a_\gamma e^{i\langle\xi,\gamma\rangle}\hat{\Phi}(\xi) \tag{5.22}$$

which clearly gives $m_f(\xi) = q^{-1}\sum_{\gamma\in\mathbb{Z}^d} a_\gamma e^{i\langle\xi,\gamma\rangle}$. From this we see that

$$\int_{[0,2\pi]^d} |m_f(\xi)|^2\, d\xi = \frac{(2\pi)^d}{q^2}\sum_{\gamma\in\mathbb{Z}^d}|a_\gamma|^2$$

and from 5.21 we get

$$\sum_{\gamma\in\mathbb{Z}^d}|a_\gamma|^2 = \int_{\mathbb{R}^d}|f(A^{-1}x)|^2\, dx = q\int_{\mathbb{R}^d}|f(x)|^2\, dx$$

so 5.20 follows. □

Let us now fix the notation to be used throughout the rest of this chapter.

Let $E_0, E_1, \ldots, E_{q-1}$ be different cosets of $A(\mathbb{Z}^d)$ in $\mathbb{Z}^d$. Assume that we have a function $G \in V_1$. From Lemma 5.8 we get a $2\pi\mathbb{Z}^d$-periodic function

$$m_G(\xi) = \sum_{\gamma\in\mathbb{Z}^d} b(\gamma)e^{i\langle\xi,\gamma\rangle}$$

such that $\hat{G}(A^*\xi) = m_G(\xi)\hat{\Phi}(\xi)$. By $m_G^r(\xi)$ for $r = 0, 1, \ldots, q-1$ let us denote the function $\sum_{\gamma\in E_r} b(\gamma)e^{i\langle\xi,\gamma\rangle}$. Clearly $m_G(\xi) = \sum_{r=0}^{q-1} m_G^r(\xi)$. Since each coset E_r can be written as $\Gamma_r + A(\mathbb{Z}^d)$ for appropriate $\Gamma_r \in \mathbb{Z}^d$ we get

$$\begin{aligned} m_G^r(\xi) &= e^{i\langle\xi,\Gamma_r\rangle}\cdot\sum_{\gamma\in A(\mathbb{Z}^d)}\tilde{b}(\gamma)e^{i\langle\xi,\gamma\rangle} \\ &= e^{i\langle\xi,\Gamma_r\rangle}\cdot\sum_{\gamma\in\mathbb{Z}^d} c(\gamma)e^{i\langle A^*\xi,\gamma\rangle}. \end{aligned}$$

Writing $\mu_G^r(\xi) = \sum_{\gamma\in\mathbb{Z}^d} c(\gamma)e^{i\langle\xi,\gamma\rangle}$ we have

$$m_G^r(\xi) = e^{i\langle\xi,\Gamma_r\rangle}\mu_G^r(A^*\xi).$$

Proposition 5.9 *Let $G_0, G_1, \ldots, G_{q-1}$ be functions in V_1 and let us shorten the above notation to $m_{G_k}(\xi) = m_k(\xi)$ and $\mu_{G_k}^r(\xi) = \mu_k^r(\xi)$. With the notation established above we have*

(i) *The system* $\{G_0(t-\gamma)\}_{\gamma\in\mathbb{Z}^d}$ *is an orthonormal system if and only if*

$$\sum_{r=0}^{q-1} |\mu_0^r(\xi)|^2 = 1 \ a.e. \tag{5.23}$$

(ii) *The system* $\{G_j(t-\gamma)\}_{\gamma\in\mathbb{Z}^d,\ j=0,1,\ldots,s}$ *is orthonormal if and only if the vectors* $v_j(\xi) = (\mu_j^r(\xi))_{r=0}^{s}$ *are orthonormal in* $\mathbb{C}^q$ *for almost all* $\xi \in \mathbb{R}^d$

(iii) *The system* $\{G_j(t-\gamma)\}_{\gamma\in\mathbb{Z}^d,\ j=0,1,\ldots,q-1}$ *is an orthonormal basis in* V_1 *if and only if the matrix*

$$U(\xi) = \left[\mu_j^r(\xi)\right]_{r,j=0,1,\ldots,q-1}$$

is unitary for almost all $\xi \in \mathbb{R}^d$.

REMARK 5.5. Observe that condition (i) of the above proposition corresponds to Lemma 2.12 while condition (iii) generalizes 2.40.

Proof The main part of the proof is the computation of the scalar product. Fix $j_1, j_2 \in \{0, 1, \ldots, q-1\}$ and $\gamma_1, \gamma_2 \in \mathbb{Z}^d$. We are interested in

$$I =: \langle G_{j_1}(x-\gamma_1), G_{j_2}(x-\gamma_2)\rangle .$$

From Plancherel's theorem A1.2–IV and A1.2–III we get

$$I = \int_{\mathbb{R}^d} \hat{G}_{j_1}(\xi)\overline{\hat{G}_{j_2}(\xi)} e^{i\langle\xi,\gamma_1-\gamma_2\rangle}\, d\xi.$$

Substituting $\xi = A^*\eta$ and using Lemma 5.8 we get

$$I = q\int_{\mathbb{R}^d} m_{j_1}(\eta)\overline{m_{j_2}(\eta)}\hat{\Phi}(\eta)\overline{\hat{\Phi}(\eta)} e^{i\langle A^*\eta,\gamma_1-\gamma_2\rangle}\, d\eta.$$

Since $m_{j_1}(\eta)$ and $m_{j_2}(\eta)$ are $2\pi\mathbb{Z}^d$-periodic we get

$$I = q\int_{[0,2\pi]^d} m_{j_1}(\eta)\overline{m_{j_2}(\eta)} e^{i\langle\eta,A(\gamma_1-\gamma_2)\rangle} \sum_{l\in\mathbb{Z}^d} |\hat{\Phi}(\eta+2\pi l)|^2\, d\eta.$$

Since Φ is a scaling function, from Definition 5.4 and formula 5.19 we obtain

$$\begin{aligned} I &= \frac{q}{(2\pi)^d}\int_{[0,2\pi]^d} m_{j_1}(\eta)\overline{m_{j_2}(\eta)} e^{i\langle\eta,A(\gamma_1-\gamma_2)\rangle}\, d\eta \\ &= \frac{q}{(2\pi)^d}\int_{[0,2\pi]^d} \left(\sum_{r=0}^{q-1} m_{j_1}^r(\eta)\right)\overline{\left(\sum_{r=0}^{q-1} m_{j_2}^r(\eta)\right)} e^{i\langle\eta,A(\gamma_1-\gamma_2)\rangle}\, d\eta \end{aligned}$$

$$= \frac{q}{(2\pi)^d} \int_{[0,2\pi]^d} \left(\sum_{r=0}^{q-1} e^{i\langle \eta, A(\gamma_1 - \gamma_2)\rangle} m_{j_1}^r(\eta) \right) \overline{\left(\sum_{r=0}^{q-1} m_{j_2}^r(\eta) \right)} d\eta.$$

Since

$$m_{j_1}^r(\eta) = \sum_{\gamma \in E_r} b(\gamma) e^{i\langle \eta, \gamma \rangle}$$

and E_r is a coset of $A(\mathbb{Z}^d)$ in $\mathbb{Z}^d$ we infer that

$$e^{i\langle \eta, A(\gamma_1 - \gamma_2)\rangle} m_{j_1}^r(\eta) = \sum_{\gamma \in E_r} c(\gamma) e^{i\langle \eta, \gamma \rangle}.$$

Since also

$$m_{j_2}^r(\eta) = \sum_{\gamma \in E_r} \tilde{b}(\gamma) e^{i\langle \eta, \gamma \rangle}$$

and cosets are disjoint, using orthogonality of exponentials we get

$$\begin{aligned} I &= \frac{q}{(2\pi)^d} \int_{[0,2\pi]^d} \sum_{r=0}^{q-1} e^{i\langle \eta, A(\gamma_1 - \gamma_2)\rangle} m_{j_1}^r(\eta) \overline{m_{j_2}^r(\eta)}\, d\eta \\ &= \frac{q}{(2\pi)^d} \int_{[0,2\pi]^d} e^{i\langle \eta, A(\gamma_1 - \gamma_2)\rangle} \sum_{r=0}^{q-1} m_{j_1}^r(\eta) \overline{m_{j_2}^r(\eta)}\, d\eta \\ &= \frac{q}{(2\pi)^d} \int_{[0,2\pi]^d} e^{i\langle A^*(\eta), (\gamma_1 - \gamma_2)\rangle} \sum_{r=0}^{q-1} \mu_{j_1}^r(A^*\eta) \overline{\mu_{j_2}^r(A^*\eta)}\, d\eta \\ &= \frac{1}{(2\pi)^d} \int_{[0,2\pi]^d} e^{i\langle \xi, (\gamma_1 - \gamma_2)\rangle} \sum_{r=0}^{q-1} \mu_{j_1}^r(\xi) \overline{\mu_{j_2}^r(\xi)}\, d\xi. \end{aligned}$$

Thus putting all the above together we get

$$\begin{aligned} &\langle G_{j_1}(x - \gamma_1), G_{j_2}(x - \gamma_2) \rangle \\ &\qquad = \frac{1}{(2\pi)^d} \int_{[0,2\pi]^d} e^{i\langle \xi, (\gamma_1 - \gamma_2)\rangle} \sum_{r=0}^{q-1} \mu_{j_1}^r(\xi) \overline{\mu_{j_2}^r(\xi)}\, d\xi. \end{aligned} \tag{5.24}$$

Now let us prove (i). For $j_1 = j_2 = 0$ we interpret the right hand side of 5.24 as an appropriate Fourier coefficient of the function $\sum_{r=0}^{q-1} |\mu_0^r(\xi)|^2$. Thus if $\{G_0(t-\gamma)\}_{\gamma \in \mathbb{Z}^d}$ is orthonormal we get $\sum_{r=0}^{q-1} |\mu_0^r(\xi)|^2 = 1$ a.e. and conversely if $\sum_{r=0}^{q-1} |\mu_0^r(\xi)|^2 = 1$ a.e we get orthogonality of the system $\{G_0(t-\gamma)\}_{\gamma \in \mathbb{Z}^d}$. We infer (ii) from 5.24 in a similar fashion. To prove (iii) let us take q functions $G_0, G_1, \ldots, G_{q-1}$ in V_1 such that the system $\{G_j(t-\gamma)\}_{\gamma \in \mathbb{Z}^d,\ j=0,1,\ldots,q-1}$ is orthonormal and suppose that it is not a basis in V_1. Thus we can take a function, call it G_q, in V_1 such that

$$G_q \perp \{G_j(t-\gamma)\}_{\gamma\in\mathbb{Z}^d,\ j=0,1,\dots,q-1}\,.$$

From 5.24 with $j_1 = q$ we get that

$$\sum_{r=0}^{q-1} \mu_q^r(\xi)\overline{\mu_j^r(\xi)} = 0 \quad \text{a.e.}$$

for all $j = 0,1,2,\dots,q-1$. This gives us $(q+1)$ orthogonal vectors in $\mathbb{C}^q$, which is impossible. This contradiction shows that the vector $(\mu_q^0(\xi),\dots,\mu_q^{q-1}(\xi)) = 0$ a.e. so $G_q = 0$. □

Now let us see what is needed to construct a wavelet set associated with a multiresolution analysis. Let us write $W_j = V_{j+1} \ominus V_j$. Then we have $L_2(\mathbb{R}^d) = \oplus\sum_{j\in\mathbb{Z}} W_j$. From condition (iv) of Definition 5.4 we see that $W_j = U_A^j W_0$. Thus in order to find a wavelet set associated with a multiresolution analysis it clearly suffices to construct a finite set of functions $\Psi^s(x) \in W_0$ for $s \in C$ such that $\{\Psi^s(t-\gamma)\}_{\gamma\in\mathbb{Z}^d,\ s\in C}$ is an orthonormal basis in W_0. But then

$$\{\Psi^s(t-\gamma)\}_{\gamma\in\mathbb{Z}^d, s\in C} \cup \{\Phi(t-\gamma)\}_{\gamma\in\mathbb{Z}^d}\,,$$

where Φ is a scaling function, is an orthonormal basis in V_1. Now Proposition 5.9 tells us that such a wavelet set has to have cardinality $(q-1)$ and gives us a way to construct it. We simply take functions $\mu_0^r(\xi) = \mu_\Phi^r(\xi)$ for $r = 0,1,\dots,q-1$ as described before Proposition 5.9 and choose $2\pi\mathbb{Z}^d$-periodic functions $\mu_j^r(\xi)$ for $r = 0,1,\dots,q-1$ and $j = 1,2,\dots,q-1$ so that the matrix

$$U(\xi) = \left[\mu_j^r(\xi)\right]_{r,j=0,1,\dots,q-1}$$

is unitary. There is no difficulty finding such measurable μ_j^r's. This simply requires building a unitary matrix whose first row is given. Having done this we define functions $\Psi^s(x)$ for $s = 1,2,\dots,q-1$ by the condition

$$\hat{\Psi}^s(A^*\xi) = \sum_{r=0}^{q-1} e^{i\langle\xi,\Gamma_r\rangle}\mu_s^r(A^*\xi)\hat{\Phi}(\xi) \tag{5.25}$$

where the Γ_r's are representatives of different cosets of $A(\mathbb{Z}^d)$ in $\mathbb{Z}^d$. It follows directly from Proposition 5.9 (iii) and our earlier discussion that $(\Psi^s)_{s=1}^{q-1}$ is a wavelet set associated with the given multiresolution analysis. We can summarize our considerations in the following theorem.

Theorem 5.10 *For every multiresolution analysis on $\mathbb{R}^d$ associated with a dilation matrix A there exists an associated wavelet set (consisting of $q-1$ functions).*

REMARK 5.6. The above theorem says nothing about the smoothness and decay of the wavelet set obtained. Since the functions $\mu_s^r(\xi)$ are bounded, formula 5.25 shows that the decay of $\hat{\Psi}^s$ is the same as the decay of $\hat{\Phi}(\xi)$, so (see A1.2–VIII and A1.2–IX) the wavelet set is basically as smooth as the scaling function. The question of the decay of Ψ^s translates into the question of the smoothness of $\hat{\Psi}^s$. Thus formula 5.25 tells us that we need smooth μ_s^r's. Here however is the problem. Suppose that the function $v(\xi) = (\mu_0^r)_{r=0}^{q-1}$ maps $\mathbb{R}^d$ onto the unit sphere in $\mathbb{C}^q$. This is perfectly possible for continuous v. If $2q-1 \leq d$ we can have such situation even for v of class C^∞. But then it is impossible to build a unitary matrix $U(\xi)$ with the first row equal $v(\xi)$ in a continuous fashion, except for very special values of q.

We will present only a very limited discussion of the regularity of multivariable wavelets. It will be done in the framework of the following definitions.

DEFINITION 5.11 *The function F on $\mathbb{R}^d$ is r-regular, if F is of class C^r, $r = -1, 0, 1, \ldots$ and*

$$\left| \frac{\partial^\alpha}{\partial x^\alpha} F(x) \right| \leq \frac{C_k}{(1+|x|)^k}$$

for each $k = 0, 1, 2, \ldots$ and each multiindex α with $|\alpha| \leq \max(r, 0)$ and some constant C_k. As usual class C^{-1} means measurable function and class C^0 means continuous function.

DEFINITION 5.12 *A multiresolution analysis on $\mathbb{R}^d$ is called r-regular if it has an r-regular scaling function.*

Lemma 5.13 *If $F(x)$ is an r-regular function on $\mathbb{R}^d$ and we define $G(x)$ by the condition $\hat{G}(\xi) = m(\xi)\hat{F}(\xi)$ for a C^∞, $2\pi\mathbb{Z}^d$-periodic function $m(\xi)$, then $G(x)$ is r-regular.*

Proof Since $m(\xi)$ is $2\pi\mathbb{Z}^d$-periodic it can be written as a Fourier series $m(\xi) = \sum_{\gamma \in \mathbb{Z}^d} a(\gamma) e^{i\langle \xi, \gamma \rangle}$, and because it is C^∞ we have

$$|a(\gamma)| \leq \frac{C_k}{(1+|\gamma|)^k} \quad \text{for each } k = 0, 1, 2, \ldots. \tag{5.26}$$

Since we have $G(x) = \sum_{\gamma\in\mathbb{Z}^d} a(\gamma)F(x-\gamma)$, from 5.26 and Definition 5.11 we infer that for each multiindex α with $|\alpha| \le \max(r,0)$ and for any natural number $k > d+1$ we have

$$\begin{aligned} \left|\frac{\partial^\alpha}{\partial x^\alpha} G(x)\right| &\le C_k \sum_{\gamma\in\mathbb{Z}^d} \frac{1}{(1+|\gamma|)^k} \frac{1}{(1+|x-\gamma|)^k} \qquad (5.27) \\ &\le C_k \sum_{|\gamma|>2|x|} + \sum_{|\gamma|\le 2|x|} \frac{1}{(1+|\gamma|)^k} \frac{1}{(1+|x-\gamma|)^k}. \end{aligned}$$

Note that there are at most $C|x|^d$ elements $\gamma \in \mathbb{Z}^d$ with $|\gamma| < 2|x|$ so we get

$$\begin{aligned} \left|\frac{\partial^\alpha}{\partial x^\alpha} G(x)\right| &\le C_k \sum_{|\gamma|>2|x|} \frac{1}{(1+|\gamma|)^k} \\ &\quad + C|x|^d \max_{|\gamma|\le 2|x|} \frac{1}{(1+|\gamma|)^k} \frac{1}{(1+|x-\gamma|)^k} \\ &\le C_k \frac{1}{(1+|x|)^{k-d}} + C|x|^d \frac{1}{(1+|x/2|)^k} \qquad (5.28) \\ &\le \frac{C}{(1+|x|)^{k-d}}. \end{aligned}$$

This shows that G is r-regular. □

REMARK 5.7. The above calculation is analogous to the proof of 3.32.

Corollary 5.14 *If $\Phi(x)$ is an r-regular function on $\mathbb{R}^d$ such that $\{\Phi(t-\gamma)\}_{\gamma\in\mathbb{Z}^d}$ is a Riesz sequence, then the function $\Phi_1(x)$ defined by*

$$\hat{\Phi}_1(\xi) = \left(\sum_{l\in\mathbb{Z}^d} |\hat{\Phi}(\xi+2\pi l)|^2\right)^{1/2} \hat{\Phi}(\xi)$$

is also r-regular.

Proof From Lemma 5.13 we see that it suffices to show that the function $\sum_{l\in\mathbb{Z}^d} |\hat{\Phi}(\xi+2\pi l)|^2$ is of class C^∞. As in Lemma 3.15 we see that

$$\sum_{l\in\mathbb{Z}^d} |\hat{\Phi}(\xi+2\pi l)|^2 = \tfrac{1}{(2\pi)^d} \sum_{\gamma\in\mathbb{Z}^d} \int_{\mathbb{R}^d} \overline{\Phi(x)}\Phi(x-\gamma)\,dx \cdot e^{i\langle\xi,\gamma\rangle}.$$

Since Φ is (-1)-regular, repeating 5.27 and 5.28 with integrals instead of sums we easily estimate

$$\left| \int_{\mathbb{R}^d} \overline{\Phi(x)} \Phi(x-\gamma)\, dx \right| \leq \frac{C}{(1+|\gamma|)^{k-d}}. \tag{5.29}$$

Since k is arbitrary we get the corollary. □

Now we are ready to state the next theorem of this section.

Theorem 5.15 *For every r-regular multiresolution analysis on $\mathbb{R}^d$ associated with a dilation matrix A, $|\det A| = q$, such that $2q-1 > d$, there exists an associated wavelet set consisting of $q-1$ r-regular functions.*

REMARK 5.8. Note that always $q \geq 2$ so for $d = 1$ and $d = 2$ there exists an r-regular wavelet set associated with any r-regular multiresolution analysis associated with any dilation. In the most important case of dyadic dilations we have $q = 2^d$ so the r-regular wavelet set always exists.

As we know from the proof of Theorem 5.10 and the remark following it, the main obstacle in the construction of r-regular wavelets is the necessity of building a smooth unitary matrix-function given its first row. So let us first address this question.

Proposition 5.16 *Let $\mathbb{S}$ be the unit sphere in $\mathbb{C}^s$ or $\mathbb{R}^s$ and let $B \subset \mathbb{S}$ be an open subset. There exists a map $F : \mathbb{S} \setminus B \longrightarrow \mathcal{U}(s)$ (unitary $s \times s$ matrices) such that*

(i) *F is of class C^∞*
(ii) *for each $x \in \mathbb{S} \setminus B$ the first row of the matrix $F(x)$ equals x.*

Proof Without loss of generality we can assume that $(1, 0, \ldots, 0) \in B$. Let us define s functions from $\mathbb{S}$ into $\mathbb{C}^s$ (or $\mathbb{R}^s$) as follows:

$$\begin{aligned}
v_1(x) &= (x_1, x_2, \ldots, x_s) \\
v_2(x) &= (\bar{x}_2, \alpha, 0, \ldots, 0) \\
v_3(x) &= (\bar{x}_3, 0, \alpha, 0, \ldots, 0) \\
&\vdots \\
v_s(x) &= (\bar{x}_s, 0, \ldots, 0, \alpha)
\end{aligned}$$

where $x = (x_1, x_2, \ldots, x_s)$ and α is a positive real number to be fixed later. For α sufficiently small, the vectors $(v_j(x))_{j=1}^s$ are linearly independent for each $x \in \mathbb{S} \setminus B$. To see this, note that $(v_j(x))_{j=2}^s$ are always linearly independent. If we attempt to write $v_1(x)$ as a linear combination of them we get $x_1 = \frac{1}{\alpha}\sum_{j=2}^{s}|x_j|^2$. This forces x_1 to be real, and since x is separated from $(1, 0, \ldots, 0)$ we get $x_1^2 < 1 - \delta$ for some positive δ (depending on B). This implies that $\sum_{j=2}^{s}|x_j|^2 \geq \delta$. For $\alpha < \frac{2}{\delta}$ we get $x_1 > 2$ which is a clear contradiction.

Now we perform the Schmidt orthonormalization of the vectors

$$v_1(x), v_2(x), \ldots, v_s(x)$$

(in this order) to get s orthonormal vectors $v_1(x), u_2(x), \ldots, u_s(x)$. Since the vectors $v_1(x)$, $v_2(x)$, ..., $v_s(x)$ were C^∞ functions of x we infer that vectors $v_1(x)$, $u_2(x), \ldots, u_s(x)$ are also C^∞ functions of x. Thus the matrix

$$F(x) = \begin{bmatrix} v_1(x) \\ u_2(x) \\ \vdots \\ u_s(x) \end{bmatrix}$$

is the desired unitary matrix-function. □

Proof of Theorem 5.15 The argument follows exactly the argument given for Theorem 5.10. We start with the r-regular scaling function $\Phi(x)$. From estimate 5.29 we infer that the functions $\mu_0^r(\xi) = \mu_\Phi^r(\xi)$ for $r = 0, 1, \ldots, q-1$ are $2\pi\mathbb{Z}^d$-periodic C^∞ functions. Let μ be the map from $\mathbb{R}^d$ into the unit sphere $\mathbb{S} \subset \mathbb{C}^q$, given by

$$\mu(\xi) = \left(\mu_0^0(\xi), \ldots, \mu_0^{q-1}(\xi)\right).$$

Since μ is C^∞ and $2\pi\mathbb{Z}^d$-periodic and $2q - 1 > d$ we infer that $\mu(\mathbb{R}^d)$ is a proper compact subset of $\mathbb{S}$. So we can apply Proposition 5.16 to conclude that

$$U(\xi) = F(\mu(\xi)) =: \left[\mu_j^r(\xi)\right]_{r,j=0,1,\ldots,q-1}$$

is a C^∞ unitary $q \times q$ matrix whose first row is $[\mu_0^0(\xi), \ldots, \mu_0^{q-1}(\xi)]$. We define the wavelet set by the formula 5.25 and we infer from Lemma 5.13 that they are r-regular functions. □

The above argument shows that a crucial role is played by the set $S \setminus \mu(\mathbb{R}^d)$, where as above $\mu(\xi) = (\mu_0^0(\xi), \ldots, \mu_0^{q-1}(\xi))$. If this set contains an open set then the above argument can be applied and there exists a wavelet set with basically the same decay as the scaling function. One particular instance of this idea is when we start with a scaling function Φ such that $\hat{\Phi}(\xi)$ is real. Then (using the notation established before Proposition 5.9) $m_\Phi(\xi) = \sum_{\gamma \in \mathbb{Z}^d} b(\gamma) e^{i\langle \xi, \gamma \rangle}$ is also real and $b(\gamma) = \overline{b(-\gamma)}$ for $\gamma \in \mathbb{Z}^d$. This implies that $\mu_\Phi^r(\xi) = \sum_{\gamma \in \mathbb{Z}^d} c(\gamma) e^{i\langle \xi, \gamma \rangle}$ with $c(\gamma) = \overline{c(-\gamma)}$. This implies that $\mu_\Phi^r(\xi)$ is real for $r = 0, 1, \ldots, q-1$. Thus we have the following corollary.

Corollary 5.17 *Assume that we have a multiresolution analysis on $\mathbb{R}^d$ associated with a dilation matrix A, $|\det A| = q$. Assume also that this multiresolution analysis has an r-regular scaling function $\Phi(x)$ such that $\hat{\Phi}(\xi)$ is real. Then there exists a wavelet set associated with this multiresolution analysis consisting of $q-1$ r-regular functions.*

5.3 Examples of multiresolution analyses

Our aim in this section is to produce examples, hopefully interesting, of multiresolution analyses in $\mathbb{R}^d$. In a sense we want in this section to reproduce on $\mathbb{R}^d$ the program carried out in Sections 1.1, 1.2 and 3.3. More precisely we start with discussion of multiresolution analyses whose scaling functions are characteristic functions. It turns out that this seemingly easy case is already very interesting. Next we use convolution powers of the above scaling functions to produce r-regular multiresolution analyses.

In this section we will often be manipulating subsets of $\mathbb{R}^d$. Generally speaking all our statements and equations involving sets are to be understood *almost everywhere*. If we want to stress that something actually holds for *every point* we will use the word 'exactly'. We will use the notation $\equiv$ to indicate 'exactly equal'.

Multiresolution analyses whose scaling function is a characteristic function of a set are characterized by the following proposition.

Proposition 5.18 *Let Q be a measurable subset of $\mathbb{R}^d$. Suppose that the function $\mathbf{1}_Q$ is a scaling function of a multiresolution analysis associated with a dilation A. Then*

(i) *Q and $(Q+\gamma)$ are non-overlapping for all $\gamma \in \mathbb{Z}^d$, $\gamma \neq 0$*
(ii) *there exists a set of digits, say $k_1, \ldots, k_q$, such that $A(Q) = \bigcup_{i=1}^q (Q + k_i)$*
(iii) *$\bigcup_{\gamma \in \mathbb{Z}^d} (Q + \gamma) = \mathbb{R}^d$;*

and conversely, if Q satisfies (i)–(iii) then $\mathbf{1}_Q$ is a scaling function of multiresolution analysis associated with A.

REMARK 5.9. Observe that we can directly construct a wavelet set associated with any multiresolution analysis described in the above Proposition 5.18. From (ii) we see that $Q = \bigcup_{i=1}^{q} Q_i$ where $Q_i = A^{-1}(Q+k_i)$. Clearly $|Q_i| = \frac{1}{q}$. Thus to get a wavelet set it suffices to fix any $q-1$ functions $\Psi^s(x)$ for $s = 1, 2, \ldots, q-1$ such that

- $\Psi^s(x)$ are orthonormal
- $\int_{\mathbf{R}^d} \Psi^s(x)\, dx = 0$ for $s = 1, 2, \ldots, q-1$
- $\Psi^s(x) = \sum_{l=1}^{q} \alpha_l^s \mathbf{1}_{Q_l}(x)$.

There is a clear analogy between this construction and the definition of the Haar wavelet (cf. Definition 1.1)

Proof Condition (i) follows directly from the orthogonality of translates; if $c\mathbf{1}_Q(x)$ is orthogonal to $c\mathbf{1}_Q(x+\gamma)$ then $Q \cap (Q-\gamma)$ has measure zero. Condition (ii) is a consequence of the scaling equation. We have $\mathbf{1}_Q(A^{-1}x) \in V_{-1} \subset V_0$ so we can write

$$\mathbf{1}_{A(Q)}(x) = \mathbf{1}_Q(A^{-1}x) = \sum_{\gamma \in \mathbb{Z}^d} a_\gamma c\mathbf{1}_Q(x-\gamma) = \sum_{\gamma \in \mathbb{Z}^d} a_\gamma c\mathbf{1}_{Q+\gamma}(x). \quad (5.30)$$

Since we already know that the sets $\{Q+\gamma\}_{\gamma\in\mathbb{Z}^d}$ do not overlap we infer from 5.30 that ca_γ equals either 0 or 1 and that

$$A(Q) = \bigcup_{i=1}^{q} (Q + k_i) \quad (5.31)$$

for some $k_i \in \mathbb{Z}^d$. The fact that we have exactly q summands follows from the fact that $|A(Q)| = q|Q|$. If two k_j's, say k_1 and k_2, are in the same coset, then $k_1 = k_2 + A(\gamma)$ for some non-zero $\gamma \in \mathbb{Z}^d$. In this case

$$A(Q+\gamma) = A(Q) + A(\gamma) = \bigcup_{i=1}^{q} (Q + k_i + A(\gamma)) \supset Q + k_1.$$

This compared with 5.31 shows that

$$A(Q) \cap A(Q+\gamma) \supset Q + k_1.$$

Since A is 1–1 this implies that $|Q \cap Q+\gamma| > 0$. This contradiction shows that $k_1, \ldots, k_q$ are representatives of different cosets of $A(\mathbb{Z}^d)$ in $\mathbb{Z}^d$ so they form a set of digits. To show (iii) let $S = \bigcup_{\gamma\in\mathbb{Z}^d}(Q+\gamma)$

and let $L_2(S)$ denote the subspace of $L_2(\mathbb{R}^d)$ consisting of all functions whose support is contained in S. Using 5.31 we have

$$\begin{aligned} A(S) &= A(\bigcup_{\gamma\in\mathbf{Z}^d} Q+\gamma) = \bigcup_{\gamma\in\mathbf{Z}^d} A(Q)+A(\gamma) \\ &= \bigcup_{\gamma\in\mathbf{Z}^d}\bigcup_{i=1}^{q} Q+k_i+A(\gamma). \end{aligned} \tag{5.32}$$

Since $(k_i)_{i=1}^q$ is a set of digits we obtain $A(S)=S$. Clearly $V_0\subset L_2(S)$, and since $V_j=U_A^j(V_0)$ we obtain

$$V_j\subset U_A^j(L_2(S))=L_2(A^{-j}S)=L_2(S).$$

From condition (iii) of Definition 5.4 we get $S=\mathbb{R}^d$.

To prove the converse statement observe that Lemma 5.6 implies that $|Q|=1$, so $\{\mathbf{1}_Q(t-\gamma)\}_{\gamma\in\mathbf{Z}^d}$ is an orthonormal system. We define $V_0=\operatorname{span}\{\mathbf{1}_Q(t-\gamma)\}_{\gamma\in\mathbf{Z}^d}$ and $V_j=U_A^j(V_0)$. It is a routine exercise to check that we have defined a multiresolution analysis. □

Note that this easy proposition links our problem with some interesting geometric questions. The set Q satisfying (i) and (iii) of Proposition 5.18 is called a tile. It is very natural terminology since the sets $\{Q+\gamma\}_{\gamma\in\mathbf{Z}^d}$, i.e. all integer translates of Q, cover the whole space once, i.e. tile the space. Condition (ii) tells us that the tile Q is self-similar using the dilation A; the image of a tile is a union of translates of the tile.

Since our aim is to produce examples of sets Q satisfying (i)–(iii) of Proposition 5.18 it is natural to start with sets satisfying only (ii). Here are natural candidates. Let us fix a set of digits $S=\{k_1,\ldots,k_q\}$ and define the set

$$Q=\left\{x\in\mathbb{R}^d\ :\ x=\sum_{j=1}^{\infty}A^{-j}s_j\ \text{ where } s_j\in S\right\}. \tag{5.33}$$

Before we proceed let us observe that the series appearing in 5.33 is always absolutely convergent. This clearly follows from the fact that for all $x\in\mathbb{R}^d$

$$|A^{-j}x|\le C\alpha^j|x|\ \text{ for some } \alpha,\quad 0<\alpha<1 \tag{5.34}$$

with some constant C. This is a consequence of the fact that all eigenvalues of A^{-1} have absolute value strictly less than 1 (use 5.14), which implies (either by the spectral radius formula or by direct calculation using the Jordan canonical form) that there exists an r such that $|A^{-r}|<1$. From this our claim follows.

Proposition 5.19 *Let $S = \{k_1, \ldots, k_q\}$ be the set of digits and let Q be the set defined by 5.33. Then*

(i) *Q is a compact subset of $\mathbb{R}^d$,*
(ii) $A(Q) = \bigcup_{i=1}^{q} Q + k_i$
(iii) $\bigcup_{\gamma \in \mathbb{Z}^d} (Q + \gamma) \equiv \mathbb{R}^d$
(iv) *Q contains an open set.*

Proof The proof of (i) is a standard exercise in metric spaces; if $x_n = \sum_{j=1}^{\infty} A^{-j} s_j^n$ is a sequence of points in Q we pass to a subsequence (we call it (x_n) again) using a diagonal procedure and the fact that S is finite and obtain a sequence $(\delta_j)_{j=1}^{\infty} \subset S$ such that $s_j^n = \delta_j$ for $j \leq n$. Let $z = \sum_{j=1}^{\infty} A^{-j} \delta_j \in Q$. It easily follows from 5.34 that $x_n \to z$. To get (ii) we simply calculate:

$$\begin{aligned} A(Q) &= \left\{ x \in \mathbb{R}^d \ : \ x = s_1 + \sum_{j=1}^{\infty} A^{-j} s_{j+1} \ \text{ where } \ s_j \in S \right\} \\ &= \bigcup_{j=1}^{q} (Q + k_j). \end{aligned}$$

To show (iii) we put $K = \bigcup_{\gamma \in \mathbb{Z}^d} (Q + \gamma)$ and note that calculation 5.32 gives that $A(K) \equiv K$. K is a closed subset of $\mathbb{R}^d$. To show this take $x_n + \gamma_n \to z$ with $x_n \in Q$ and $\gamma_n \in \mathbb{Z}^d$. Since Q is bounded the γ_n's are also bounded so we can assume (taking a subsequence) that they are constant $= \gamma$. This implies that x_n converges (on the same subsequence) to $x_0 \in Q$, so $z = x_0 + \gamma \in K$. From the definition of K we infer that there exists a constant C such that for every $x \in \mathbb{R}^d$ there exists $x_0 \in K$ with $|x - x_0| \leq C$. Now let us fix an arbitrary $y \in \mathbb{R}^d$ and let x_n for $n = 1, 2, \ldots$ be elements of K such that $|A^n y - x_n| \leq C$. Since $A(K) \equiv K$ we infer that $x_n = A^n y_n$ for some $y_n \in K$. It follows from 5.34 that

$$|y - y_n| = |A^{-n}(A^n y - x_n)| \leq C\alpha^n$$

so $y_n \to y$, and since K is closed we get $y \in K$. This shows that $K \equiv \mathbb{R}^d$. Condition (iv) is a direct consequence of Baire's theorem. □

REMARK 5.10. The set Q defined in 5.33 is essentially the only set satisfying (ii) of Proposition 5.18 for a given set of digits. It does happen, however, that different sets of digits give very different sets Q.

Corollary 5.20 *Let Q be the set defined in 5.33. The following conditions are equivalent:*

(i) $\mathbf{1}_Q$ *is a scaling function of a multiresolution analysis*
(ii) $|Q| = 1$
(iii) $|Q \cap (Q+\gamma)| = 0$ *for every* $\gamma \in \mathbb{Z}^d$, $\gamma \neq 0$,

Proof This Corollary follows directly from Proposition 5.19, Lemma 5.6 and Proposition 5.18. □

Conditions (ii) and (iii) above seem very easy to check. Sometimes, however, this is not so. Thus we will give a sufficient condition for the set Q to satisfy (i)–(iii) above, expressed in terms of the Fourier transform. Given the set of digits $S = \{k_1, \ldots, k_q\}$ we form a trigonometric polynomial

$$m(\xi) = \frac{1}{q} \sum_{k \in S} e^{-i\langle \xi, k \rangle}. \tag{5.35}$$

Our aim now is to prove the following proposition.

Proposition 5.21 *Let $m(\xi)$ be given by 5.35. Suppose that there exists a compact set $K \subset \mathbb{R}^d$ such that*

(i) K *contains a neighborhood of* 0
(ii) $\bigcup_{\gamma \in \mathbb{Z}^d} (K + 2\pi\gamma) = \mathbb{R}^d$
(iii) $|K \cap (K + 2\pi\gamma)| = 0$ *for all* $\gamma \in \mathbb{Z}^d$, $\gamma \neq 0$
(iv) $m((A^*)^{-j}\xi) \neq 0$ *for all* $\xi \in K$ *and all integers* $j > 0$.

Then for the set Q defined by 5.33 the function $\mathbf{1}_Q$ is a scaling function of a multiresolution analysis.

REMARK 5.11. The sufficient condition for $\mathbf{1}_Q$ to be a scaling function of a multiresolution analysis, expressed in this proposition, is called Cohen's condition. Actually it is also necessary, (Exercise 5.4).

REMARK 5.12. We are dealing here with special compactly supported scaling functions on $\mathbb{R}^d$, so naturally our arguments are parallel to some arguments of Chapter 4. In particular the above proposition is an analog of Lemma 4.2 with the set K replacing the interval $[-\frac{\pi}{2}, \frac{\pi}{2}]$ from condition 4.3. See also Exercise 4.8.

Before we embark on the proof let us consider a sequence of 'approximations' to the set Q. Let us put $Q_0 = [-\frac{1}{2}, \frac{1}{2}]^d$ and define inductively

$$Q_{N+1} = \bigcup_{k \in S} A^{-1}(Q_N + k).$$

Lemma 5.22 *For each $N = 0, 1, 2, \dots$ the sets Q_N satisfy*

(i) $|Q_N| = 1$
(ii) $\bigcup_{\gamma \in \mathbb{Z}^d} (Q_N + \gamma) = \mathbb{R}^d$
(iii) $|Q_N \cap (Q_N + \gamma)| = 0$ *for all non-zero* $\gamma \in \mathbb{Z}^d$.

Proof We use induction. Clearly Q_0 satisfies (i)–(iii). If Q_N satisfies (i)–(iii) we use the fact that S is the set of digits and obtain

$$A(\bigcup_{\gamma \in \mathbb{Z}^d} Q_{N+1} + \gamma) = \bigcup_{\gamma \in \mathbb{Z}^d} \bigcup_{k \in S} Q_N + k + A(\gamma) = \bigcup_{\gamma \in \mathbb{Z}^d} Q_N + \gamma = \mathbb{R}^d$$

so

$$\bigcup_{\gamma \in \mathbb{Z}^d} Q_{N+1} + \gamma = A^{-1}(\mathbb{R}^d) = \mathbb{R}^d.$$

Thus (ii) holds for Q_{N+1}. From (iii) for Q_N we infer that the sets $A^{-1}(Q_N + k)$ are disjoint, so

$$|Q_{N+1}| = |\det A^{-1}| \cdot q \cdot |Q_N| = 1$$

so we see that Q_{N+1} satisfies (i). An application of Lemma 5.6 completes the proof. □

For the set Q given by 5.33 we have

$$\mathbf{1}_Q(x) = \sum_{k \in S} \mathbf{1}_Q(Ax - k)$$

so taking the Fourier transform (and denoting $\hat{\mathbf{1}}_Q(\xi)$ by $\varphi(\xi)$ and A^* by B) we get

$$\varphi(\xi) = m(B^{-1}\xi)\varphi(B^{-1}\xi). \tag{5.36}$$

When we iterate 5.36 we get for each $N = 1, 2, \dots$

$$\varphi(\xi) = \prod_{j=1}^{N} m(B^{-j}\xi)\varphi(B^{-N}\xi) =: \nu_N(\xi)\varphi(B^{-N}\xi). \tag{5.37}$$

Since $0 < |Q| < \infty$ we see that φ is continuous and $\varphi(0) = (2\pi)^{-d/2}|Q|$. From 5.34 we infer that $B^{-N}\xi \to 0$ for each $\xi \in \mathbb{R}^d$, so

$$\nu_N(\xi) \text{ tends pointwise to } \frac{(2\pi)^{d/2}}{|Q|}\varphi(\xi). \tag{5.38}$$

Analogously for the sets Q_N we have

$$\mathbf{1}_{Q_{N+1}}(x) = \sum_{k \in S} \mathbf{1}_{Q_N}(Ax - k)$$

so taking the Fourier transform and denoting $\hat{\mathbf{1}}_{Q_N}(\xi)$ by $\varphi_N(\xi)$ and (as before) A^* by B we get

$$\varphi_{N+1}(\xi) = m(B^{-1}\xi) \cdot \varphi_N(B^{-1}\xi).$$

Iterating this we get

$$\varphi_N(\xi) = \nu_N(\xi)\varphi_0(B^{-N}\xi). \tag{5.39}$$

Proof of Proposition 5.21 We know from Proposition 5.7 that

$$\sum_{\gamma \in \mathbb{Z}^d} |\varphi_0(\xi - 2\pi\gamma)|^2 = \frac{1}{(2\pi)^d} \quad \text{a.e.} \tag{5.40}$$

Let us denote $\Omega = [-\pi, \pi]^d$. From 5.39 we get

$$\begin{aligned} \int_{\mathbb{R}^d} |\varphi_N(\xi)|^2 \, d\xi &= \int_{\mathbb{R}^d} |\nu_N(\xi)|^2 |\varphi_0(B^{-N}\xi)|^2 \, d\xi \\ &= \sum_{\gamma \in \mathbb{Z}^d} \int_{B^N\Omega + B^N(\gamma)2\pi} |\nu_N(\xi)|^2 |\varphi_0(B^{-N}\xi)|^2 d\xi. \end{aligned}$$

Since $\nu_N(\xi)$ is $2\pi B^N(\mathbb{Z}^d)$-periodic, this equals

$$\int_{B^N\Omega} |\nu_N(\xi)|^2 \cdot \sum_{\gamma \in \mathbb{Z}^d} |\varphi_0(B^{-N}\xi - 2\pi\gamma)|^2 d\xi$$

which by 5.40 equals

$$\frac{1}{(2\pi)^d} \int_{B^N\Omega} |\nu_N(\xi)|^2 d\xi.$$

Using once more the fact that $\nu_N(\xi)$ is $2\pi B^N(\mathbb{Z}^d)$-periodic and properties (ii) and (iii) of the set K we see that

$$\int_{B^N\Omega} |\nu_N(\xi)|^2 d\xi = \int_{B^N K} |\nu_N(\xi)|^2 d\xi = \int_{\mathbb{R}^d} |\nu_N(\xi)|^2 \mathbf{1}_{B^N K}(\xi) \, d\xi$$

so putting things together we get

$$\int_{\mathbb{R}^d} |\varphi_N(\xi)|^2 d\xi = \frac{1}{(2\pi)^d} \int_{\mathbb{R}^d} |\nu_N(\xi)|^2 \cdot \mathbf{1}_{B^N K}(\xi) \, d\xi. \tag{5.41}$$

Observe that (iv) of Proposition 5.21 implies that $\nu_N(\xi) \neq 0$ for $\xi \in K$ and $N = 1, 2, \ldots$. Thus from 5.37 we get that $\varphi(\xi) \neq 0$ for $\xi \in K$. Since K is compact, it follows that there exists a constant $c > 0$ such that $|\varphi(\xi)| > c$ for $\xi \in K$. Thus for $\xi \in B^N K$ we have $|\varphi(B^{-N}\xi)| > c$. Using 5.37 once more we get

$$|\nu_N(\xi)| \le c^{-1} |\varphi(\xi)| \quad \text{for } \xi \in B^N K.$$

This means that for all $\xi \in \mathbb{R}^d$

$$|\nu_N(\xi) \cdot \mathbf{1}_{B^N K}(\xi)| \leq c^{-1} |\varphi(\xi)|. \tag{5.42}$$

Recall that K contains a neighborhood of 0, so $\bigcup_{N=1}^{\infty} B^N K = \mathbb{R}^d$. This observation together with 5.38 shows that $\nu_N(\xi) \cdot \mathbf{1}_{B^N K}(\xi)$ converges pointwise to $\frac{(2\pi)^{d/2}}{|Q|} \varphi(\xi)$. Since $\varphi(\xi) \in L_2(\mathbb{R}^d)$ we infer from 5.42 and Lebesgue's dominated convergence theorem that

$$\int_{\mathbb{R}^d} |\nu_N(\xi) \cdot \mathbf{1}_{B^N K}(\xi)|^2 d\xi \to \frac{(2\pi)^d}{|Q|^2} \int_{\mathbb{R}^d} |\varphi(\xi)|^2 d\xi.$$

This together with 5.41 gives that

$$\int_{\mathbb{R}^d} |\varphi_N(\xi)|^2 d\xi \to \frac{1}{|Q|^2} \int_{\mathbb{R}^d} |\varphi(\xi)|^2 d\xi.$$

But from Lemma 5.22 and Plancherel's theorem (A1.2–IV) we infer that

$$\int_{\mathbb{R}^d} |\varphi_N(\xi)|^2 d\xi = 1$$

so $\int_{\mathbb{R}^d} |\varphi(\xi)|^2 d\xi = |Q|^2$. But on the other hand we know from Plancherel's theorem that $\int_{\mathbb{R}^d} |\varphi(\xi)|^2 d\xi = |Q|$. Since $|Q| > 0$, we get $|Q| = 1$, so the proposition follows from Corollary 5.20. □

Now we are ready to present some examples.

EXAMPLE 5.2. Let us take $d = 1$ and consider the dilation $Ax = 3x$. As a set of digits let us take $S = \{0, 1, 2\}$. Then, as is easily seen,

$$Q = \left\{ x = \sum_{j=1}^{\infty} 3^{-j} s_j \quad \text{with } s_j = 0, 1, 2 \right\} = [0, 1]$$

which gives a multiresolution analysis. When we take $S = \{1, 2, 3\}$ we get using the above

$$Q = \left\{ x = \sum_{j=1}^{\infty} 3^{-j}(s_j - 1) + \sum_{j=1}^{\infty} 3^{-j} \quad \text{with } s_j = 1, 2, 3 \right\} = [\tfrac{1}{2}, 1\tfrac{1}{2}].$$

This also gives a multiresolution analysis, but different from (although rather similar to) the previous one.

When we take $S = \{0, 1, 5\}$ the corresponding set Q is depicted in Figure 5.2. It is disconnected. To see this formally, note that each point $x = \sum_{j=1}^{\infty} 3^{-j} s_j$ with $s_1 = 0$ or 1 is at most $\frac{1}{3} + \sum_{j=2}^{\infty} 5 \cdot 3^{-j} = \frac{7}{6}$, while each x with $s_1 = 5$ is at least $\frac{5}{3}$. Nevertheless the set Q also gives a scaling function of a multiresolution analysis. This follows from

0 2.5

Fig. 5.2. The 1-dimensional disconnected set giving the scaling function

Proposition 5.21. The polynomial $m(\xi)$ equals $\frac{1}{3}(1+e^{-i\xi}+e^{-5i\xi})$ and we take $K=[-\pi,\pi]$. Since $B(x)=\frac{1}{3}x$ it suffices to show that $m(\xi)\neq 0$ for $|\xi|\leq\frac{\pi}{3}$. But $3\Re m(\xi)=1+\cos\xi+\cos 5\xi\geq\cos\xi$ so $m(\xi)\neq 0$ for $|\xi|<\frac{\pi}{2}$. •

EXAMPLE 5.3. Now let us take $d=2$ and the simplest dilation $A=2Id$. Taking the set of digits S as $\{(0,0),\,(0,1),\,(1,0),\,(1,1)\}$ we get $Q=[0,1]^2$. This clearly gives a scaling function of a multiresolution analysis. This is the natural 2-dimensional Haar multiresolution analysis described in Example 5.1. Choosing the set S as

$$\{(0,0),\,(1,1),\,(0,1),\,(1,2)\}$$

we obtain as the set Q the parallelogram with vertices from the set S. This also gives a scaling function of a multiresolution analysis. If we take the set S as

$$\{(0,0),\,(1,0),\,(0,1),\,(-1,-1)\}$$

then we get the set Q represented in Figure 5.3. To check that it gives a scaling function we will use Proposition 5.21. The polynomial $m(\xi)$ is given by

$$m(\xi)=\tfrac{1}{4}(1+e^{-i\xi_1}+e^{-i\xi_2}+e^{i(\xi_1+\xi_2)}).$$

As the set K we take $[-\pi,\pi]^2$. Thus we need to check that $m(\xi)\neq 0$ for $\xi\in[-\frac{\pi}{2},\frac{\pi}{2}]^2$. This we get immediately because

$$\begin{aligned}\Re m(\xi) &= \frac{1}{4}(1+\cos\xi_1+\cos\xi_2+\cos(\xi_1+\xi_2))\\ &\geq \frac{1}{4}(\cos\xi_1+\cos\xi_2).\end{aligned}$$

•

EXAMPLE 5.4. Let us take $d=2$ and the dilation given by the matrix

$$\begin{bmatrix}1 & -1\\ 1 & 1\end{bmatrix}.$$

Fig. 5.3. The set Q described in Example 5.3 corresponding to the dilation $A = 2Id$

Geometrically speaking this dilation is a rotation by 45° and expansion by the factor $\sqrt{2}$. The nice thing about this dilation is that $\det A = 2$, so we obtain *one* wavelet generating an orthonormal basis in $L_2(\mathbb{R}^2)$ (provided we have a multiresolution analysis associated with this dilation). Taking $S = \{(0,0),\ (1,0)\}$ we get the set Q depicted in Figure 5.4. This is the fractal set known as the 'twin dragon'. To check that it gives a scaling function we use Proposition 5.21. The polynomial $m(\xi) = \frac{1}{2}(1 + e^{-i\xi_1})$ has zeros at the lines $\xi_1 = (2n+1)\pi$ for $n \in \mathbb{Z}$. The appropriate set K and the set $B^{-1}K$ (note that $B^{-1} = \frac{1}{2}A$) are depicted in Figure 5.5. •

We want to conclude this section with the construction of an r-regular multiresolution analysis and associated wavelet sets. The general scheme is similar to the construction of a spline multiresolution analysis. The main step in the argument is the following proposition.

Proposition 5.23 *Let Q be the set given by 5.33. Assume that $\mathbf{1}_Q(x)$ is a scaling function of a multiresolution analysis. There exists an $\varepsilon > 0$ and C such that*

$$|\hat{\mathbf{1}}_Q(\xi)| \leq C|\xi|^{-\varepsilon}. \tag{5.43}$$

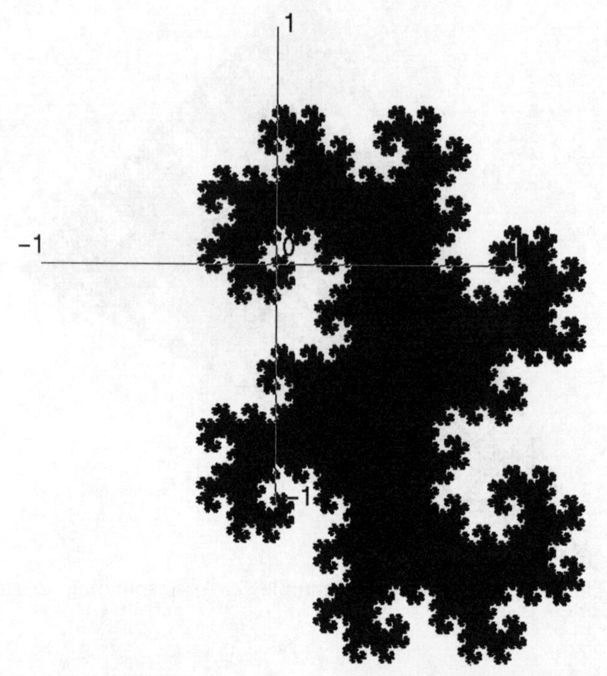

Fig. 5.4. The 'twin dragon' set Q obtained in Example 5.4

Let us start with some simplifying assumptions. If the set Q is given by 5.33 with the set of digits $S = \{k_1, k_2, \ldots, k_q\}$ with $k_1 \in A(\mathbb{Z}^d)$, then the set given by 5.33 using the set $S' = \{0, k_2 - k_1, \ldots, k_q - k_1\}$ is a translation of the set Q. This shows that it also has measure 1, so its characteristic function is also a scaling function of a multiresolution analysis (perhaps a different one, cf. Example 5.2). Thus (see A1.2–III) for our purpose we can and will assume that $0 \in S$, so also $0 \in Q$.

Observe also that if Q is given by 5.33 with a set of digits S, then for each integer k we also have

$$Q = \left\{ x \in \mathbb{R}^d \ : \ x = \sum_{j=1}^{\infty} A_k^{-j} s_j \ \text{ where } s_j \in S_k \right\}$$

where the dilation matrix A_k equals A^k and the set of digits S_k is given by

$$S_k =: \{s_1 + As_2 + \ldots + A^{k-1}s_{k-1} \ : \ s_1, \ldots, s_{k-1} \in S\}. \tag{5.44}$$

One can easily check that S_k really is a set of digits for A_k. Thus we can

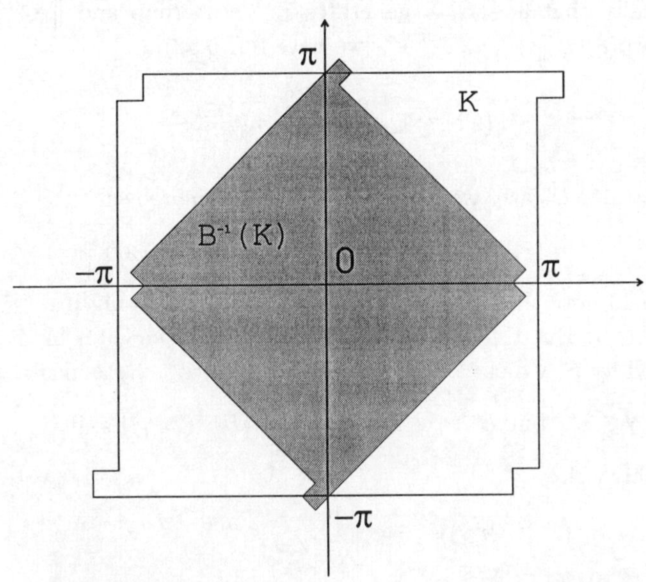

Fig. 5.5. The set K used to show that the 'twin dragon' gives a scaling function

replace the dilation matrix A by the matrix A^k for a suitable integer k. This gives us two advantages, First we can assume that $\|A^{-1}\| < 1$ (see the argument for 5.34). Second we can assume that the additive group generated by S equals $\mathbb{Z}^d$. This is not so obvious but can be seen as follows.

First observe that for any $x_0 \in \mathbb{R}^d$ the set of all $\gamma \in \mathbb{Z}^d$ such that $|x_0 - \gamma| \leq 2d$ generates the group $\mathbb{Z}^d$. From Proposition 5.19(iv) we know that Q contains a ball. This implies (see the argument for 5.34) that there exists a natural number k such that $A^k Q$ contains a ball $B(x_0, r)$ with radius $r > 2\max(2d, \text{diam } Q)$. For $z \in \mathbb{Z}^d \cap B(x_0, r/2)$ we have $z + Q \subset B(x_0, r) \subset A^k(Q)$. Since we easily infer from 5.44 and 5.33 that $A^k(Q) = \bigcup_{s \in S_{k+1}} s + Q$ and we know that Q tiles the space we obtain $z \in S_{k+1}$. Thus $\mathbb{Z}^d \cap B(x_0, r/2) \subset S_{k+1}$, so in particular S_{k+1} generates $\mathbb{Z}^d$.

Note that in Example 5.4 we saw a set of digits S giving a scaling function but such that the group generated by S is not $\mathbb{Z}^d$. Thus we really have to increase k.

Proof of Proposition 5.23 Using the above comments we will assume additionally that $0 \in S$, S generates $\mathbb{Z}^d$ as a group and $\|A^{-1}\| < 1$. Writing $m(\xi) = \frac{1}{q}\sum_{\gamma\in S} e^{-i\langle\xi,\gamma\rangle}$ we have (cf. 5.38)

$$(2\pi)^{d/2}\hat{\mathbf{1}}_Q(\xi) = \prod_{k=1}^{\infty} m(B^{-k}\xi) \tag{5.45}$$

with $B = A^*$. Clearly we have $|m(\xi)| \le 1$ and moreover

$$|m(\xi)| = 1 \iff m(\xi) = 1 \iff \xi \in 2\pi\mathbb{Z}^d. \tag{5.46}$$

To see 5.46 note that since $0 \in S$, $|m(\xi)| = 1$ if and only if $e^{-i\langle\xi,\gamma\rangle} = 1$ for all $\gamma \in S$. But this is equivalent to $e^{i\langle\xi,\gamma\rangle} = 1$ for all γ in the group generated by S, which is $\mathbb{Z}^d$. This forces $\xi \in 2\pi\mathbb{Z}^d$. Note also that

$$\text{if } \gamma \in \mathbb{Z}^d \text{ but } B^{-1}\gamma \notin \mathbb{Z}^d, \text{ then } m\big(B^{-1}(2\pi\gamma)\big) = 0. \tag{5.47}$$

To see this write

$$\begin{aligned} m\big(B^{-1}(2\pi\gamma)\big)^2 &= \frac{1}{q^2}\sum_{r,s\in S} e^{-i\langle B^{-1}(2\pi\gamma),r+s\rangle} \\ &= \frac{1}{q^2}\sum_{r,s\in S} e^{-i\langle 2\pi\gamma,A^{-1}(r+s)\rangle}. \end{aligned} \tag{5.48}$$

Observe that the individual exponentials in 5.48 depend only on the coset to which $(r+s)$ belongs. Since S is a set of digits, we infer that for each coset there are exactly q pairs (r, s) with $r, s \in S$ such that $r + s$ belongs to this coset. Thus we get from 5.48

$$m\big(B^{-1}(2\pi\gamma)\big)^2 = \frac{1}{q}\sum_{s\in S} e^{-i\langle 2\pi\gamma,A^{-1}(s)\rangle} = m\big(B^{-1}(2\pi\gamma)\big).$$

This shows that $m\big(B^{-1}(2\pi\gamma)\big)$ equals either 0 or 1. But $B^{-1}(\gamma) \notin \mathbb{Z}^d$, so we see from 5.46 that it cannot equal 1. Now let us fix certain numbers. We put

$$\delta = (\max|\text{grad } m(\xi)|)^{-1}\,\|A\|^{-1}. \tag{5.49}$$

Since m is a non-zero trigonometric polynomial δ is a well defined positive number. We also fix $\rho < 1$ such that for every ξ

$$\text{either } |m(\xi)| \le \rho \text{ or } |\xi - \gamma| < \delta \text{ for some } \gamma \in 2\pi\mathbb{Z}^d. \tag{5.50}$$

The existence of such a ρ follows from 5.46. Now we fix ϵ, $0 < \epsilon < 1$, such that

$$\epsilon < \frac{\log\|A^{-1}\|^{-1}}{\log\|A\|} \quad \text{so} \quad \|A\|^{\epsilon}\cdot\|A^{-1}\| \le 1 \tag{5.51}$$

and

$$\epsilon < \frac{\log \rho^{-1}}{\log \|A\|} \quad \text{so} \quad \rho \|A\|^\epsilon \leq 1. \tag{5.52}$$

This is the ε for which 5.43 holds.

We will prove 5.43 by induction. When $|\xi| \leq 1 = \|A^{-1}\|^{-0}$ then 5.43 holds for an appropriate C. Now assume that 5.43 holds (with *the same* C) for $|\xi| < \|A^{-1}\|^{-N}$ and take ξ such that $|\xi| < \|A^{-1}\|^{-N-1}$. We want to show that $|\hat{\mathbf{1}}_Q(\xi)| \leq C|\xi|^{-\epsilon}$. Let us denote $\zeta = B^{-1}\xi$. We will consider two cases according to 5.50.

Case 1: $|m(\zeta)| \leq \rho$. Then we see from 5.45 that

$$|\hat{\mathbf{1}}_Q(\xi)| = |m(\zeta)| \cdot |\hat{\mathbf{1}}_Q(B^{-1}\xi)| \leq \rho |\hat{\mathbf{1}}_Q(B^{-1}\xi)| \tag{5.53}$$

Since $|B^{-1}\xi| \leq \|B^{-1}\| \, |\xi| \leq \|A^{-1}\|^{-N}$ we can apply the inductive hypothesis to obtain

$$|\hat{\mathbf{1}}_Q(\xi)| \leq \rho C |B^{-1}\xi|^{-\epsilon}.$$

Since $|\xi| \leq \|B\| \, |B^{-1}\xi|$ we see that

$$\frac{1}{|B^{-1}\xi|} \leq \frac{\|B\|}{|\xi|}$$

so using 5.52 we get

$$|\hat{\mathbf{1}}_Q(\xi)| \leq \rho C \frac{\|A\|^\epsilon}{|\xi|^\epsilon} \leq C|\xi|^{-\epsilon}.$$

Case 2: *there exists* $\eta \in 2\pi\mathbb{Z}^d$ *such that* $|\zeta - \eta| < \delta$. Let k be the smallest integer such that $B^{-k}\eta \notin 2\pi\mathbb{Z}^d$. Then from 5.47 we see that $m(B^{-k}\eta) = 0$. So we have

$$\begin{aligned}
|m(B^{-k-1}\xi)| &= |m(B^{-k-1}\xi) - m(B^{-k}\eta)| \\
&\leq \max|\text{grad}\, m| \cdot |B^{-k-1}\xi - B^{-k}\eta| \\
&\leq \max|\text{grad}\, m| \cdot \|B^{-k}\| \, |B^{-1}\xi - \eta| \\
&= \max|\text{grad}\, m| \cdot \|B^{-k}\| \, |\zeta - \eta| \\
&\leq \max|\text{grad}\, m| \cdot \|B^{-k}\| \cdot \delta \\
&= \|A\|^{-1} \, \|B^{-k}\|.
\end{aligned}$$

Thus we see from 5.45, 5.46 and the choice of k that

$$\begin{aligned}
|\hat{\mathbf{1}}_Q(\xi)| &\leq |m(B^{-k-1}\xi)| \, |\hat{\mathbf{1}}_Q(B^{-k-1}\xi)| \\
&\leq \|A\|^{-1} \, \|B^{-k}\| \, |\hat{\mathbf{1}}_Q(B^{-k-1}\xi)|.
\end{aligned}$$

Since

$$\begin{aligned} |B^{-k-1}\xi| &\leq \|A^{-1}\|^{k+1} |\xi| \leq \|A^{-1}\|^{k+1} \|A^{-1}\|^{-N-1} \\ &= \|A^{-1}\|^{-N+k} < \|A^{-1}\|^{-N} \end{aligned}$$

we can apply the inductive hypothesis in order to obtain

$$|\hat{\mathbf{1}}_Q(\xi)| \leq \|A\|^{-1} \|A^{-1}\|^k C|B^{-k-1}\xi|^{-\epsilon}.$$

But

$$|\xi| = |B^{k+1}B^{-k-1}\xi| \leq \|A\|^{k+1} |B^{-k-1}\xi|$$

so

$$\frac{1}{|B^{-k-1}\xi|} \leq \frac{\|A\|^{k+1}}{|\xi|}.$$

Using 5.51 and the above estimates we obtain

$$\begin{aligned} |\hat{\mathbf{1}}_Q(\xi)| &\leq \|A\|^{-1} \|A^{-1}\|^k C \|A\|^{\epsilon(k+1)} |\xi|^{-\epsilon} \\ &= C \left(\|A^{-1}\| \|A\|^{\epsilon}\right)^k \|A\|^{\epsilon-1} |\xi|^{-\epsilon} \leq C|\xi|^{-\epsilon}. \end{aligned}$$

Since by 5.50 the above two cases cover all possibilities, the inductive argument is complete. □

Now let us define the function $g_1(x) =: \mathbf{1}_Q * \mathbf{1}_{-Q}$ and for $n = 2, 3, \ldots$ define the function g_n as the n-th convolution power of g_1, or equivalently $g_{n+1} = g_n * g_1$. The properties of the functions g_n are summarized in the following proposition.

Proposition 5.24

(i) *g_n is a compactly supported, positive $L_1(\mathbb{R}^d)$ function and*

$$\int_{\mathbb{R}^d} g_n(x)\, dx = 1,$$

so $\hat{g}_n$ is continuous, $|\hat{g}_n(\xi)| \leq \frac{1}{(2\pi)^{d/2}}$ and $\hat{g}_n(0) = \frac{1}{(2\pi)^{d/2}}$.

(ii) *g_n is of class C^k for $k < \frac{2n\epsilon}{d} - 1$ where ϵ is given by Proposition 5.23.*

(iii) *the system $\{g_n(t-\gamma)\}_{\gamma \in \mathbb{Z}^d}$ is a Riesz sequence.*

Proof Property (i) follows directly from the definition of convolution A1.2–X, the definition of Fourier transform A1.2–I and from A1.2–II.

Property (ii) follows from A1.2–XI and A1.2–IX. To prove (iii) we check condition (i) of Proposition 5.7. Let us consider the series

$$\sum_{\gamma\in\mathbb{Z}^d} |\hat{\mathbf{1}}_Q(\xi+2\pi\gamma)|^2. \tag{5.54}$$

Since $\{\mathbf{1}_Q(t-\gamma)\}_{\gamma\in\mathbb{Z}^d}$ is an orthonormal system we see from Proposition 5.7 that this series converges a.e. to $(2\pi)^{-d}$. Since

$$|\mathcal{F}_1 g_n(\xi)| = |\mathcal{F}_1 g_1(\xi)|^n = |\mathcal{F}_1 \mathbf{1}_Q(\xi)|^{2n}$$

(cf. A1.2–XI) and $|\mathcal{F}_1 g_1(\xi)| \leq 1$ we see that

$$\sum_{\gamma\in\mathbb{Z}^d} |\hat{g}_n(\xi+2\pi\gamma)|^2 \leq C. \tag{5.55}$$

Observe also that for each multiindex α the function $x^\alpha \mathbf{1}_Q(x)$ also has orthogonal translates, so the series

$$\sum_{\gamma\in\mathbb{Z}^d} |D^\alpha \hat{\mathbf{1}}_Q(\xi+2\pi\gamma)|^2$$

converges a.e. to a non-zero constant. This implies that the series 5.54 converges uniformly on $[0,2\pi]^d$. Thus there exists a finite set $\Gamma\subset\mathbb{Z}^d$ such that

$$\sum_{\gamma\in\Gamma} |\hat{\mathbf{1}}_Q(\xi+2\pi\gamma)|^2 \geq \tfrac{1}{2}(2\pi)^{-d} \tag{5.56}$$

for all $\xi\in[0,2\pi]^d$. Since $|\hat{g}_n(\xi)| = (2\pi)^{d(n-1)/2}|\hat{\mathbf{1}}_Q(\xi)|^{2n}$ we see from 5.56 that for each n there exists a constant c_n such that

$$\sum_{\gamma\in\Gamma} |\hat{g}_n(\xi+2\pi\gamma)|^2 \geq c_n$$

for $\xi\in[0,2\pi]^d$. This clearly implies that $\sum_{\gamma\in\mathbb{Z}^d} |\hat{g}_n(\xi+2\pi\gamma)|^2 \geq c_n$ for all $\xi\in\mathbb{R}^d$, so from 5.55 we get (iii). □

The functions g_n, $n=1,2,\ldots$ are important for our purposes because they are scaling functions of a multiresolution analysis with

$$V_0 = \operatorname{span}\{g_n(t-\gamma)\}_{\gamma\in\mathbb{Z}^d}.$$

From this definition and Propositions 5.7(iii) and 5.24 we see that (v) and (vi) of Definition 5.4 hold. To ensure (i) and (iv) it suffices to check that

$$g_n(x) = \sum_{\gamma\in S} a_\gamma g_n(Ax-\gamma). \tag{5.57}$$

First let us check it for $n = 1$. Since $\mathbf{1}_Q$ is a scaling function we have

$$\mathbf{1}_Q(x) = \sum_{\gamma \in S} a_\gamma \mathbf{1}_Q(Ax - \gamma)$$

and also

$$\mathbf{1}_{-Q}(x) = \mathbf{1}_Q(-x) = \sum_{\gamma \in S} a_\gamma \mathbf{1}_Q(-Ax - \gamma) = \sum_{\gamma \in S} a_\gamma \mathbf{1}_{-Q}(Ax + \gamma).$$

Since

$$f(Ax - \gamma_1) * g(Ax - \gamma_2) = (f * g)(Ax - \gamma_1 - \gamma_2) \tag{5.58}$$

we infer that $g_1(x) = \mathbf{1}_Q * \mathbf{1}_{-Q}(x)$ can be written as a finite sum of the form $\sum_\gamma a(\gamma) g_1(Ax - \gamma)$. Using 5.58 we inductively get 5.57 for $n = 2, 3, \ldots$.

Conditions (ii) and (iii) of Definition 5.4 follow from Proposition 5.24 using standard arguments (cf. Section 2.3 or Theorem 8.4 and Proposition 8.5).

Note also that $\hat{\mathbf{1}}_{-Q}(\xi) = \overline{\hat{\mathbf{1}}_Q(\xi)}$, so $\hat{g}_n(\xi)$ is real and positive for $n = 1, 2, \ldots$. From Proposition 5.24 and Corollary 5.14 we infer that for $r < \frac{2n\varepsilon}{d} - 1$ the multiresolution analyses described above have r-regular scaling functions with real Fourier transforms. Applying Corollary 5.17 we get the following theorem.

Theorem 5.25 *Suppose that A is a dilation matrix such that for some set of digits the set Q given by 5.33 has measure 1, i.e. the function $\mathbf{1}_Q$ is a scaling function of a multiresolution analysis. Then for each natural number $r = 1, 2, \ldots$ there exists an r-regular wavelet set (consisting of $|\det A| - 1$ functions) associated with the dilation matrix A.*

Sources and comments

It was clear from the very beginning that various aspects of the theory of wavelets generalize to several variables and/or to more general dilations. The tensoring process is very common in mathematics. The procedure indicated in Example 5.1 was already applied in [107] to construct spline wavelet sets on $\mathbb{R}^d$. The general construction of multi-variable wavelets associated with the dyadic dilation was given by K. Gröchenig in [47] and reproduced in [85]. Multiresolution analyses with general matrix dilation appear e.g. in [76], [50] and [21]. Our presentation of the general theory in Section 5.2 is an adaptation of one-dimensional theory and multivariable theory for dyadic dilation to this more general context. The notion of r-regular multiresolution analysis was used extensively at the very beginning of the theory cf. [83], and our treatment of r-regular multiresolution analyses and wavelets closely follows [85], [83] or [24]. Our examples of multiresolution analyses with the scaling function of the form $\Phi(x) = \mathbf{1}_Q$ and connections with tilings follow the paper [50]. Similar two-dimensional examples are considered in detail in [21]. From the pictures we present it is clear that the sets Q can be very irregular. Such sets are called fractals. More about such sets can be found e.g. in [38]. Connections with wavelets are discussed e.g. in [80]. Proposition 5.23 and Theorem 5.25 are due to R. S. Strichartz [106]. Our construction of wavelets is slightly different from his.

It should be pointed out that estimate 5.43 is really a special property of the set Q. It follows from Exercise 3 on page 103 of [57] that there exists a bounded set $E \subset \mathbb{R}$ such that

$$\limsup_{|\xi| \to \infty} \left| \hat{\mathbf{1}}_E(\xi) \ln |\xi| \right| = \infty.$$

The geometric theory of self-affine tilings of $\mathbb{R}^d$ is discussed e.g. in [9] or [49].

It is an interesting question for which dilation matrix A there exists a set of digits such that the function $\mathbf{1}_Q$, where Q is given by 5.33, is a scaling function of a multiresolution analysis. An equivalent form of this question is when $|Q| = 1$, cf. Proposition 5.20. This is interesting in itself but also is of some practical importance since given such a set Q we can construct smooth wavelets, cf. Proposition 5.23 and Theorem 5.25. This question has also close connections with some problems of algebraic number theory. The series of papers by J. C. Lagarias and Y. Wang [59]–[62] deal with various aspects of this problem. The answer is known in dimension $d = 1$ and $d = 2$ when every dilation matrix A admits a set of digits yielding a scaling function. For $d = 1$ this was shown in [50] and for $d = 2$ in [62].

It is interesting to note that if we do not wish to work in the framework of multiresolution analyses then there is no necessity to increase the number of wavelets. It was shown in [23] that for each dilation matrix A in $\mathbb{R}^d$ there exists a function Ψ such that $\left\{ |\det A|^{j/2} \Psi(A^j x - \gamma) \right\}_{j \in \mathbb{Z}, \gamma \in \mathbb{Z}^d}$ is an orthonormal basis in $L_2(\mathbb{R}^d)$. This Ψ has quite bad decay, since $\hat{\Psi} = \mathbf{1}_K$ for some set $K \subset \mathbb{R}^d$.

The general presentation in this chapter was influenced by [8].

About the exercises. Exercise 5.7 is taken from [76] section 3. Exercise 5.8 is a special case of some results of [59].

Exercises

5.1 Let $\Phi \in L_1(\mathbb{R}^d) \cap L_2(\mathbb{R}^d)$ be a scaling function of a multiresolution analysis associated with dyadic dilations. Modify the proofs of Propositions 2.16 and 2.17 to show that

$$\int_{\mathbb{R}^d} \Phi(x)\, dx = 1 \text{ and } \sum_{\gamma \in \mathbb{Z}^d} \Phi(x - \gamma) = 1.$$

Modify the arguments further to show it for scaling functions associated with arbitrary dilations.

5.2 Prove in detail both assertions made in the third paragraph of Section 5.1.

5.3 Show that if Ψ is an r-regular wavelet on $\mathbb{R}^d$ (associated with any dilation) and $p(x)$ is a polynomial on $\mathbb{R}^d$ of degree $\leq r$ then $\int_{\mathbb{R}^d} p(x)\Psi(x)\, dx = 0$.

5.4 Show the converse implication in Proposition 5.21.

5.5 Construct explicitly a Haar type (i.e. supp $\Psi \subset [0, 1]$) and a Strömberg type (i.e. continuous, piecewise linear) wavelet set for the dilation $x \mapsto 3x$ on $\mathbb{R}$. Show that the Strömberg type wavelets can have exponential decay. Construct periodic versions of these wavelet sets.

5.6 Suppose that $\Phi(x)$ is a scaling function of a multiresolution analysis and satisfies $\Phi(a + x) = \Phi(a - x)$ for some $a \in \mathbb{R}^d$.

(a) Show that $\hat{\Phi}(\xi) = e^{i\langle a,\xi\rangle}\varphi(\xi)$ where $\varphi(\xi)$ is a purely imaginary function.

(b) Show that each of the functions μ^r_Φ defined before Proposition 5.9 on page 112 is of the form $e^{i\langle b_r,\xi\rangle}\varphi_r(\xi)$ for some $b_r \in \mathbb{R}^d$ and a purely imaginary function $\varphi_r(\xi)$.

(c) Assume additionally that Φ is r-regular. Show that there exists an r-regular wavelet set associated with this multiresolution analysis.

5.7 Let $\ldots \subset V_{-1} \subset V_0 \subset V_1 \ldots$ be a multiresolution analysis on $\mathbb{R}^d$ associated with a dilation A.

(a) Show that if V_0 is translation invariant, i.e. for every $f \in V_0$ and every $h \in \mathbb{R}^d$ the function $f(x - h)$ is also in V_0, then $\mathcal{F}(V_0) = L_2(\Omega)$ for an appropriate $\Omega \subset \mathbb{R}^d$.

(b) Show that the set Ω above satisfies the following conditions, where $B = A^*$.

1. $\Omega \subset B\Omega$
2. $|\Omega \cap (\Omega + 2\pi\gamma)| = 0$ for all $\gamma \in \mathbb{Z}^d$
3. $\bigcup_{\gamma \in \mathbb{Z}^d} (\Omega + 2\pi\gamma) = \mathbb{R}^d$ a.e.
4. $\lim_{j\to\infty} \frac{1}{|B^{-j}Q|} \int_{B^{-j}Q} \mathbf{1}_\Omega(\xi)\, d\xi = 1$ for every cube Q of finite diameter.

(c) Show by exhibiting appropriate examples that conditions 1.–4. above are independent

(d) Find an example of a set Ω satisfying 1.–4. above which does not contain a neighborhood of 0.

5.8 Let Q be the set defined by 5.33 for some set of digits S and dilation A. Show that

$$\sum_{\gamma \in \mathbb{Z}^d} \mathbf{1}_Q(x - \gamma) = |Q| \quad \text{a.e.}$$

so in particular $|Q|$ is an integer.

5.9 Let $Ax = 3x$ be a dilation on $\mathbb{R}$. Show that $\{0, 2, 4\}$ is a set of digits for A. Show that the function $\mathbf{1}_Q$ where Q is given by 5.33 for this set of digits is not a scaling function of a multiresolution analysis.

5.10 Find explicitly an interval contained in the set considered in Example 5.2 and depicted in Figure 5.2

6
Function spaces

The aim of this chapter is to introduce some important function spaces on $\mathbb{R}^d$ and basic tools to investigate operators on those spaces.

6.1 L_p-spaces

Let us recall that for a measurable function f on $\mathbb{R}^d$ and $1 \leq p \leq \infty$ we define the L_p norm of f as

$$\|f\|_p = \left(\int_{\mathbb{R}^d} |f(x)|^p \, dx \right)^{1/p} \qquad \text{if } 1 \leq p < \infty$$

and $\|f\|_\infty = \operatorname{supess} |f(x)|$.

By $L_p(\mathbb{R}^d)$ we mean the space of all functions such that $\|f\|_p < \infty$. Actually we identify functions equal almost everywhere and then $L_p(\mathbb{R}^d)$ becomes a Banach space. Everywhere in this chapter we can think about complex-valued or real-valued functions as well. Some facts will be proved only for real-valued functions. Then the corresponding fact for complex-valued functions can always be established by considering real and imaginary parts separately.

As a first and easy observation let us note that a standard change of variables argument gives

$$\|T_h f\|_p = \|f\|_p \tag{6.1}$$

for $1 \leq p \leq \infty$ and every $h \in \mathbb{R}^d$, where T_h is the translation operator defined in 5.4. For a dilation operator $J_s(f)(x) =: f(2^s x)$ (cf. 5.2) we have

$$\|J_s f\|_p = 2^{-sd/p} \|f\|_p \tag{6.2}$$

so the operator $2^{sd/p} J_s$ is an isometry on $L_p(\mathbb{R}^d)$, $1 \leq p \leq \infty$.

Intuitively speaking the L_p norm measures the size of the function. The other natural way to measure the size of the function is to use its distribution, i.e. the function

$$d_f(t) = |\{x \in \mathbb{R}^d : |f(x)| > t\}| \tag{6.3}$$

defined for $t \geq 0$. Clearly $d_f(t)$ is a decreasing function on $[0, \infty)$. There is a very close connection between $d_f(t)$ and $\|f\|_p$, namely we have:

Proposition 6.1 *If f is a function on $\mathbb{R}^d$ and $1 \leq p < \infty$ then*

$$d_f(t) \leq t^{-p} \|f\|_p^p \tag{6.4}$$

and

$$\|f\|_p^p = p \int_0^\infty t^{p-1} d_f(t)\, dt. \tag{6.5}$$

Proof Clearly

$$t^p \cdot d_f(t) = \int_{\{x \in \mathbb{R}^d : |f(x)| > t\}} t^p\, dx \leq \int_{\mathbb{R}^d} |f(x)|^p\, dx$$

so we get 6.4. To obtain 6.5 let us introduce the set $A \subset \mathbb{R}^d \times [0, \infty)$ defined as $A = \{(x, s) : s < |f(x)|^p\}$. Using Fubini's theorem we get

$$\begin{aligned} \int_{\mathbb{R}^d} |f(x)|^p\, dx &= \int_{\mathbb{R}^d} \int_0^{|f(x)|^p} 1\, ds dx = \int_{\mathbb{R}^d \times [0,\infty)} \mathbf{1}_A(x, s)\, ds dx \\ &= \int_{\mathbb{R}^d \times [0,\infty)} \mathbf{1}_A(x, s)\, dx ds \\ &= \int_0^\infty |\{x : |f(x)|^p > s\}|\, ds. \end{aligned}$$

Making the substitution $s = t^p$ in the last integral we see that it equals

$$p \int_0^\infty t^{p-1} |\{x : |f(x)| > t\}|\, dt$$

which is 6.5. □

REMARK 6.1. One easily checks that for $f(x) = |x|^{-1/p}$ on $\mathbb{R}$ we have $d_f(t) = 2^p t^{-p}$ but $f \notin L_p(\mathbb{R})$, so the condition $d_f(t) \leq ct^{-p}$ is really weaker than the condition $\|f\|_p < \infty$. Inequality 6.4 is a well known inequality usually called Markov's inequality.

By a cube in $\mathbb{R}^d$ we mean a set $I_1 \times I_2 \times \ldots \times I_d$ where the I_j are closed intervals in $\mathbb{R}$ all of the same length. In other words we assume that the cube has sides parallel to the coordinate axes. Generally in this chapter we will reserve the letter Q to denote a cube in $\mathbb{R}^d$. For a positive number c and a cube Q, by $c \diamond Q$ we will mean the cube with the same center as Q but whose sides are c times longer than sides of Q. More formally, if $Q = I_1 \times \ldots \times I_d$ with $I_j = [a_j - h, a_j + h]$ then $c \diamond Q = J_1 \times \ldots \times J_d$ where $J_j = [a_j - ch, a_j + ch]$.

REMARK 6.2. The notation $c \diamond Q$ makes sense only when Q is a cube. The reader should be warned that it is an entirely ad hoc notation. In harmonic analysis papers and books (cf. e.g. [42] or [108]) the notation cQ is used. This however flatly contradicts another well established usage, namely $cQ = \{cx : x \in Q\}$ which we also use in this book. To avoid possible collisions and misunderstandings I decided to use some other notation. The symbol $\diamond$ is to remind the reader that we mean expanded cube.

DEFINITION 6.2 *For a function f on $\mathbb{R}^d$ we define its Hardy–Littlewood maximal function $Mf(x)$ by the formula*

$$Mf(x) = \sup\left\{\frac{1}{|Q|}\int_Q |f(t)|dt \ : \ Q \subset \mathbb{R}^d \text{ is a cube and } x \in Q\right\}.$$

It is clear from this definition that $Mf(x)$ has the following properties

$$0 \le Mf(x) \le \infty \tag{6.6}$$

$$M(f+g)(x) \le Mf(x) + Mg(x) \tag{6.7}$$

$$M(\lambda f)(x) = |\lambda| Mf(x). \tag{6.8}$$

One can easily find a function f such that $Mf(x) = \infty$ for each $x \in \mathbb{R}^d$ (see Exercise 6.1). We will be interested in Mf for $f \in L_p(\mathbb{R}^d)$. For $f \in L_\infty(\mathbb{R}^d)$ we immediately see that $Mf \in L_\infty(\mathbb{R}^d)$ and that for each $x \in \mathbb{R}^d$

$$Mf(x) \le \|f\|_\infty . \tag{6.9}$$

The analogous statement is not true for $f \in L_1(\mathbb{R}^d)$. For example for $f = \mathbf{1}_{[0,1]}$ we easily obtain that

$$Mf(x) = \begin{cases} \frac{1}{x} & \text{if } x \ge 1 \\ 1 & \text{if } 0 < x < 1 \\ \frac{1}{1-x} & \text{if } x \le 0 \end{cases}$$

so $Mf \notin L_1(\mathbb{R})$. Nevertheless something remains, namely we have:

Theorem 6.3 *For each function f on $\mathbb{R}^d$ we have*

$$|\{x \ : \ Mf(x) > t\}| \le 6^d t^{-1} \|f\|_1 .$$

Proof If $\|f\|_1 = \infty$ there is nothing to prove, so assume that $\|f\|_1$ is finite. It follows from 6.8 that without loss of generality we can assume $\|f\|_1 = 1$. For a fixed $t \geq 0$ let $E_t = \{x : Mf(x) > t\}$. It follows from Definition 6.2 that for each $x \in E_t$ there is a cube Q_x such that $x \in Q_x \subset E_t$ and

$$\frac{1}{|Q_x|}\int_{Q_x} |f(y)|\,dy > t. \tag{6.10}$$

This in particular gives $|Q_x| \leq \frac{\|f\|_1}{t} = \frac{1}{t}$. Let $\alpha_1 = \max\{|Q_x| : x \in E_t\}$, so $\alpha_1 \leq t^{-1}$, and let us fix a cube Q_x, call it Q_1, such that $|Q_1| > \frac{1}{2}\alpha_1$. Consider all cubes Q_x such that $Q_x \cap Q_1 = \emptyset$. If there are no such cubes we stop, if there are we put

$$\alpha_2 = \max\{|Q_x| : x \in E_t \text{ and } Q_x \cap Q_1 = \emptyset\}$$

and we fix such a cube Q_x, call it Q_2, satisfying $|Q_2| > \frac{1}{2}\alpha_2$. We continue in this way and we get a sequence of cubes $Q_1, Q_2, \ldots$ (possibly finite) such that:

(i) the cubes Q_i are disjoint
(ii) $|Q_j| > \frac{1}{2}\max\{|Q_x| :$ and $Q_x \cap Q_s = \emptyset$ for $s = 1, 2, \ldots, j-1\}$.
(iii) if $Q_x \cap Q_s = \emptyset$ for $s = 1, 2, \ldots, j-1$ then $|Q_x| \leq 2|Q_j|$.

From (i) above and 6.10 we get

$$\left|\bigcup_{i=1} Q_i\right| = \sum_{i=1} |Q_i| \leq \frac{1}{t}\sum_{i=1}\int_{Q_i} |f(y)|\,dy \leq \frac{1}{t}. \tag{6.11}$$

The important observation we have to make is that *each* Q_x intersects some Q_i. To see this note that if there exists a Q_x disjoint from all Q_i's, then our process was infinite, so from 6.11 we see that $|Q_j| \to \infty$, but this contradicts (iii) above.

Now, given Q_x let Q_s be the first Q_i that intersects Q_x. Condition iii) above implies that $|Q_x| \leq 2|Q_s|$, so easy geometry shows that $Q_x \subset 6 \diamond Q_s$. Thus from 6.11 we get

$$\begin{aligned} |\{x : Mf(x) > t\}| &= \left|\bigcup_x Q_x\right| \leq \left|\bigcup_i 6 \diamond Q_i\right| \\ &\leq \sum_{i=1} |6 \diamond Q_i| \leq 6^d \sum_{i=1} |Q_i| \leq 6^d \frac{1}{t}. \end{aligned}$$

This proves the theorem. □

As our first application of the above theorem we will present the following classical result.

Theorem 6.4 (Lebesgue differentiation theorem) *Let $f \in L_1(\mathbb{R}^d)$ be given. For almost all $x \in \mathbb{R}^d$ and for every decreasing sequence of cubes $(Q_j)_{j=1}^{\infty}$ such that $\bigcap_{j=1}^{\infty} Q_j = \{x\}$ we have*

$$\lim_{j\to\infty} \frac{1}{|Q_j|} \int_{Q_j} f(y)\,dy = f(x). \tag{6.12}$$

Proof If $f \in L_1(\mathbb{R}^d) \cap C(\mathbb{R}^d)$ then 6.12 holds for all $x \in \mathbb{R}^d$. Given $f \in L_1(\mathbb{R}^d)$ we take ϵ such that $0 < \epsilon < 1$ and write $f = g + h$ with $g \in L_1(\mathbb{R}^d) \cap C(\mathbb{R}^d)$ and $\|h\|_1 < \epsilon$. From the above observation and the definition of Hardy–Littlewood maximal function (Definition 6.2) we have

$$\begin{aligned}
&\limsup_{j\to\infty} \left| \frac{1}{|Q_j|} \int_{Q_j} f(y)\,dy - f(x) \right| = \\
&= \limsup_{j\to\infty} \left| \frac{1}{|Q_j|} \int_{Q_j} g(y)\,dy - g(x) + \frac{1}{|Q_j|} \int_{Q_j} h(y)\,dy - h(x) \right| \\
&= \limsup_{j\to\infty} \left| \frac{1}{|Q_j|} \int_{Q_j} h(y)\,dy - h(x) \right| \\
&\leq |h(x)| + \limsup_{j\to\infty} \frac{1}{|Q_j|} \int_{Q_j} |h(y)|\,dy \\
&\leq |h(x)| + Mh(x).
\end{aligned}$$

Note however that Theorem 6.3 gives

$$|\{x : Mh(x) > \sqrt{\epsilon}\}| \leq 6^d \frac{\|h\|_1}{\sqrt{\epsilon}} \leq 6^d \sqrt{\epsilon}$$

and 6.4 gives

$$|\{x : |h(x)| > \sqrt{\epsilon}\}| \leq \frac{\|h\|_1}{\sqrt{\epsilon}} \leq \sqrt{\epsilon}.$$

This shows that outside the set $\{x : Mh(x) > \sqrt{\epsilon}\} \cup \{x : |h(x)| > \sqrt{\epsilon}\}$ which has measure at most $C\sqrt{\epsilon}$ we have

$$\limsup_{j\to\infty} \left| \frac{1}{|Q_j|} \int_{Q_j} f(y)\,dy - f(x) \right| \leq 2\sqrt{\epsilon}.$$

Since ϵ can be as small as we wish we get 6.12. □

Motivated by the properties of the Hardy–Littlewood maximal function we introduce the following general definitions.

DEFINITION 6.5 *An operator T defined on some class of measurable functions and mapping it into measurable functions is called sublinear if*

$$|T(f+g)(x)| \leq |T(f)(x)| + |T(g)(x)|$$

and

$$|T(\lambda f)(x)| \leq |\lambda||T(f)(x)|$$

for all admissible functions f and g and all scalars λ. The above inequalities are to be understood as holding almost everywhere.

DEFINITION 6.6 *A sublinear operator T defined on $L_1(\mathbb{R}^d)$ is of weak type (1,1) if there exists a constant C such that for each $f \in L_1(\mathbb{R}^d)$ and each $t > 0$ we have*

$$|\{x : |Tf(x)| > t\}| \leq \tfrac{C}{t} \|f\|_1 . \tag{6.13}$$

It is clear that each linear operator is sublinear. It follows immediately from 6.4 that if a sublinear operator T satisfies $\|Tf\|_1 \leq C \|f\|_1$ then T is of weak type (1,1). The usefulness of the above notions stems from the following interpolation theorem.

Theorem 6.7 (Marcinkiewicz) *Suppose that T is a sublinear operator defined on $L_1(\mathbb{R}^d) + L_\infty(\mathbb{R}^d)$ which is of weak type (1,1) and for some C satisfies*

$$\|Tf\|_\infty \leq C \|f\|_\infty . \tag{6.14}$$

Then for each p, $1 < p < \infty$, there exists a constant $C(p)$ such that $\|Tf\|_p \leq C(p) \|f\|_p$.

Proof Multiplying the operator T by an appropriate constant we can (and will) assume that it satisfies 6.13 and 6.14 with constant $C = 1$. For $f \in L_p(\mathbb{R}^d)$ and given $t > 0$ we write $f(x) = f_t(x) + f^t(x)$ where $f_t(x) = f(x) \cdot \mathbf{1}_{\{s \,:\, |f(s)| \leq t\}}$. Since T is sublinear

$$\{x : |Tf(x)| > t\} \subset \{x : |Tf_{t/2}(x)| > \tfrac{t}{2}\} \cup \{x : |Tf^{t/2}(x)| > \tfrac{t}{2}\}$$

so from 6.5 we get

$$\begin{aligned}
\|Tf\|_p^p &= \int_{\mathbb{R}^d} |Tf(x)|^p\,dx \\
&= p\int_0^\infty t^{p-1}|\{x : |Tf(x)| > t\}|dt \qquad (6.15)\\
&\le p\int_0^\infty t^{p-1}|\{x : |Tf_{t/2}(x)| > \tfrac{t}{2}\}|dt \\
&\quad + p\int_0^\infty t^{p-1}|\{x : |Tf^{t/2}(x)| > \tfrac{t}{2}\}|dt.
\end{aligned}$$

But we have assumed that T satisfies 6.14 with constant 1, and this implies $\|Tf_{t/2}\|_\infty \le \frac{t}{2}$ and thus the second integral in the above is zero. So using 6.13, 6.15 and Fubini's theorem we obtain

$$\begin{aligned}
\|Tf\|_p^p &\le p\int_0^\infty t^{p-1}\frac{2}{t}\left\|f_{t/2}\right\|_1\,dt \\
&= 2p\int_0^\infty t^{p-2}\int_{\{x : |f(x)|>t/2\}} |f(x)|\,dx\,dt \\
&= 2p\int_{\mathbb{R}^d} |f(x)|\int_0^{2|f(x)|} t^{p-2}dt\,dx \\
&= 2p\int_{\mathbb{R}^d} |f(x)|\frac{1}{p-1}\left(2|f(x)|\right)^{p-1}dx \\
&= \frac{2^p p}{(p-1)}\int_{\mathbb{R}^d} |f(x)|^p\,dx.
\end{aligned}$$

□

Immediately from Theorem 6.7, 6.9 and Theorem 6.3 we obtain:

Corollary 6.8 *For each p, $1 < p \le \infty$, there exists a constant $C(p)$ such that $\|Mf\|_p \le C(p)\|f\|_p$ where M is the Hardy–Littlewood maximal function.*

The heart of the proof of Theorem 6.7 is the splitting of the function into a 'small' and a 'large' part. In the future we will need a more refined decomposition of this type. In order to obtain it we need to introduce the family of dyadic cubes in $\mathbb{R}^d$. This is a direct generalization of the family of dyadic intervals introduced in Section 1.1. By a dyadic cube in $\mathbb{R}^d$ we mean any cube $Q = I_1 \times \ldots \times I_d$ where $I_j = [k_j 2^s, (k_j+1)2^s]$ for an integer s and some integers k_j. If $Q = [a_1, a_1+h] \times \ldots \times [a_d, a_d+h]$ is any cube in $\mathbb{R}^d$ then by a dyadic subcube of Q we will mean any cube of the form

$$\left[a_1 + \frac{k_1 h}{2^s}, a_1 + \frac{(k_1+1)h}{2^s}\right] \times \ldots \times \left[a_d + \frac{k_d h}{2^s}, a_d + \frac{(k_d+1)h}{2^s}\right]$$

where $s = 0, 1, 2, \ldots$ and $k_i = 0, 1, \ldots, 2^s - 1$. All this simply means that we take the cube Q itself, then we subdivide it naturally into 2^d equal cubes, next each of those we subdivide into 2^d equal subcubes, etc. It is an easy but very important observation that *two dyadic cubes (dyadic subcubes of a given cube) are either non-overlapping or one is contained in the other.*

Theorem 6.9 (Calderón–Zygmund decomposition) *Suppose we are given a function $f \in L_1(\mathbb{R}^d)$ (or $f \in L_1(\mathbf{Q})$ for some cube $\mathbf{Q} \subset \mathbb{R}^d$) and a number $t > 0$ ($t > \frac{1}{|\mathbf{Q}|}\int_{\mathbf{Q}} |f(x)|\,dx$). There exists a family $C(f,t)$ of non-overlapping dyadic cubes (dyadic subcubes of $\mathbf{Q}$) such that*

(i) *each $Q \in C(f,t)$ is a maximal, in the sense of inclusion, dyadic cube (dyadic subcube of $\mathbf{Q}$) such that*

$$\frac{1}{|Q|}\int_Q |f(x)|\,dx > t \tag{6.16}$$

(ii) *for each cube $Q \in C(f,t)$ we have*

$$\frac{1}{|Q|}\int_Q |f(x)|\,dx \leq 2^d t$$

(iii) *the total measure covered by cubes from $C(f,t)$ can be estimated as*

$$\sum_{Q \in C(f,t)} |Q| \leq \frac{1}{t}\int_{\{x\,:\,|f(x)|>t\}} |f(y)|\,dy \leq \frac{1}{t}\|f\|_1$$

(iv) *if $x \notin \bigcup_{Q\in C(f,t)} Q$ then $|f(x)| \leq t$.*

The above Calderón–Zygmund decomposition is a very powerful tool. We can use it as stated in the above Theorem to get a collection of cubes (in this way we will prove Theorem 6.16) but also for each t it yields a very useful decomposition of functions. For each t we define the 'bad' part of a function f to be

$$b_t(x) = \sum_{Q \in C(f,t)} (f(x) - f_Q)\,\mathbf{1}_Q(x)$$

where $f_Q = \frac{1}{|Q|}\int_Q f(y)\,dy$, and the 'good' part to be $f - b_t$. This decomposition will be used to prove Theorem 6.23 which is the improved version of the Marcinkiewicz interpolation theorem.

Proof Since the argument is almost exactly the same for $\mathbb{R}^d$ as for $\mathbf{Q}$, we will consider both cases simultaneously. We define $C(f,t)$ to be the family of all maximal dyadic cubes satisfying 6.16. This definition obviously ensures that (i) holds. Our choice of t ensures that $\mathbf{Q} \notin C(f,t)$. For $Q \in C(f,t)$ let Q' be the smallest dyadic cube strictly containing it. By the maximality of Q we have $Q' \notin C(f,t)$ so

$$t \geq \frac{1}{|Q'|}\int_{Q'} |f(x)|\,dx \geq \frac{1}{|Q'|}\int_Q |f(x)|\,dx = \frac{1}{2^d|Q|}\int_Q |f(x)|\,dx$$

which gives (ii). If $x \notin \bigcup_{Q\in C(f,t)} Q$ then for each dyadic cube Q such that $x \in Q$ we have

$$\left|\frac{1}{|Q|}\int_Q f(x)\,dx\right| \leq \frac{1}{|Q|}\int_Q |f(x)|\,dx \leq t.$$

So taking a sequence of decreasing dyadic cubes Q_j such that $\bigcap_j Q_j = \{x\}$ we get from Theorem 6.4 that $|f(x)| \leq t$ for almost all x. This gives (iv). To see (iii) note that $C(f,t)$ consists of non-overlapping cubes, so using 6.16 and condition (iv) which is already proven, we get

$$\begin{aligned}
\sum_{Q\in C(f,t)} |Q| &\leq \sum_{Q\in C(f,t)} \frac{1}{t}\int_Q |f(x)|\,dx \\
&= \frac{1}{t}\int_{\bigcup\{Q\,:\,Q\in C(f,t)\}} |f(x)|\,dx \\
&\leq \frac{1}{t}\int_{\{x\,:\,|f(x)|>t\}} |f(x)|\,dx \leq \frac{1}{t}\,\|f\|_1\,.
\end{aligned}$$

□

6.2 *BMO* and H_1

In this section we will introduce and investigate two new spaces of functions: $BMO(\mathbb{R}^d)$ and $H_1(\mathbb{R}^d)$. We start with two series of spaces $BMO_p(\mathbb{R}^d)$ and $H_1^q(\mathbb{R}^d)$ and after some effort we will show that each of these series collapses into one space and that these spaces are in duality.

In this and in the next section we will very often use the notation f_Q

to denote the mean value of the function f on the cube Q, i.e.

$$f_Q =: \frac{1}{|Q|}\int_Q f(x)\,dx.$$

DEFINITION 6.10 *We say that the function f on $\mathbb{R}^d$ has bounded p-mean oscillation, $1 \le p < \infty$ if*

$$\|f\|_{*,p} =: \sup_Q \left(\frac{1}{|Q|}\int_Q |f(x) - f_Q|^p\,dx\right)^{1/p} < \infty. \tag{6.17}$$

The set of all functions of bounded p–mean oscillation will be denoted by $BMO_p(\mathbb{R}^d)$.

Let us make some observations about these definitions.

- The quantity $\|.\|_{*,p}$ has the following properties:

$$\begin{aligned}\|f+g\|_{*,p} &\le \|f\|_{*,p} + \|g\|_{*,p}\\ \|\lambda f\|_{*,p} &= |\lambda|\cdot\|f\|_{*,p}\\ \|f\|_{*,p} &= 0 \text{ if and only if } f = \text{const. a.e}\end{aligned}$$

  This shows that $\|.\|_{*,p}$ is a quasi–norm and it becomes a norm when we identify functions which differ by a constant. With this identification $BMO_p(\mathbb{R}^d)$ becomes a normed space.
- Each bounded function has bounded p-mean oscillation, in other words $L_\infty(\mathbb{R}^d) \subset BMO(\mathbb{R}^d)$ for $1 \le p < \infty$.
- In 6.17 we can replace f_Q by other numbers, more precisely we have the following: *Suppose that for each cube $Q \subset \mathbb{R}^d$ there exists a number c_Q such that*

$$\sup_Q \left(\frac{1}{|Q|}\int_Q |f(x) - c_Q|^p\,dx\right)^{1/p} < \infty.$$

  Then $f \in BMO_p(\mathbb{R}^d)$. To see this, observe that

$$\begin{aligned}&\left(\frac{1}{|Q|}\int_Q |f(x) - f_Q|^p\,dx\right)^{1/p}\\ &\quad\le \left(\frac{1}{|Q|}\int_Q |f(x) - c_Q|^p\,dx\right)^{1/p} + |f_Q - c_Q|\end{aligned}$$

  and

$$|f_Q - c_Q| = \left|\frac{1}{|Q|}\int_Q (f(x) - c_Q)\,dx\right| \le \frac{1}{|Q|}\int_Q |f(x) - c_Q|\,dx$$

$$\leq \left(\frac{1}{|Q|}\int_Q |f(x)-c_Q|^p\,dx\right)^{1/p}.$$

- There are unbounded functions in $BMO(\mathbb{R}^d)$. In particular the function $\ln|x|$ on $\mathbb{R}^d$ is in $BMO_1(\mathbb{R}^d)$. In order to avoid cumbersome calculations let us check this only for $d=1$. Since $\ln|x|$ is even it suffices to assume that $Q=[a,b]$ with $|a|\leq b$. We have

$$\begin{aligned}\frac{1}{b-a}\int_a^b |\ln|x|-\ln b|\,dx &= \frac{1}{b-a}\int_a^b \ln b-\ln|x|\,dx \\ &= \frac{1}{b-a}\left[(b-a)\ln b - x(\ln|x|-1)|_a^b\right] \\ &= \frac{1}{b-a}\left[b-a+a(\ln|a|-\ln b)\right].\end{aligned}$$

For $a=-b$ this equals 1 while for $|a|<b$ it is at most

$$1+|a|\frac{\ln b-\ln|a|}{b-|a|}.$$

From the mean value theorem we infer that it equals $1+|a|\eta^{-1}$ for some η, $|a|\leq\eta\leq b$, so finally we see that $\|\ln|x|\|_{*,1}\leq 2$.
- Note also that $f(x)=\operatorname{sgn} x\cdot\ln|x|\notin BMO_1(\mathbb{R})$. To see this take $Q=[-a,a]$, with $a>0$. Since f is odd $f_Q=0$ so

$$\frac{1}{|Q|}\int_Q |f(x)-f_Q|\,dx = \frac{1}{2a}\int_{-a}^a |f(x)|\,dx \geq \min_{x\in Q}|f(x)| = |\ln a|$$

which is unbounded.
- Note that the above two observations point to an important difference between $BMO_p(\mathbb{R}^d)$ and $L_p(\mathbb{R}^d)$ spaces. Namely there are functions f and g such that $|f(x)|\leq|g(x)|$ and $g\in BMO_p(\mathbb{R}^d)$ while $f\notin BMO_p(\mathbb{R}^d)$.

Now we will introduce the second series of spaces, namely $H_1^q(\mathbb{R}^d)$ for $1<q\leq\infty$. We start with the following

DEFINITION 6.11 *Let q, $1<q\leq\infty$, be given. A q-atom is a function $a(x)$ defined on $\mathbb{R}^d$ such that there exists a cube $Q\subset\mathbb{R}^d$ such that*

(i) supp $a(x)\subset Q$
(ii) $\int_{\mathbb{R}^d} a(x)\,dx=0$
(iii) $\|a\|_q\leq|Q|^{1/q-1}$ *that is* $\int_{\mathbb{R}^d}|a(x)|^q\,dx\leq|Q|^{1-q}$. *Note that for $q=\infty$ we mean $\|a\|_\infty\leq|Q|^{-1}$.*

Let us start with some remarks about this definition. We observe that *if $1 < q_1 < q_2 \le \infty$ then each q_2-atom is a q_1-atom.* The only thing to check is condition (iii). Applying Hölder's inequality for $s = \frac{q_2}{q_1} > 1$ and r, $s^{-1} + r^{-1} = 1$ we have

$$\begin{aligned}
\int_{\mathbb{R}^d} |a(x)|^{q_1}\, dx &= \int_{\mathbb{R}^d} |a(x)|^{q_1} \cdot \mathbf{1}_Q(x)\, dx \\
&\le \left(\int_{\mathbb{R}^d} |a(x)|^{q_1 s}\, dx \right)^{1/s} \cdot \left(\int_{\mathbb{R}^d} |\mathbf{1}_Q(x)|^r\, dx \right)^{1/r} \\
&= |Q|^{1/r} \left(\int_{\mathbb{R}^d} |a(x)|^{q_2} \right)^{1/s} \\
&\le |Q|^{1/r} |Q|^{(1-q_2)/s} = |Q|^{1-q_1}.
\end{aligned}$$

Note also that the same argument for $q_1 = 1$ gives: *if $a(x)$ is a q-atom for $1 < q \le \infty$ then $\|a\|_1 \le 1$.*

DEFINITION 6.12 *The space $H_1^q(\mathbb{R}^d)$ is the space of all functions f on $\mathbb{R}^d$ (identified when equal almost everywhere) which can be written as*

$$f = \sum_j \lambda_j a_j \quad \textit{with} \; \sum_j |\lambda_j| < \infty \quad \textit{and } a_j \; q\textit{-atoms} \tag{6.18}$$

so the series converges in $L_1(\mathbb{R}^d)$. We define $|\!|\!| f |\!|\!|_{1,q} = \inf \sum_j |\lambda_j|$ where the infimum is taken over all representations of f as the series 6.18.

It is clear from the above definition that the following useful fact holds:

Proposition 6.13 *If T is a linear map from $H_1^q(\mathbb{R}^d)$ into a Banach space X such that $\|Ta\| \le C$ for each q-atom a then T is a continuous linear operator and $\|T\| \le C$.*

Our next proposition summarizes some functional analytic properties of spaces $H_1^q(\mathbb{R}^d)$.

Proposition 6.14

(i) *For $1 < q_1 \le q_2 \le \infty$ we have $\|f\|_1 \le |\!|\!| f |\!|\!|_{1,q_1} \le |\!|\!| f |\!|\!|_{1,q_2}$ so the identity is a continuous embedding*

$$H_1^{q_2}(\mathbb{R}^d) \hookrightarrow H_1^{q_1}(\mathbb{R}^d) \hookrightarrow L_1(\mathbb{R}^d)$$

(ii) *For each q, $1 < q \leq \infty$, $\|\cdot\|_{1,q}$ is a norm and $H_1^q(\mathbb{R}^d)$ equipped with this norm is a Banach space*

(iii) *If $f \in H_1^q(\mathbb{R}^d)$ then $\int_{\mathbb{R}^d} f(x)\,dx = 0$*

(iv) *If Q is a cube in $\mathbb{R}^d$ and $f \in L_q(Q)$ with $\int_Q f(x)\,dx = 0$ then $f \in H_1^q(\mathbb{R}^d)$.*

Proof Condition (i) follows directly from our observations about q-atoms. It implies that if $\|f\|_{1,q} = 0$ then $f = 0$ a.e., because $\|f\|_1 = 0$. It is routine to check directly from the definition that $\|\cdot\|_{1,q}$ satisfies the triangle inequality and that $\|\lambda f\|_{1,q} = |\lambda| \cdot \|f\|_{1,q}$. Thus $\|\cdot\|_{1,q}$ is a norm. To check that $H_1^q(\mathbb{R}^d)$ is complete, it suffices to check that if $f_n \in H_1^q(\mathbb{R}^d)$ and $\sum_n \|f_n\|_{1,q} < \infty$ then $\sum_n f_n$ is in $H_1^q(\mathbb{R}^d)$. But we can represent each f_n as $\sum_j \lambda_j^n a_j^n$ where a_j^n are q-atoms and $\sum_j |\lambda_j^n| \leq 2\|f_n\|_{1,q}$. Thus the representation $\sum_n \sum_j \lambda_j^n a_j^n$ shows that $\sum_n f_n \in H_1^q(\mathbb{R}^d)$. Condition (iii) is clear because it holds for each atom and the integral exists because $f \in L_1(\mathbb{R}^d)$. To see (iv) note that each such f is an appropriate multiple of a q-atom. □

Proposition 6.15 *Let g be a function on $\mathbb{R}^d$. The formula $\varphi(f) = \int_{\mathbb{R}^d} f(x)g(x)\,dx$ considered only when f is a finite sum of q-atoms gives (extends to) a bounded linear functional on $H_1^q(\mathbb{R}^d)$ if and only if $g \in BMO_p(\mathbb{R}^d)$ with $q^{-1}+p^{-1} = 1$. The norms are equivalent, i.e. there exist two constants $0 < C_1 \leq C_2$ such that $C_1\|\varphi\| \leq \|g\|_{*,p} \leq C_2\|\varphi\|$.*

REMARK 6.3. Since $BMO_p(\mathbb{R}^d)$ may contain unbounded functions, we have to be a bit careful in the formulation of the above proposition. It is not true (cf. Exercise 6.6) that if $f \in H_1^q(\mathbb{R}^d)$ and $g \in BMO_p(\mathbb{R}^d)$ then $fg \in L_1(\mathbb{R}^d)$.

Proof Let us take $g \in BMO_p(\mathbb{R}^d)$. For $f = \sum_j \lambda_j a_j$ with a_j q-atoms supported on cubes Q_j we have

$$\left|\int_{\mathbb{R}^d} f(x)g(x)\,dx\right| \leq \sum_j |\lambda_j| \left|\int_{Q_j} a_j(x)g(x)\,dx\right|$$

$$= \sum_j |\lambda_j| \left|\int_{Q_j} a_j(x)(g(x) - g_{Q_j})\,dx\right|$$

$$\leq \sum_j |\lambda_j| \left(\int_{Q_j} |a_j(x)|^q\,dx\right)^{1/q} \left(\int_{Q_j} |g(x) - g_{Q_j}|^p\,dx\right)^{1/p}$$

$$\le \sum_j |\lambda_j||Q_j|^{1/q-1}\left(\int_{Q_j}|g(x)-g_{Q_j}|^p\,dx\right)^{1/p}$$
$$=\sum_j|\lambda_j|\left(\frac{1}{|Q_j|}\int_{Q_j}|g(x)-g_{Q_j}|^p\,dx\right)^{1/p}$$
$$\le \|g\|_{*,p}\sum_j|\lambda_j|.$$

This clearly implies

$$\left|\int_{\mathbb{R}^d} f(x)g(x)\,dx\right| \le \|f\|_{1,q}\cdot\|g\|_{*,p}. \tag{6.19}$$

In other words g gives a linear functional on $H_1^q(\mathbb{R}^d)$ with norm at most $\|g\|_{*,p}$.

On the other hand for a function g on $\mathbb{R}^d$ and a cube Q we have (because $L_p^*=L_q$)

$$\left(\frac{1}{|Q|}\int_Q|g(x)-g_Q|^p\,dx\right)^{1/p}=\frac{1}{|Q|}\int_Q(g(x)-g_Q)b(x)\,dx \tag{6.20}$$

for some $b(x)$ on Q such that

$$\left(\frac{1}{|Q|}\int_Q|b(x)|^q\right)^{1/q}=1. \tag{6.21}$$

Since $\big(g(x)-g_Q\big)$ has mean zero we can continue 6.20 as follows:

$$\begin{aligned}\left(\frac{1}{|Q|}\int_Q|g(x)-g_Q|^p\,dx\right)^{1/p} &= \frac{1}{|Q|}\int_Q\big(g(x)-g_Q\big)\big(b(x)-b_Q\big)\,dx\\ &= \int_Q g(x)\frac{1}{|Q|}\big(b(x)-b_Q\big)\,dx. \end{aligned}\tag{6.22}$$

Note that

$$|b_Q|\le\frac{1}{|Q|}\int_Q|b(x)|\,dx\le 1. \tag{6.23}$$

From 6.21 and 6.23 we get

$$\begin{aligned}&\left(\int_Q\left|\tfrac{1}{|Q|}(b(x)-b_Q)\right|^q dx\right)^{1/q}\\ &\quad=|Q|^{1/q-1}\left(\frac{1}{|Q|}\int_Q|b(x)-b_Q|^q\,dx\right)^{1/q}\le 2|Q|^{1/q-1}\end{aligned}$$

which means that $\frac{1}{|Q|}(b(x) - b_Q)\mathbf{1}_Q$ equals twice a q-atom. Thus using 6.22 we get

$$\begin{aligned}\|g\|_{H_1^q(\mathbb{R}^d)^*} &\geq \sup\left\{\int_{\mathbb{R}^d} g(x)a(x)\,dx \;:\; a(x) \text{ is a } q\text{-atom}\right\} \\ &\geq \frac{1}{2}\sup_Q\left(\frac{1}{|Q|}\int_Q |g(x) - g_Q|^p\right)^{1/p} = \frac{1}{2}\|g\|_{*,p}. \qquad (6.24)\end{aligned}$$

Comparing 6.24 and 6.19 we see that a function on $\mathbb{R}^d$ gives a functional on $H_1^q(\mathbb{R}^d)$ if and only if $g \in BMO_p(\mathbb{R}^d)$ and the norms are equivalent. □

The fundamental result about $BMO(\mathbb{R}^d)$ is the following theorem.

Theorem 6.16 (John–Nierenberg) *There exist two constants C_1 and C_2 (depending only on d) such that for every $f \in BMO_1(\mathbb{R}^d)$ and every cube $Q \subset \mathbb{R}^d$ and every $t \geq 0$*

$$|\{x \in Q \;:\; |f(x) - f_Q| > t\}| \leq C_1|Q| \exp -\frac{C_2 t}{\|f\|_{*,1}}. \qquad (6.25)$$

Proof It suffices to show 6.25 for f with $\|f\|_{*,1} = 1$, so we will assume this. Our basic tool will be the Calderón–Zygmund decomposition described in Theorem 6.9. We apply it to the function $f - f_Q$ on the cube Q with $t = e$. We obtain a family of non-overlapping dyadic subcubes Q_j^1 of Q such that

$$e \leq \frac{1}{|Q_j^1|}\int_{Q_j^1} |f(x) - f_Q|\,dx \leq 2^d e \qquad (6.26)$$

$$\text{for } x \in Q \setminus \bigcup_j Q_j^1 \text{ we have } |f(x) - f_Q| \leq e \qquad (6.27)$$

$$\sum_j |Q_j^1| \leq \frac{1}{e}\int_Q |f(x) - f_Q|\,dx \leq \frac{1}{e}|Q|. \qquad (6.28)$$

Note that 6.26 gives

$$2^d e \geq \left|\frac{1}{|Q_j^1|}\int_{Q_j^1} (f(x) - f_Q)\,dx\right| \geq |f_{Q_j^1} - f_Q|. \qquad (6.29)$$

Now for each cube Q_j^1, exactly as before, we apply Theorem 6.9 to the function $f(x) - f_{Q_j^1}$ considered on the cube Q_j^1 with $t = e$. We get non-overlapping dyadic subcubes of Q_j^1, which we call $Q_{j,s}^1$, such that (see 6.27, 6.28, 6.29)

$$\text{for } x \in Q_j^1 \setminus \bigcup_s Q_{j,s}^1 \text{ we have } |f(x) - f_{Q_j^1}| \leq e \tag{6.30}$$

$$\sum_s |Q_{j,s}^1| \leq \frac{1}{e}|Q_j^1| \tag{6.31}$$

$$|f_{Q_{j,s}^1} - f_{Q_j^1}| \leq 2^d e. \tag{6.32}$$

Let us reindex all cubes $Q_{j,s}^1$ and denote them by Q_j^2. This is a family of non-overlapping dyadic subcubes of Q such that

$$\sum_k |Q_k^2| = \sum_j \sum_s |Q_{j,s}^1| \leq \frac{1}{e} \sum_j |Q_j^1| \leq \left(\frac{1}{e}\right)^2 |Q|$$

and

$$\text{for } x \in Q \setminus \bigcup_k Q_k^2 \text{ we have } |f(x) - f_Q| \leq 2 \cdot 2^d e. \tag{6.33}$$

To verify 6.33 note that if x belongs to no Q_j^1 then 6.27 implies 6.33 and when $x \in Q_j^1$ then from 6.27 and 6.32 we have

$$|f(x) - f_Q| \leq |f(x) - f_{Q_j^1}| + |f_{Q_j^1} - f_Q| \leq e + 2^d e \leq 2 \cdot 2^d e.$$

Now we apply Theorem 6.9 to each function $f(x) - f_{Q_j^2}$ considered on the cube Q_j^2 and with $t = e$. Continuing in this manner we get for each $N = 1, 2, \ldots$ a family of non-overlapping dyadic subcubes of Q which we call $\{Q_j^N\}$ such that

$$\text{each } Q_j^{N+1} \text{ is a dyadic subcube of some cube } Q_s^N \tag{6.34}$$

$$\sum_j |Q_j^N| \leq e^{-N}|Q| \tag{6.35}$$

and

$$\text{for } x \in Q \setminus \bigcup_j Q_j^N \text{ we have } |f(x) - f_Q| \leq N \cdot 2^d e. \tag{6.36}$$

Note that condition follows 6.34 directly from the construction, condition 6.35 is simply a repetitive application of 6.28 while 6.36 follows by induction exactly like 6.33. Now we are ready to show 6.25. Let us start with $t \geq 2^d e$ and fix $N = 1, 2, \dots$ such that $N 2^d e \leq t < (N+1) 2^d e$. Using 6.36 and 6.34 we obtain

$$\begin{aligned} |\{x \in Q : |f(x) - f_Q| > t\}| &\leq |\{x \in Q : |f(x) - f_Q| > N 2^d e\}| \\ &\leq \Big| \bigcup_j Q_j^N \Big| \leq e^{-N} |Q|. \end{aligned}$$

Taking $C_2 = e^{-1} 2^{-d-1}$ this gives 6.25 for all $t \geq 2^d e$ with $C_1 = 1$. It is clear that we can increase C_1 to get 6.25 for all $t \geq 0$ (with the same C_2). □

The above theorem shows that unbounded functions in $BMO_1(\mathbb{R}^d)$ are only 'slightly' unbounded. The sets where they differ very much from the mean value take up only an exceedingly small part of each cube. In particular it implies that it does not matter in what L_p norm we measure oscillation. This easily implies that the spaces $BMO_p(\mathbb{R}^d)$ are the same for all p, $1 \leq p < \infty$. This is proved in the following corollary.

Corollary 6.17 *For each p, $1 \leq p < \infty$, there exists a constant C_p such that for each $f \in BMO_p(\mathbb{R}^d)$ we have*

$$\|f\|_{*,1} \leq \|f\|_{*,p} \leq C_p \|f\|_{*,1} . \tag{6.37}$$

Proof The left hand side inequality is clear. To get the right hand side note that Proposition 6.1 and 6.25 yield

$$\begin{aligned} \int_Q |f(x) - f_Q|^p \, dx &= p \int_0^\infty t^{p-1} |\{x \in Q : |f(x) - f_Q| > t\}| \, dt \\ &\leq C_1 p |Q| \int_0^\infty t^{p-1} \exp\Big(- \frac{C_2 t}{\|f\|_{*,1}} \Big) \, dt. \end{aligned}$$

Substituting $u = C_2 t \|f\|_{*,1}^{-1}$ we get

$$\begin{aligned} \int_Q |f(x) - f_Q|^p \, dx &\leq |Q| \, \|f\|_{*,1}^p \, \frac{C_1 p}{C_2^p} \int_0^\infty u^{p-1} e^{-u} \, du \\ &= |Q| \, \|f\|_{*,1}^p \, C_p^p. \end{aligned}$$

□

In view of the above corollary we will drop the subscript p and use the notation $BMO(\mathbb{R}^d)$. We will also talk about functions of bounded mean oscillation to denote functions satisfying any of the equivalent conditions 6.17. Any of the equivalent BMO norms $\|.\|_{*,p}$ will be denoted by $\|.\|_*$.

Since the spaces $BMO_p(\mathbb{R}^d)$ turn out to be just one space $BMO(\mathbb{R}^d)$, looking at Proposition 6.15 it is natural to expect that the spaces $H_1^q(\mathbb{R}^d)$ are really one space. This is actually the case, namely we have:

Theorem 6.18 *For $1 < q \leq \infty$ the spaces $H_1^q(\mathbb{R}^d)$ coincide and the norms $|\!|\!|.|\!|\!|_{1,q}$ are equivalent. This one space will be denoted by $H_1(\mathbb{R}^d)$ and any of the norms $|\!|\!|.|\!|\!|_{1,q}$ will be denoted by $|\!|\!|.|\!|\!|$. We have also $H_1(\mathbb{R}^d)^* = BMO(\mathbb{R}^d)$ in the sense that for each continuous linear functional φ on $H_1(\mathbb{R}^d)$ there exists a unique (up to a constant) function $g \in BMO(\mathbb{R}^d)$ such that if f is any finite sum of atoms we have*

$$\varphi(f) = \int_{\mathbb{R}^d} f(x)g(x)\,dx.$$

The BMO norm $\|g\|_*$ and the functional norm of φ are equivalent.

Proof Let us start with the case $q < \infty$. By $L_q^0(Q)$ we denote the subspace of $L_q(Q)$ consisting of all functions of mean zero. This is easily seen to be a closed subspace, and since $L_q(Q)^* = L_p(Q)$ where $\frac{1}{p}+\frac{1}{q} = 1$, one can easily check that each linear, continuous functional φ on $L_q^0(Q)$ is given by a function $f \in L_p(Q)$. This function f however is not unique. Two such functions f_1 and f_2 give the same functional if and only if $f_1 - f_2$ is constant. Let us take cubes $Q_n = [-2^n, 2^n]^d$. From Proposition 6.14(iv) we know that $L_q^0(Q_n) \subset H_1^q(\mathbb{R}^d)$. This implies that a functional φ on $H_1^q(\mathbb{R}^d)$ induces a functional on each $L_q^0(Q_n)$. Thus there exists a function f_n on Q_n which on $L_q^0(Q_n)$ induces the same linear functional as φ. This implies that the function $f_{n+1}|Q_n$ gives $\varphi|L_q^0(Q_n)$, so f_n and $f_{n+1}|Q_n$ differ by a constant. Since adding the constant does not change the functional we can assume that $f_{n+1}|Q_n = f_n$. Doing this inductively we find a function f_∞ defined on $\mathbb{R}^d$ by $f_\infty|Q_n = f_n$ which gives the functional φ. This shows that φ is given by a function, so we can apply Proposition 6.15 to see that $f_\infty \in BMO_p(\mathbb{R}^d)$. This means that for $q < \infty$ we have $H_1^q(\mathbb{R}^d)^* = BMO_p(\mathbb{R}^d)$.

For $q = \infty$ the above proof does not work, because functionals on L_∞^0 are not functions. To treat this case let us denote by $I_{s,r}$, $1 < r \leq s \leq \infty$, the identity acting as a map from $H_1^s(\mathbb{R}^d)$ into $H_1^r(\mathbb{R}^d)$. It follows from Proposition 6.14(i) that $I_{s,r}$ is a continuous, linear, 1–1 map. Thus $I_{s,r}^* : H_1^r(\mathbb{R}^d)^* \to H_1^s(\mathbb{R}^d)^*$. When $1 < r \leq s < \infty$ we infer from the above that $I_{s,r}^*$ is the identity between $BMO_{r'}(\mathbb{R}^d)$ and $BMO_{s'}(\mathbb{R}^d)$, that is an isomorphism (see Corollary 6.17). This implies that $I_{s,r}$ is an isomorphism, so $H_1^r(\mathbb{R}^d)$ coincides with $H_1^s(\mathbb{R}^d)$ for $1 < r \le s < \infty$. For $1 < r < s = \infty$ we see from Proposition 6.15 that $I_{s,r}^*$ is an isomorphic embedding (possibly there are some functionals on $H_1^\infty(\mathbb{R}^d)$ not given by a function). But by a standard duality result, this implies that $I_{s,r}$ is onto. Since it is also 1–1, we infer that $I_{s,r}$ is an isomorphism, in other words $H_1^\infty(\mathbb{R}^d) = H_1^s(\mathbb{R}^d)$. □

Corollary 6.19 *The space $BMO(\mathbb{R}^d)$ is complete, so it is a Banach space.*

Proof It is a general fact that the dual of a normed space is a Banach space. Actually it is almost obvious that a norm limit of continuous linear functionals is a continuous linear functional. □

Now let us see how the dyadic dilations act on BMO and H_1 spaces. Recall that J_s is defined by 5.2, i.e. $J_s f(x) = f(2^s x)$. Suppose that $a(x)$ is a q-atom supported on the cube Q. Then $J_s a$ is supported on the cube $2^{-s}Q$ and $\int_{\mathbb{R}^d} J_s a(x)\,dx = 0$. Also

$$\begin{aligned}
\left(\int_{\mathbb{R}^d} |J_s a(x)|^q\,dx\right)^{1/q} &= \left(\int_{\mathbb{R}^d} |a(2^s x)|^q\,dx\right)^{1/q} \\
&= \left(\int_{\mathbb{R}^d} |a(x)|^q 2^{-sd}\,dx\right)^{1/q} \\
&\le 2^{-sd/q}\,\|a\|_q \le 2^{-\frac{sd}{q}} |Q|^{1/q-1} \\
&= 2^{-sd} 2^{-sd(1/q-1)} |Q|^{1/q-1} \\
&= 2^{-sd} |2^{-s} \circ Q|^{1/q-1}.
\end{aligned}$$

This calculation tells us that J_s maps atoms into fixed multiples of atoms, so we get

$$\|J_s f\|_{1,q} = 2^{-sd}\,\|f\|_{1,q}\,. \tag{6.38}$$

A similarly straightforward calculation gives for each p, $1 \le p < \infty$, that

$$\|J_s f\|_{*,p} = \|f\|_{*,p} \tag{6.39}$$

Clearly

Translates are isometries of $H_1(\mathbb{R}^d)$ and of $BMO(\mathbb{R}^d)$. (6.40)

One of the difficulties when working with $H_1(\mathbb{R}^d)$ is that it may be difficult to show that a given function is in $H_1(\mathbb{R}^d)$. Let us offer one easy but very useful result in this direction.

Proposition 6.20 *Suppose that f is a function on $\mathbb{R}^d$ such that $\int_{\mathbb{R}^d} f(x)\,dx = 0$ and $\left(\int_{\mathbb{R}^d} |f(x)|^p\,dx\right)^{1/p} \leq C$ for some $p > 1$ and $|f(x)| \leq C|x|^{-\alpha}$ for $|x| > \beta > 0$ with $\alpha > d$. Then $f \in H_1(\mathbb{R}^d)$ and $\|f\|$ can be estimated by a constant depending only on C, α and β.*

Proof Clearly we have to write f as a series of atoms. Let $Q_n =: [-2^n, 2^n]^d$ for $n = 1, 2, \ldots$ and let $L_n =: Q_n \backslash Q_{n-1}$ with $n = 2, 3, \ldots$. Fix r such that $Q_r \supset \{x \in \mathbb{R}^d : |x| < \beta\}$. Let us put $g_n =: (f - f_{Q_n})\mathbf{1}_{Q_n}$. Note that $\mathbf{1}_{Q_n}$ is, as always, the indicator function of the cube Q_n. Since

$$\|f - f\mathbf{1}_{Q_n}\|_1 = \int_{\mathbb{R}^d \backslash Q_n} |f(x)|\,dx \leq C \int_{\mathbb{R}^d \backslash Q_n} |x|^{-\alpha}\,dx$$

we see that $f \cdot \mathbf{1}_{Q_n}$ converges to f in $L_1(\mathbb{R}^d)$. Also

$$\begin{aligned} f_{Q_n}\mathbf{1}_{Q_n} &= \int_{Q_n} f(x)\,dx \cdot \frac{1}{|Q_n|}\mathbf{1}_{Q_n} \\ &= \int_{\mathbb{R}^d \backslash Q_n} f(x)\,dx \cdot \frac{1}{|Q_n|}\mathbf{1}_{Q_n} \end{aligned}$$

so $f_{Q_n}\mathbf{1}_{Q_n}$ converges to 0 in $L_1(\mathbb{R}^d)$. This implies that $g_n \to f$ in $L_1(\mathbb{R}^d)$ so we can write

$$f = g_r + \sum_{n=r}^{\infty}(g_{n+1} - g_n). \tag{6.41}$$

Note that g_r is a multiple of a p-atom with the multiplicity constant depending only on C and β. Clearly

$$\int_{\mathbb{R}^d} (g_{n+1}(x) - g_n(x))\,dx = 0$$

and supp $(g_{n+1} - g_n) \subset Q_{n+1}$. Writing $g_{n+1} - g_n$ explicitly as

$$g_{n+1}(x) - g_n(x) = f \cdot \mathbf{1}_{L_{n+1}} - f_{Q_{n+1}}\mathbf{1}_{Q_{n+1}} + f_{Q_n}\mathbf{1}_{Q_n}$$

we can estimate

$$\|g_{n+1} - g_n\|_\infty \leq C2^{-\alpha n} + |f_{Q_{n+1}}| + |f_{Q_n}|.$$

Since

$$|f_{Q_n}| = \frac{1}{|Q_n|}\left|\int_{Q_n} f(x)\,dx\right| \leq 2^{-(n+1)d} \int_{\mathbb{R}^d \backslash Q_n} |f(x)|\,dx$$

$$\leq \quad 2^{-(n+1)d} C \int_{\mathbb{R}^d \setminus Q_n} |x|^{-\alpha}\, dx \leq C 2^{-nd} 2^{n(d-\alpha)}$$
$$= \quad C 2^{-n\alpha}$$

we get

$$\|g_{n+1} - g_n\|_\infty \leq C 2^{-\alpha n}.$$

This means that $C^{-1} 2^{n(\alpha-d)}(g_{n+1} - g_n)$ is an ∞-atom. Since the series $\sum_{n=r}^\infty 2^{(d-\alpha)n}$ converges, formula 6.41 gives a representation of f as an absolutely convergent series of atoms. □

Putting together 6.38 and the above Proposition 6.20 we obtain:

Corollary 6.21 *If we have a function f on $\mathbb{R}^d$ such that $\int_{\mathbb{R}^d} f(x)\, dx = 0$, $\|f\|_2 \leq C 2^{dN/2}$ and for $|x| > \beta 2^{-N}$ we have for some $\alpha > d$*

$$|g(x)| \leq C 2^{(d-\alpha)N} |x|^{-\alpha}$$

then $\|f\| \leq$ const.

One of the reasons for introducing the spaces $H_1(\mathbb{R}^d)$ and $BMO(\mathbb{R}^d)$ is that quite often they successfully replace the spaces L_1 and L_∞. More precisely, many natural operators which are unbounded on L_1 are bounded on H_1 or operators which are unbounded on L_∞ are bounded on BMO. To emphasize this point and for future use we will conclude this section with the interpolation theorem which generalizes Theorem 6.7. Let us start with the following definition, which extends ideas of Definition 6.6.

DEFINITION 6.22 *A sublinear operator T is of weak type $(H_1, 1)$ if there exists a constant C such that*

$$|\{x \,:\, |Tf(x)| > t\}| \leq \frac{C}{t} \|f\| \tag{6.42}$$

and is of weak type (p,p), $1 \leq p < \infty$, if there exists a constant C such that

$$|\{x \,:\, |Tf(x)| > t\}| \leq \left(\frac{C\|f\|_p}{t}\right)^p. \tag{6.43}$$

REMARK 6.4. We can repeat the remarks made after Definition 6.6. If T is continuous on $H_1(\mathbb{R}^d)$, i.e. $\|Tf\| \leq C\|f\|$, or if T maps $H_1(\mathbb{R}^d)$ into $L_1(\mathbb{R}^d)$ i.e. $\|Tf\|_1 \leq c\|f\|$, then T is of weak type $(H_1, 1)$. Also if T maps $L_p(\mathbb{R}^d)$ into itself i.e. $\|Tf\|_p \leq c\|f\|_p$ then T is of weak type (p,p). When the above does not hold the condition 6.42 is not easy to check. In particular it cannot be checked on atoms only, cf. Exercise 6.7.

Now we can formulate:

Theorem 6.23 *Suppose that T is a sublinear operator defined on $H_1(\mathbb{R}^d)+L_q(\mathbb{R}^d)$ for some $q>1$ and assume that it is of weak type $(H_1,1)$ and of weak type (q,q). Then for all p, $1<p<q$, the operator T is continuous from $L_p(\mathbb{R}^d)$ into itself, i.e. there exists a constant $C(p)$ such that $\|Tf\|_p \leq C(p)\|f\|_p$ for all $f\in L_p(\mathbb{R}^d)$. Besides p this constant depends only on the constants appearing in 6.42 and 6.43.*

REMARK 6.5. It follows from the proof given below that it suffices instead of 6.42 to assume only the following weaker condition:

> there exists a constant C such that if $f=\sum_j \lambda_j a_j$ where the a_j's are atoms supported on *dyadic* cubes then
> $$|\{x : |Tf(x)|>t\}| \leq Ct^{-1}\sum_j |\lambda_j|. \tag{6.44}$$

Proof Let us fix a number s, $1<s<p$. Let us consider a function $f\in L_p(\mathbb{R}^d)$ such that $|f(x)|^s\in L_1(\mathbb{R}^d)$. Since such functions are dense in $L_p(\mathbb{R}^d)$ it suffices to consider only such functions. We apply the Calderón–Zygmund decomposition (Theorem 6.9) to the function $|f(x)|^s$ with the number t^s. We obtain a family of non-overlapping dyadic cubes Q_j (depending on t) such that

$$t\leq\left(\frac{1}{|Q_j|}\int_{Q_j}|f(x)|^s\,dx\right)^{1/s}\leq 2^{d/s}t. \tag{6.45}$$

Writing $E_t=\bigcup_j Q_j$ we have $|E_t|\leq t^{-s}\|f\|_s^s$ and for $x\in\mathbb{R}^d\setminus E_t$ we have $|f(x)|\leq t$. The heart of the proof is the splitting of the function $f(x)$ as $f(x)=g_t(x)+b_t(x)$ with

$$g_t(x)=f(x)\cdot\mathbf{1}_{\mathbb{R}^d\setminus E_t}+\sum_j f_{Q_j}\mathbf{1}_{Q_j}$$

and

$$b_t(x)=\sum_j\left(f(x)-f_{Q_j}\right)\mathbf{1}_{Q_j}.$$

It follows from 6.45 that $|f_{Q_j}|\leq 2^{d/s}t$ so

$$|g_t(x)|\leq 2^{d/s}t. \tag{6.46}$$

One also immediately observes that $b_t(x)$ looks like a sum of atoms.

More precisely we get from 6.45 that

$$\left(\frac{1}{|Q_j|}\int_{Q_j} |f(x) - f_{Q_j}|^s\,dx\right)^{1/s} \leq 2\left(\frac{1}{|Q_j|}\int_{Q_j} |f(x)|^s\,dx\right)^{1/s} \leq 2\cdot 2^{d/s}t$$

so

$$\left(\int_{Q_j} |f(x) - f_{Q_j}|^s\,dx\right)^{1/s} \leq 2\cdot 2^{d/s}t|Q_j|^{1/s}$$

which shows that

$$(f(x) - f_{Q_j})\mathbf{1}_{Q_j} = 2\cdot 2^{d/s}t|Q_j|a_j$$

where a_j is an s-atom. Thus we obtain

$$|\!|\!|b_t|\!|\!| \leq \sum_j 2\cdot 2^{d/s}t|Q_j| = 2\cdot 2^{d/s}t|E_t|. \tag{6.47}$$

Now we will estimate $\|Tf\|_p$ using Proposition 6.1 and the fact that T is sublinear.

$$\begin{aligned} \|Tf\|_p^p &= \int_{\mathbb{R}^d} |Tf(x)|^p\,dx \qquad (6.48)\\ &= p\int_0^\infty t^{p-1}|\{x\ :\ |Tf(x)| > t\}|\,dt\\ &\leq p\int_0^\infty t^{p-1}|\{x\ :\ |Tg_t(x)| > t/2\}|\,dt\\ &\quad +p\int_0^\infty t^{p-1}|\{x\ :\ |Tb_t(x)| > t/2\}|\,dt\\ &=: I_1 + I_2. \end{aligned}$$

Using the fact that T is of weak type (q,q) and the properties of the Calderón–Zygmund decomposition recalled at the beginning of this proof we get

$$\begin{aligned} I_1 &\leq Cp\int_0^\infty t^{p-1}t^{-q}\,\|g_t\|_q^q\,dt\\ &= Cp\int_0^\infty t^{p-q-1}\int_{\mathbb{R}^d} |g_t(x)|^q\,dx\,dt\\ &\leq C\int_0^\infty t^{p-q-1}\int_{\{x\ :\ |f(x)\leq t\}} |f(x)|^q\,dx\,dt\\ &\quad +C\int_0^\infty t^{p-q-1}\int_{E_t} |g_t(x)|^q\,dx\,dt\\ &\leq C\int_{\mathbb{R}^d}\int_{|f(x)|}^\infty t^{p-q-1}\,dt\,|f(x)|^q\,dx + C\int_0^\infty t^{p-q-1}t^q|E_t|\,dt \end{aligned}$$

$$
\begin{aligned}
&= C\int_{\mathbb{R}^d} |f(x)|^{p-q}|f(x)|^q\,dx + C\int_0^\infty t^{p-1}|E_t|\,dt \\
&= C\int_{\mathbb{R}^d} |f(x)|^p\,dx + C\int_0^\infty t^{p-1}|E_t|\,dt. \qquad (6.49)
\end{aligned}
$$

Now let us estimate the second summand in the above. Recalling the definition of Hardy–Littlewood maximal function (see Definition 6.2) we see that

$$
|E_t| \le |\{x \,:\, (M|f|^s)(x) > t^s\}|
$$

so

$$
\int_0^\infty t^{p-1}|E_t|\,dt \le \int_0^\infty t^{p-1}|\{x \,:\, (M|f|^s)(x) > t^s\}|\,dt.
$$

Making the substitution $t^s = u$, using 6.5 and Corollary 6.8 for the exponent $\frac{p}{s} > 1$ we get

$$
\begin{aligned}
\int_0^\infty t^{p-1}|E_t|\,dt &\le C\int_0^\infty u^{p/s-1}|\{x \,:\, (M|f|^s)(x) > u\}|\,du \\
&= C\int_{\mathbb{R}^d} |M(|f|^s)(x)|^{p/s}\,dx \qquad (6.50) \\
&\le C\int_{\mathbb{R}^d} (|f(x)|^s)^{p/s}\,dx \\
&= C\int_{\mathbb{R}^d} |f(x)|^p\,dx.
\end{aligned}
$$

Substituting 6.50 into 6.49 we get

$$
I_1 \le C\,\|f\|_p^p. \qquad (6.51)
$$

To estimate I_2 we use the fact that T is of weak type $(H_1, 1)$ to get

$$
I_2 \le C\int_0^\infty t^{p-1}t^{-1}\,\|b_t\|\,dt.
$$

Applying 6.47 and 6.50 we get

$$
I_2 \le C\int_0^\infty t^{p-2}t|E_t|\,dt = C\int_0^\infty t^{p-1}|E_t|\,dt \le C\,\|f\|_p^p. \qquad (6.52)
$$

From 6.48, 6.51 and 6.52 we get the theorem. □

Sources and comments

This chapter contains a quick intoduction to basic ideas and tools of harmonic analysis on $\mathbb{R}^d$. This branch of mathematics grew up from the classical theory of Fourier series and integrals. The subject is extensively presented in numerous books, e.g. [105], [104], [108], [44], [40]. In this chapter I have presented only those results which will be needed later. The reader may have noticed that almost every result in Section 6.1 has a name attached. This identifies the results according to the standard usage in this area but also indicates the fundamental nature of those results.

The space BMO was introduced by F. John and L. Nierenberg in [55] where Theorem 6.16 is proved. The space H_1 has a very long history. It appeared already in the work of G. H. Hardy and F. Riesz as a space of analytic functions on the disk. Then its real-variable version was investigated by C. Fefferman and E. Stein in [39]. In our definition of this space and in extensive use of atoms in this book we follow the ideas of R. R. Coifman and G. Weiss [22]. The fact that $H_1^* = BMO$ (cf. Theorem 6.18) is usually called the Fefferman duality theorem. This is a deep theorem when a more standard definition of H_1 is used; our definition using atoms makes it almost a tautology.

Theorem 6.7 was obtained by J. Marcinkiewicz [79]. It, as well as its generalization Theorem 6.23, is an example of the so-called real interpolation method.

There is a general theory of interpolation of operators. Its general idea is that very often Banach spaces are organized in scales indexed by parameters (like L_p-spaces or Besov spaces, to mention the examples discussed in this book) in such a way that if an operator behaves nicely at two points of the scale, it behaves nicely in between. This general theory is explained e.g. in [4] and its most important special cases also in [40], [44], [73], [105] or [108].

About the exercises. The Hilbert transform discussed in Exercise 6.9 is a classical operator. It is discussed in detail in e.g. [57], [44], [105] or [108]. The results of this exercise are classical. Hardy's inequality discussed in Exercise 6.13 is also a classical result to be found in e.g. [57] or [108]. The argument using atoms is from [22]. The operators M_m defined in Exercise 6.13 are called multipliers and are of fundamental importance in harmonic analysis (see eg. [105] or [108]). As previously the use of atoms is from [22]. The space $\delta H_1(\mathbb{R}^d)$ discussed in Exercise 6.14 has intimate connections with martingale theory.

Exercises

6.1 (a) Show that for $f(t) = |t|^\alpha$ with $\alpha > 0$ we have $Mf(x) = \infty$ for each $x \in \mathbb{R}$.

(b) Let f be a function on $\mathbb{R}$ such that $f(t) = 0$ for $x < 0$ and $f(t)$ is an increasing function with $\lim_{t\to\infty} f(t) = \infty$. Show that $Mf(t) = \infty$ for all $x \in \mathbb{R}$.

(c) Give an example of an unbounded function f on $\mathbb{R}$ such that $Mf(x) < \infty$ for all $x \in \mathbb{R}$.

(d) Find a function $f \in L_1(\mathbb{R})$ such that the support of f is in $[0,1]$ and such that $Mf \mid [0,1] \notin L_1([0,1])$.

6.2 Show that we obtain the same classes $BMO_p(\mathbb{R}^d)$ when by Q in Definition 6.10 we mean an arbitrary *ball*, not an arbitrary cube.

6.3 Show that on $\mathbb{R}$ the function $\big|\ln|x|\big|^s$ is in $BMO(\mathbb{R})$ for $s \le 1$ and is not in $BMO(\mathbb{R})$ for $s > 1$.

6.4 Show that $|x|^\alpha \notin BMO_p(\mathbb{R}^d)$ for any p and any $\alpha > 0$. Show also that the function

$$(\log|x|)_+^\alpha = \begin{cases} (\log|x|)^\alpha & \text{if } |x| > 1 \\ 0 & \text{if } |x| \le 1 \end{cases}$$

is in BMO if and only if $\alpha \le 1$.

6.5 Show that a 2π-periodic function on $\mathbb{R}$ with Fourier series $f(t) = \sum_{n\in\mathbb{Z}} a_n e^{i2^n t}$ is in $BMO(\mathbb{R})$ if and only if $\sum_{n\in\mathbb{Z}} |a_n|^2 < \infty$. Show that $f \in L_\infty(\mathbb{R})$ if and only if $\sum_{n\in\mathbb{Z}} |a_n| < \infty$.

6.6 Let a function on $\mathbb{R}$ be defined as

$$f(x) = \begin{cases} x^{-1}\big|\log|x|\big|^{-3/2} & \text{if } |x| < 1 \\ 0 & \text{otherwise.} \end{cases}$$

By considering $f = \sum_n f_n$ where $f_n = f \cdot \mathbf{1}_{D_n}$ with

$$D_n = [-2^{-n}, -2^{-n-1}] \cup [2^{-n-1}, 2^{-n}],$$

show that $f \in H_1^q(\mathbb{R})$ for all $q > 1$. Conclude that there exist functions $f \in H_1(\mathbb{R}^d)$ and $g \in BMO(\mathbb{R}^d)$ such that $fg \notin L_1(\mathbb{R}^d)$.

6.7 Show that there exists a positive constant c such that for each $N = 1, 2, \ldots$ we have $\sum_{k=1}^N |x - \frac{k}{N}|^{-1} \ge cN \log N$ on a set of measure ≥ 1. Using this, construct a sequence of functions $(f_n)_{n=1}^\infty$ such that

$$|\{t : |f_n(t)| > \lambda\}| \le \lambda^{-1}$$

for all $\lambda > 0$ and $n = 1, 2, 3, \ldots$, and a sequence of positive numbers $(a_n)_{n=1}^{\infty}$ such that $\sum_{n=1}^{\infty} a_n = 1$ and such that the function $\sum_{n=1}^{\infty} a_n f_n$ is > 1 on a set of infinite measure.

6.8 (a) Construct a compactly supported C^∞ function g on $\mathbb{R}^d$ such that $g(x) \geq 0$ and $\int_{\mathbb{R}^d} g(x)\, dx = 1$ (cf. Exercise 3.2).

(b) Let a be an atom on $\mathbb{R}^d$. Show that $a * g_n$, where $g_n(x) =: n^{-d} g(nx)$, tends to a in H_1 norm. Show also that $a * g_n = c_n b_n$ where $c_n < 2$ and the b_n's are atoms.

(c) Show that any $f \in H_1(\mathbb{R}^d)$ can be written as $\sum_j \lambda_j a_j$ with $\sum_j |\lambda_j| \leq 2 \|f\|$ and the a_j's are ∞-atoms of class C^∞.

6.9 For $f \in C_{00}^1(\mathbb{R})$ let us define its Hilbert transform as

$$Hf(x) =: \lim_{\varepsilon \to 0+} \frac{1}{\pi} \int_{\varepsilon}^{\infty} \frac{f(x-u) - f(x+u)}{u}\, du.$$

(a) Show that for $f \in C_{00}^1(\mathbb{R})$, Hf is a well defined function and that

$$\mathcal{F}(Hf)(\xi) = \operatorname{sgn}\xi\, \hat{f}(\xi).$$

(b) Show that H extends to a unitary operator on $L_2(\mathbb{R})$.

(c) Show that there exists a constant C such that $\|Ha\| \leq C$ for every ∞-atom of class C^1. Conclude that H extends to a bounded operator on $H_1(\mathbb{R})$.

(d) Show that $H : L_p(\mathbb{R}) \to L_p(\mathbb{R})$ when $1 < p < \infty$.

(e) Show that H is not continuous on $L_1(\mathbb{R})$ or on $L_\infty(\mathbb{R})$.

6.10 Suppose φ is a measurable function on $\mathbb{R}^d$ such that for every $f \in H_1(\mathbb{R}^d)$ we have $\|\varphi f\| \leq C \|f\|$. Show that $\varphi \equiv \text{const}$.

6.11 Show that the space $BMO(\mathbb{R}^d)$ is not separable. Show that there exists a function $f \in BMO(\mathbb{R})$ such that $\|f - g\|_* \geq 1$ for every $g \in L_\infty(\mathbb{R})$.

6.12 Suppose $1 \leq p < q < \infty$ and suppose that T is a sublinear operator defined on $L_p(\mathbb{R}^d) + L_q(\mathbb{R}^d)$ which is of weak type (p,p) and of weak type (q,q). Show that for $p < s < q$ there is a constant $C(s)$ such that $\|Tf\|_s \leq C(s) \|f\|_s$.

6.13 Let a be a 2-atom on $\mathbb{R}$ supported on an interval I.

(a) Show that $|\hat{a}(\xi)| \leq \min(1, |\xi||I|)$.

(b) Show that

$$\int_{-\infty}^{\infty} |\hat{a}(\xi)| \frac{d\xi}{|\xi|} \leq C.$$

(c) Show Hardy's inequality, i.e. for $f \in H_1(\mathbb{R})$ we have

$$\int_{-\infty}^{\infty} |\hat{f}(\xi)| \frac{d\xi}{|\xi|} \leq C \|f\|.$$

(d) Let $m \in L_\infty(\mathbb{R})$. Define an operator M_m on $L_2(\mathbb{R})$ by the formula $M_m(f)(x) = \mathcal{F}^{-1}(m(\xi) \cdot \hat{f}(\xi))$. Show that $\|M_m\| = \|m\|_\infty$.

(e) Suppose that $f \in L_2(\mathbb{R})$ and let $f_0(x) =: xf(x)$. Show that $\|f\|_1 \leq \left(8\|f\|_2 \cdot \|f_0\|_2\right)^{1/2}$.

(f) Assume additionally that $m \in C^1(\mathbb{R})$ and satisfy the Hörmander condition, i.e.

$$\sup_{R>0} R \int_{R \leq |x| \leq 2R} |m'(x)|^2 \, dx = C_0 < \infty.$$

Assume also that the interval I on which the atom a is supported is symmetric with respect to 0. Show that

$$\|M_m a\|_1 \leq \left(8\|m\|_\infty |I|^{-1/2} \|(m\hat{a})'\|_2\right)^{1/2}$$

Show next that

$$\|(m\hat{a})'\|_2 \leq \|m'\hat{a}\|_2 + \|m\|_\infty \|\hat{a}'\|_2.$$

Use the Hörmander condition to show that $\|\hat{a}'\|_2 \leq \sqrt{|I|}$ and also $\|m'\hat{a}\|_2 \leq C\sqrt{|I|}$.

(g) Using the above show that $M_m : H_1(\mathbb{R}) \to L_1(\mathbb{R})$. Conclude that $M_m : L_p(\mathbb{R}) \to L_p(\mathbb{R})$ for $1 < p < \infty$.

6.14 Let us define a dyadic q-atom by Definition 6.11 with the additional requirement that Q is a *dyadic* cube. Using dyadic q-atoms we define the dyadic H_1 space $\delta H_1^q(\mathbb{R}^d)$ as in Definition 6.12 but using only dyadic q-atoms.

(a) Modify the arguments given in Section 6.2 and show that the space $\delta H_1^q(\mathbb{R}^d)$ does not depend on q as a set and that the appropriate norms are equivalent.

(b) Show that translation operators are *not* continuous on $\delta H_1(\mathbb{R}^d)$.

(c) Modify the arguments given in Section 6.2 and show that the dual of $\delta H_1(\mathbb{R}^d)$ is the dyadic BMO space defined as in Definition 6.10 but with the additional condition that Q is a dyadic cube.

7
Unconditional convergence

7.1 Unconditional convergence of series

When we want to sum a family of vectors (or numbers) the order of summation is very important: see Exercise 7.1. Since in many cases no natural order is apparent, this is a rather inconvenient situation. In order to rectify this we introduce a stronger convergence of series, namely unconditional convergence.

DEFINITION 7.1 *Let $(x_n)_{n\in A}$ be a countable family of vectors in a Banach space X. We say that the series $\sum_{n\in A} x_n$ is unconditionally convergent if for each $\sigma : \mathbb{N} \to A$, a 1–1 and onto map, the series $\sum_{k=0}^{\infty} x_{\sigma(k)}$ converges.*

The map σ fixes the order in the set A, and naturally enough it will be called an *order* in the rest of this chapter. It is a nice fact, expressed precisely in the next theorem, that unconditionally convergent series have many other useful properties.

Theorem 7.2 *Let $(x_n)_{n\in A}$ be a countable family of vectors in a Banach space X. The following conditions are equivalent:*

(i) *the series $\sum_{n\in A} x_n$ is unconditionally convergent*

(ii) *there exists an order $\sigma : \mathbb{N} \to A$ such that the series*

$$\sum_{k=0}^{\infty} \varepsilon_k x_{\sigma(k)}$$

converges for every choice of numbers $\varepsilon_k = \pm 1$ for $k = 0, 1, 2, \ldots$

(iii) *for every choice of signs $(\varepsilon_n)_{n\in A}$ with $\varepsilon_n = \pm 1$ the series $\sum_{n\in A} \varepsilon_n x_n$ converges unconditionally*

(iv) *there exists an order $\sigma : \mathbb{N} \to A$ such that the series*

$$\sum_{k=0}^{\infty} \alpha_k x_{\sigma(k)}$$

converges for every bounded sequence $(\alpha_k)_{k=0}^{\infty}$

(v) *for every bounded sequence of numbers $(\alpha_n)_{n\in A}$ the series $\sum_{n\in A} \alpha_n x_n$ converges unconditionally*

(vi) *there exists an order $\sigma : \mathbb{N} \to A$ such that for every strictly increasing sequence of integers n_k the series $\sum_{k=0}^{\infty} x_{\sigma(n_k)}$ converges*

(vii) *for every 1–1 map $\gamma : \mathbb{N} \to A$ the series $\sum_{k=0}^{\infty} x_{\gamma(k)}$ converges.*

Proof We will prove this theorem by proving the following two chains of implications

$$\text{(iv)} \Longrightarrow \text{(i)} \Longrightarrow \text{(vii)} \Longrightarrow \text{(vi)} \Longrightarrow \text{(ii)} \Longrightarrow \text{(iv)}$$

and

$$\text{(vii)} \Longrightarrow \text{(iii)} \Longrightarrow \text{(v)} \Longrightarrow \text{(vi)}.$$

iv) $\Longrightarrow$ (i) Suppose that a series $\sum_{n\in A} x_n$ satisfies (iv) but does not satisfy (i). This means that there exists an order $\gamma : \mathbb{N} \to A$ and an increasing sequence of integers $n(k)$ and a $\delta > 0$ such that

$$\left\| \sum_{s=n(2k)}^{n(2k+1)} x_{\gamma(s)} \right\| \geq \delta. \tag{7.1}$$

Let $I_k = \{n(2k), n(2k)+1, \ldots, n(2k+1)\}$. We can find a subsequence of integers $(k_r)_{r=0}^{\infty}$ such that $\gamma(I_{k_r})$ will lie in increasing blocks in the order σ, which means precisely that

$$\max\{\sigma^{-1}\gamma(s) \ : \ s \in I_{k_r}\} < \min\{\sigma^{-1}\gamma(s) \ : \ s \in I_{k_{r+1}}\}$$

for $r = 0, 1, 2, \ldots$. Now we fix a bounded sequence of numbers $(\alpha_l)_{l=0}^{\infty}$ defined as

$$\alpha_l =: \begin{cases} 1 & \text{if } l \in \bigcup_{r=0}^{\infty} \sigma^{-1}\gamma(I_{k_r}) \\ 0 & \text{if } l \notin \bigcup_{r=0}^{\infty} \sigma^{-1}\gamma(I_{k_r}). \end{cases}$$

Unwrapping what we did above we see that for each $r = 0, 1, 2, \ldots$

$$\sum_{s=n(2k_r)}^{n(2k_r+1)} x_{\gamma(s)} = \sum_{l=N_r}^{M_r} \alpha_l x_{\sigma(l)}$$

for appropriate $N_r < M_r < N_{r+1}$. Thus $\sum_{l=0}^{\infty} \alpha_l x_{\sigma(l)}$ does not converge, because it violates the Cauchy condition, see 7.1. So (iv) does not hold, and this implication is proven.

(i) $\Longrightarrow$ (vii) Suppose (vii) does not hold, i.e. there exists a 1–1 map $\gamma : \mathbb{N} \to A$ and a strictly increasing sequence of integers $(n(k))_{k=0}^{\infty}$ and $\delta > 0$ such that

$$\left\| \sum_{s=n(2k)}^{n(2k+1)} x_{\gamma(s)} \right\| > \delta \quad \text{for } k = 0, 1, 2, \ldots \tag{7.2}$$

As in the previous implication, let $I_k = \{n(2k), n(2k)+1, \ldots, n(2k+1)\}$. Taking a further subsequence we can assume that

$$\mathbb{N} \setminus \bigcup_{k=0}^{\infty} I_k \quad \text{is infinite.} \tag{7.3}$$

Now we define the order $\sigma : \mathbb{N} \to A$ to be any 1–1 and onto map which satisfies $\sigma \mid I_k = \gamma \mid I_k$ for all $k = 0, 1, 2, \ldots$. This is possible because of 7.3. From 7.2 we easily see that $\sum_{n=0}^{\infty} x_{\sigma(n)}$ diverges, so the implication holds.

(vii) $\Longrightarrow$ (vi) is obvious.

(vi) $\Longrightarrow$ (ii) Take the order specified in (iv) and fix a sequence $\varepsilon_k = \pm 1$. Split the integers into two strictly increasing sequences n_k and m_k by the rule that $\varepsilon_{n_k} = 1$ and $\varepsilon_{m_k} = -1$. We infer from (vi) that both series $\sum_{k=0}^{\infty} x_{\sigma(n_k)}$ and $\sum_{k=0}^{\infty} x_{\sigma(m_k)}$ converge. Since each partial sum of the series $\sum_{k=0}^{\infty} \varepsilon_k x_{\sigma(k)}$ equals a partial sum of the series $\sum_{k=0}^{\infty} x_{\sigma(n_k)}$ minus a partial sum of the series $\sum_{k=0}^{\infty} x_{\sigma(m_k)}$ we infer that $\sum_{k=0}^{\infty} \varepsilon_k x_{\sigma(k)}$ converges.

(ii) $\Longrightarrow$ (iv) To prove this implication we need the following lemma.

Lemma 7.3 *Suppose $(\alpha_l)_{l=1}^{N}$ is a sequence of real numbers such that $|\alpha_l| \leq 1$ for $l = 1, 2, \ldots, N$. There exist sequences $(\varepsilon_l^r)_{l=1}^{N}$ such that $\varepsilon_l^r = \pm 1$ and positive numbers β_r with $\sum_r \beta_r = 1$ such that $(\alpha_l) = \sum_r \beta_r (\varepsilon_l^r)$. In other words, each sequence $(\alpha_l)_{l=1}^{N}$ with $|\alpha_l| \leq 1$ is a convex combination of sequences with entries ± 1.*

Proof of Lemma 7.3 For readers who know the Krein–Milman theorem it suffices to say that the sequences $(\varepsilon_l)_{l=1}^{N}$ with $\varepsilon_l = \pm 1$ are extreme points of the convex set of all sequences $(\alpha_l)_{l=1}^{N}$ such that $|\alpha_l| \leq 1$. For those who do not know or do not like the Krein–Milman theorem, here is a direct proof by induction on N. For $N = 1$ the claim is easy: if $|\alpha| \le 1$ then

$$\alpha = \frac{1+\alpha}{2}(1) + \frac{1-\alpha}{2}(-1). \tag{7.4}$$

If our claim holds for N and we have $(\alpha_l)_{l=1}^{N+1}$ then we apply the inductive hypothesis to get $(\alpha_l)_{l=1}^{N} = \sum_r \beta_r (\varepsilon_l^r)_{l=1}^{N}$. Then

$$(\alpha_l)_{l=1}^{N+1} = \sum_r \beta_r (\eta_l^r)_{l=1}^{N+1} \tag{7.5}$$

where $(\eta_l^r)_{l=1}^{N+1} = (\varepsilon_1^r, \dots, \varepsilon_N^r, \alpha_{N+1})$. Applying 7.4 to the last coordinate we see that for each r

$$(\eta_l^r)_{l=1}^{N+1} = \gamma_r (\varepsilon_l)_{l=1}^{N+1} + (1-\gamma_r)(\varepsilon_l')_{l=1}^{N+1} \tag{7.6}$$

with $\varepsilon_l, \varepsilon_l' = \pm 1$ and $0 \le \gamma_r \le 1$. Substituting 7.6 into 7.5 we get the Lemma. □

Now let us return to the implication (ii) $\Longrightarrow$ (iv). Suppose that (ii) holds but (iv) does not hold for some order σ, i.e. there exists a sequence $(\alpha_l)_{l=0}^{\infty}$ such that $\sum_{k=0}^{\infty} \alpha_k x_{\sigma(k)}$ diverges. This implies that there exists a $\delta > 0$ and an increasing sequence of integers $n(l)$ such that $\|\sum_{k=n(2l)}^{n(2l+1)} \alpha_k x_{\sigma(k)}\| \ge \delta$. To each sequence $(\alpha_k)_{k=n(2l)}^{n(2l+1)}$ we apply Lemma 7.3 to obtain

$$(\alpha_k)_{k=n(2l)}^{n(2l+1)} = \sum_r \beta_r (\varepsilon_k^r)_{k=n(2l)}^{n(2l+1)}$$

with $\beta_r \ge 0$ and $\sum_r \beta_r = 1$. We have

$$\begin{aligned} \delta &\le \left\| \sum_{k=n(2l)}^{n(2l+1)} \alpha_k x_{\sigma(k)} \right\| = \left\| \sum_r \beta_r \sum_{k=n(2l)}^{n(2l+1)} \varepsilon_k^r x_{\sigma(k)} \right\| \\ &\le \sum_r \beta_r \left\| \sum_{k=n(2l)}^{n(2l+1)} \varepsilon_k^r x_{\sigma(k)} \right\|. \end{aligned}$$

Since $\sum_r \beta_r = 1$ we infer that for each l there exists a sequence of ± 1's $(\varepsilon_k^l)_{k=n(2l)}^{n(2l+1)} = (\varepsilon_k^{r(l)})_{k=n(2l)}^{n(2l+1)}$ such that

$$\left\| \sum_{k=n(2l)}^{n(2l+1)} \varepsilon_k^l x_{\sigma(k)} \right\| \ge \delta. \tag{7.7}$$

Since $n(l)$ is strictly increasing we can find a sequence $\varepsilon_k = \pm 1$ for $k = 0, 1, 2, \dots$ such that for $n(2l) \le k \le n(2l+1)$ we have $\varepsilon_k = \varepsilon_l^l$. It is clear from 7.7 that $\sum_{k=0}^{\infty} \varepsilon_k x_{\sigma(k)}$ diverges.

(vii) $\Longrightarrow$ (iii) is analogous to (vi) $\Longrightarrow$ (ii).

(iii) $\Longrightarrow$ (v) is proved in exactly the same way as (ii) $\Longrightarrow$ (iv).

(v) $\Longrightarrow$ (vi) For any order σ and any increasing sequence of integers (n_k) we can write the series $\sum_{k=0}^{\infty} x_{\sigma(n_k)}$ as $\sum_{s=0}^{\infty} \alpha_s x_{\sigma(s)}$ where

$$\alpha_s = \begin{cases} 1 & \text{if } s = n_k \text{ for some } k \\ 0 & \text{otherwise.} \end{cases}$$

□

Corollary 7.4 *Suppose that the series $\sum_{n\in A} x_n$ converges unconditionally. Then there exists a constant C such that*

$$\Big\| \sum_{n\in A} \varepsilon_n x_n \Big\| \leq C \tag{7.8}$$

for all $(\varepsilon_n)_{n\in A}$ with $\varepsilon_n = \pm 1$.

Proof The fastest argument is to define an operator $T : \ell_\infty(A) \to X$ by $T(\alpha_n) = \sum_n \alpha_n x_n$. It follows from Theorem 7.2 that T has a closed graph, so the closed graph theorem implies 7.8. For those readers who do not know or do not like the closed graph theorem here is a direct proof. Let us fix an order so we can consider a series $\sum_{n=1}^{\infty} x_n$ and let $\alpha_N = \max\{\|\sum_{n=1}^{N} \varepsilon_n x_n\| \;:\; \varepsilon_n = \pm 1\}$. It is clear that if 7.8 is violated then $\alpha_N \to \infty$ as $N \to \infty$. So we can take a sequence N_s tending to ∞ and such that $\alpha_{N_{s+1}} > \alpha_{N_s} + 1$. We see that for a sequence $(\varepsilon_n)_{n=1}^{N_{s+1}}$ such that $\|\sum_{n=1}^{N_{s+1}} \varepsilon_n x_n\| = \alpha_{N_{s+1}}$ we have $\|\sum_{n=N_s+1}^{N_{s+1}} \varepsilon_n x_n\| \geq 1$. Doing this for each s and putting these pieces $(\varepsilon_n)_{n=N_s+1}^{N_{s+1}}$ together we get a sequence $(\varepsilon_n)_{n=1}^{\infty}$ such that the series $\sum_{n=1}^{\infty} \varepsilon_n x_n$ diverges. □

REMARK 7.1. Let us make one very natural thing clear. If the series $\sum_{n\in A} x_n$ is unconditionally convergent in a Banach space, then for each order σ the series $\sum_{k=0}^{\infty} x_{\sigma(k)}$ converges to the same sum, which naturally enough will be called the sum of this series. The easiest way to see this is to observe that for numerical series unconditional convergence is the same as absolute convergence so the sum is the same in any order. Next, given a vector-valued series $\sum_{n\in A} x_n$ we consider all numerical series $\sum_{n\in A} x^*(x_n)$ with $x^* \in X^*$. Since they have the same sum in each order, from the Hahn–Banach theorem we infer that the series $\sum_{n\in A} x_n$ has the same sum when summed in any order. Another argument follows from Exercise 7.3. Let us also note that when $A = \bigcup_{\alpha\in B} A_\alpha$ with the A_α's disjoint and $\sum_{n\in A} x_n$ is unconditionally convergent, then for each $\alpha \in B$ the series $\sum_{n\in A_\alpha} x_n$ converges unconditionally to some y_α and the series $\sum_{\alpha\in B} y_\alpha$

converges unconditionally to $\sum_{n\in A} x_n$. In fact Theorem 7.2 assures us that all natural manipulations on unconditionally convergent series are legitimate.

EXAMPLE 7.1. Let us give some examples of unconditionally convergent series.

- Every absolutely convergent series, i.e. a series $\sum_{n\in A} x_n$ such that $\sum_{n\in A} \|x_n\| < \infty$, is unconditionally convergent.
- If $(\psi_n)_{n\in A}$ is any orthogonal system in a Hilbert space H such that $\sum_{n\in A} \|\psi_n\|^2 < \infty$ then $\sum_{n\in A} \psi_n$ converges unconditionally in H.
- The series $\sum_{n=1}^{\infty} n^{-1} e^{int} e^{-t^2}$ is unconditionally convergent in $L_2(\mathbb{R})$. It is not absolutely convergent.

Note that if $\sum_{n\in A} x_n$ is unconditionally convergent in a Banach space X and $T : X \to Y$ is a continuous linear operator, then the series $\sum_{n\in A} Tx_n$ is unconditionally convergent in Y. In this way we can easily produce more examples. •

7.2 Unconditional bases

Our aim in this section is to discuss one way of representing all elements of a Banach space as sums of unconditionally convergent series.

DEFINITION 7.5 *A system $(x_n, x_n^*)_{n\in A}$ of elements x_n from X and functionals x_n^* from X^* is called a biorthogonal system if*

$$x_n^*(x_m) = \begin{cases} 1 & \text{if } n = m \\ 0 & \text{if } n \neq m. \end{cases} \tag{7.9}$$

Let us give some examples. First note that each orthonormal system can be thought of as a biorthogonal system. Namely if $(\psi_n)_{n\in A}$ is an orthonormal system in $L_2(\mathbb{R})$ then the system $(\psi_n, \psi_n)_{n\in A}$, where the first ψ_n in each pair is treated as an element of $L_2(\mathbb{R})$ and the second as a functional on $L_2(\mathbb{R})$, is a biorthogonal system. If e.g. $\psi_n \in L_1(\mathbb{R}) \cap L_\infty(\mathbb{R})$ then in the same way we can treat $(\psi_n)_{n\in A}$ as a biorthogonal system in $L_1(\mathbb{R})$ (or in $L_\infty(\mathbb{R})$ as well). If $(\psi_n)_{n\in A}$ is a Riesz sequence in a Hilbert space H then there are biorthogonal functionals $(\psi_n^*)_{n\in A}$ (cf. Lemma 2.7) so that $(\psi_n, \psi_n^*)_{n\in A}$ is a biorthogonal system. Let us note that a biorthogonal system allows us to associate with each $x \in X$ the series $\sum_{n\in A} x_n^*(x) x_n$. This is an entirely formal operation. In order to ensure proper convergence we must impose additional restrictions. Thus we are ready to define an unconditional basis. This is not the definition one is likely to find in books on functional analysis but it is equivalent to the standard one and it will be easy to check later.

DEFINITION 7.6 *A biorthogonal system $(x_n, x_n^*)_{n\in A}$ with a countable index set A is an unconditional basis in X if*

$$\overline{\operatorname{span}}\{x_n\}_{n\in A} = X \tag{7.10}$$

and there exists a constant C such that for every $x \in X$ and every finite set $B \subset A$

$$\Big\|\sum_{n\in B} x_n^*(x)x_n\Big\| \le C\|x\|. \tag{7.11}$$

As of now we can only give a few and rather easy examples.

- Every Riesz basis (in particular an orthonormal basis) in a Hilbert space is an unconditional basis.
- For $X = \ell_p$ with $1 \le p < \infty$, vectors $e_n = (0, \ldots, 0, 1, 0 \ldots)$ with 1 in the n-th place together with the biorthogonal functionals e_n^* (where naturally $e_n^*(x)$ is the n-th coordinate of the sequence x) form an unconditional basis in X.

It should be noted however that one of the aims of the next two chapters is to show that good wavelet bases are unconditional bases in many function spaces.

REMARK 7.2. If the biorthogonal system $(x_n, x_n^*)_{n\in A}$ is an unconditional basis in X then it follows from Definition 7.5 and formula 7.10 that the functionals $(x_n^*)_{n\in A}$ are determined by the vectors $(x_n)_{n\in A}$. Thus we will often abuse the notation and say that the system of vectors $(x_n)_{n\in A}$ is an unconditional basis in X.

The following theorem shows why unconditional bases can be useful. They provide an efficient way to represent an arbitrary function (element of X) in terms of known functions (elements x_n).

Theorem 7.7 *Suppose that $(x_n, x_n^*)_{n\in A}$ is an unconditional basis in X. Then*

(i) *for each $x \in X$ the series $\sum_{n\in A} x_n^*(x)x_n$ converges unconditionally to x*

(ii) *there exists a constant C such that for every $x \in X$ and every bounded sequence of numbers $(\alpha_n)_{n\in A}$ we have*

$$\Big\|\sum_{n\in A} \alpha_n x_n^*(x)x_n\Big\| \le C \sup_{n\in A}|\alpha_n| \cdot \|x\|. \tag{7.12}$$

Proof To prove (i) let us fix any order σ. Given $x \in X$ and $\varepsilon > 0$, using 7.10 we can find $y = \sum_{k=1}^{N} \beta_k x_{\sigma(k)}$ such that $\|x - y\| \leq \varepsilon$. It follows from 7.9 that for $M > N$ we have $\sum_{k=1}^{M} x^*_{\sigma(k)}(y) x_{\sigma(k)} = y$. Thus taking $B = \sigma(\{1, 2, \ldots, M\})$ we get from 7.11 that

$$\Big\|\sum_{k=1}^{M} x^*_{\sigma(k)}(x - y) x_{\sigma(k)}\Big\| = \Big\|\sum_{k=1}^{M} x^*_{\sigma(k)}(x) x_{\sigma(k)} - y\Big\| \leq C\varepsilon,$$

which implies that

$$\Big\|\sum_{k=1}^{M} x^*_{\sigma(k)}(x) x_{\sigma(k)} - x\Big\| \leq (C+1)\varepsilon.$$

Since ε and σ were arbitrary (i) follows. To obtain (ii) note first that from 7.11 we infer that for each finite $B \subset A$ and each $\varepsilon_n = \pm 1$ we have

$$\Big\|\sum_{n \in B} \varepsilon_n x^*_n(x) x_n\Big\| \leq 2C\|x\|.$$

Using Lemma 7.3 this yields

$$\Big\|\sum_{n \in B} \alpha_n x^*_n(x) x_n\Big\| \leq 2C \sup_n |\alpha_n| \cdot \|x\| \tag{7.13}$$

for finite $B \subset A$. Since for a fixed bounded sequence $(\alpha_n)_{n \in A}$ the series $\sum_{n \in A} \alpha_n x^*_n(x) x_n$ converges unconditionally (use (i) and Theorem 7.2(v)) condition 7.12 follows from 7.13. □

7.3 Unconditional convergence in L_p spaces

After the general discussion of the previous sections let us concentrate on a more concrete problem of unconditional convergence in $L_p(\mathbb{R}^d)$ with $1 \leq p < \infty$. Let me note at the very beginning that the results of this section do not depend in any way on the fact that we consider functions on $\mathbb{R}^d$. Any other decent measure space is as good.

At the basis of our considerations lies the classical Khintchine's inequality. In order to formulate it we need to define Rademacher functions. These are a sequence of functions $\big(r_n(t)\big)_{n=1}^{\infty}$ defined on $[0, 1]$ by the formula

$$r_n(t) =: \operatorname{sgn} \sin 2^n t\pi.$$

An alternative, perhaps more explicit description is that we define a function $\varphi(x)$ on $\mathbb{R}$ by

$$\varphi(x) =: \begin{cases} 1 & \text{if } n \le t < n + \frac{1}{2} \text{ for some } n \in \mathbb{Z} \\ -1 & \text{if } n + \frac{1}{2} \le t < n \text{ for some } n \in \mathbb{Z} \end{cases}$$

and we put $r_n(t) =: \varphi(2^{n-1}t) \mid [0,1]$. We can also use probabilistic language and say that $\big(r_n(t)\big)_{n=1}^{\infty}$ is a sequence of independent random variables each taking value 1 with probability $\frac{1}{2}$ and value -1 with probability $\frac{1}{2}$.

Whatever the description, we easily see that $\big(r_n(t)\big)_{n=1}^{\infty}$ is an orthonormal system in $L_2[0,1]$. Another important observation is that given an arbitrary sequence $(\varepsilon_j)_{j=1}^{N}$ with $\varepsilon_j = \pm 1$ we have $r_j(t) = \varepsilon_j$ for $j =, 1, \ldots, N$ exactly on the set of measure 2^{-N}. The basic result about Rademacher functions is:

Proposition 7.8 (Khintchine's inequality) *There exist constants A_p and B_p, $1 \le p < \infty$, such that for all (finite) sequences of scalars $(a_n)_{n=1}^{\infty}$ and every p, $1 \le p < \infty$, we have*

$$A_p \Big\| \sum_{n=1}^{\infty} a_n r_n \Big\|_p \le \Big(\sum_{n=1}^{\infty} |a_n|^2 \Big)^{1/2} \le B_p \Big\| \sum_{n=1}^{\infty} a_n r_n \Big\|_p. \tag{7.14}$$

This is a classical inequality and its proof can be found in many places, e.g. [72] Theorem 2.6.3 or [56] Chapter II Theorem 6.

Lemma 7.9 *There exist constants $0 < c \le C$ such that for each sequence of functions $(f_n)_{n=1}^{\infty}$ from $L_p(\mathbb{R}^d)$, $1 \le p < \infty$, we have*

$$c \Big\| \Big(\sum_{n=1}^{\infty} |f_n|^2 \Big)^{\frac{1}{2}} \Big\|_p \le \Big(\int_0^1 \Big\| \sum_{n=1}^{\infty} r_n(t) f_n \Big\|_p^p dt \Big)^{\frac{1}{p}} \le C \Big\| \Big(\sum_{n=1}^{\infty} |f_n|^2 \Big)^{\frac{1}{2}} \Big\|_p. \tag{7.15}$$

Proof Writing the norm inside the central integral in terms of integrals and changing the order of integration we get

$$\left(\int_{\mathbb{R}^d} \int_0^1 \Big| \sum_{n=1}^{\infty} r_n(t) f_n(x) \Big|^p dt\, dx \right)^{1/p}.$$

Applying Khintchine's inequality to the inner integral (for each point $x \in \mathbb{R}^d$ separately) we get 7.15. □

Corollary 7.10 *Suppose that the series $\sum_{n\in A} f_n$ converges unconditionally in $L_p(\mathbb{R}^d)$, $1 \le p < \infty$. Then $\left\|(\sum_{n\in A}|f_n|^2)^{1/2}\right\|_p < \infty$.*

Proof It follows from Theorem 7.2 that it suffices to consider series in one fixed order, so we can write our series as $\sum_{n=1}^{\infty} f_n$. It follows from Corollary 7.4 that there exists a constant C such that $\left\|\sum_{n=1}^{\infty} r_n(t)f_n\right\|_p \le C$ for every number $t \in [0,1]$. This implies that $C^p \ge \int_0^1 \|\sum_{n=1}^{\infty} r_n(t)f_n\|_p^p\,dt$, so the claim follows directly from 7.15. □

Corollary 7.11 *Suppose that $(f_n)_{n\in A}$ is an unconditional basis in $L_p(\mathbb{R}^d)$, $1 < p < \infty$. Then there are constants $0 < c \le C$ such that for all sequences of scalars $(a_n)_{n\in A}$ we have*

$$c\Big\|\sum_{n\in A} a_n f_n\Big\|_p \le \Big\|\Big(\sum_{n\in A}|a_n|^2|f_n|^2\Big)^{1/2}\Big\|_p \le C\Big\|\sum_{n\in A} a_n f_n\Big\|_p. \tag{7.16}$$

Proof As usual we can fix an order and assume that we have $(f_n)_{n=1}^{\infty}$. It follows from 7.12 that for each $t \in [0,1]$

$$\Big\|\sum_{n=1}^{\infty} r_n(t)a_n f_n\Big\|_p \le C\Big\|\sum_{n=1}^{\infty} a_n f_n\Big\|_p \tag{7.17}$$

so in particular $y_t =: \sum_{n=1}^{\infty} r_n(t)a_n f_n \in L_p(\mathbb{R}^d)$. Applying 7.12 for an element y_t and the same $t \in [0,1]$ we get

$$\Big\|\sum_{n=1}^{\infty} a_n f_n\Big\|_p \le C\Big\|\sum_{n=1}^{\infty} r_n(t)a_n f_n\Big\|_p. \tag{7.18}$$

Clearly 7.16 follows from 7.15 using 7.17 and 7.18. □

As the last observation in this section let us note the following duality result.

Proposition 7.12 *Suppose that $(f_n, f_n^*)_{n\in A}$ with $f_n \in L_p(\mathbb{R}^d)$ and $f_n^* \in L_q(\mathbb{R}^d)$, $1 < p < \infty$ and $\frac{1}{p}+\frac{1}{q} = 1$, is an unconditional basis in $L_p(\mathbb{R}^d)$. Then $(f_n^*)_{n\in A}$ is an unconditional basis in $L_q(\mathbb{R}^d)$.*

Proof First we need to show that $\operatorname{span}(f_n^*)_{n\in A}$ is dense in $L_q(\mathbb{R}^d)$. But if not then by the Hahn–Banach theorem there exists an $h \in L_p(\mathbb{R}^d)$,

$h \neq 0$, such that $\int_{\mathbb{R}^d} h(t)f_n^*(t)\,dt = 0$ for all $n \in A$. But Theorem 7.7(i) implies that

$$h = \sum_{n\in A} \int_{\mathbb{R}^d} h(t)f_n^*(t)\,dt \cdot f_n = 0$$

so we have a contradiction, i.e. $\operatorname{span}(f_n^*)_{n\in A}$ is dense in $L_q(\mathbb{R}^d)$.

For any $B \subset A$ consider the operator

$$P_B(g) = \sum_{n\in B} \int_{\mathbb{R}^d} g(t)f_n^*(t)\,dt \cdot f_n.$$

Since $(f_n)_{n\in A}$ is an unconditional basis in $L_p(\mathbb{R}^d)$ we can reformulate 7.11 to say that $\|P_B\| \leq C$. This implies that the adjoint operator $P_B^* : L_q(\mathbb{R}^d) \to L_q(\mathbb{R}^d)$ also satisfies $\|P_B^*\| \leq C$. An easy calculation shows that

$$P_B^*(g) = \sum_{n\in B} \int_{\mathbb{R}^d} g(t)f_n(t)\,dt \cdot f_n^*$$

which shows that $(f_n^*)_{n\in A}$ is an unconditional basis in $L_q(\mathbb{R}^d)$. □

REMARK 7.3. It is known that $L_1(\mathbb{R}^d)$ does not have an unconditional basis, cf. [116] II.D.10. This partly explains our restriction on p in Corollary 7.11.

Sources and comments

The concept of unconditional convergence is one of the classical topics in functional analysis. Everything in this chapter is classical and well known. A more detailed exposition together with some further references can be found in [116] or in [72] and [73].

About the exercises. Exercise 7.5 is a classical theorem of Orlicz which was at the beginning of the theory of unconditionally convergent series in Banach spaces. The fact that every unconditional basis in a Hilbert space is a Riesz basis is a well known theorem of N. Bari.

Exercises

7.1 Consider the family of numbers $\{(-1)^n n^{-1}\}_{n=1}^{\infty}$ and look at three orders to sum this family:

- $\sum_{n=1}^{\infty}(-1)^n n^{-1}$
- $\sum_{k=1}^{\infty}(\frac{1}{4k} + \frac{1}{4k-2} - \frac{1}{2k-1})$
- $\sum_{k=1}^{\infty}\left(\sum_{s=k^2}^{(k+1)^2-1} \frac{1}{2s}\right) - \frac{1}{2k-1}.$

Show that the first two series converge but to different sums while the third series diverges.

7.2 Let X be a finite-dimensional Banach space. Show that the series $\sum_{n=1}^{\infty} x_n$ converges unconditionally in X if and only if $\sum_{n=1}^{\infty} \|x_n\| < \infty$. Let $r_n(t)$ be Rademacher functions. Show that the series

$$\sum_{n=1}^{\infty} n^{-1} r_n(t)$$

converges unconditionally in $L_p[0,1]$ for each p, $1 \leq p < \infty$, but $\sum_{n=1}^{\infty} \|n^{-1} r_n\|_p = \infty$.

7.3 Suppose that $\sum_{n \in A} x_n$ converges unconditionally. Show that there exists a constant C such that for each sequence $(\alpha_n)_{n \in A}$ with $|\alpha_n| \leq 1$ we have $\|\sum_{n \in A} \alpha_n x_n\| \leq C$. Give an example of elements $(x_n)_{n=1}^{\infty}$ in a Banach space and a constant $C < \infty$ such that $\|\sum_{n=1}^{N} \alpha_n x_n\| \leq C \sup |\alpha_n|$ for each integer N but such that the series $\sum_{n=1}^{\infty}$ is not convergent.

7.4 Show that for each $\alpha > 0$ the series $\sum_{n=1}^{\infty} n^{-1} e^{int} e^{-\alpha|t|}$ is unconditionally convergent in $L_2(\mathbb{R})$. Show that it is unconditionally convergent in $L_p(\mathbb{R})$ for $1 \leq p \leq 2$.

7.5 [Orlicz's theorem] Suppose that the series $\sum_{n=1}^{\infty} f_n$ converges unconditionally in $L_p(\mathbb{R}^d)$, $1 \leq p \leq 2$. Show that

$$\sum_{n=1}^{\infty} \|f_n\|_p^2 < \infty.$$

7.6 Let $(e_n)_{n \in \mathbb{Z}}$ be the unit vector basis in $\ell_1(\mathbb{Z})$ and let $(e_n^*)_{n \in \mathbb{Z}}$ be coordinate functionals. Show that $(e_n^*)_{n \in \mathbb{Z}}$ is not an unconditional basis in $\ell_1(\mathbb{Z})^* = \ell_\infty(\mathbb{Z})$. This shows that Proposition 7.12 fails in general.

7.7 Show that if X has an unconditional basis then X is separable. Thus $L_\infty(\mathbb{R}^d)$ does not have any unconditional basis.

7.8 Let $\{h_{jk}\}$ with $j = 0, 1, \ldots$ and $k = 0, \ldots, 2^j - 1$, be periodic Haar wavelet basis. Let $f =: 2^N \mathbf{1}_{[0, 2^{-N}]}$. Calculate

$$\sum_{j=0}^{\infty} \sum_{k=0}^{2^j - 1} \langle f, h_{jk} \rangle \, h_{jk}.$$

Consider $\sum_{j=0}^{\infty} \sum_{k=0}^{2^{2j+1}-1} \langle f, h_{2j,k} \rangle \, h_{2j,k}$ and show that (h_{jk}) is not an unconditional basis in $L_1[0,1]$.

7.9 For $f_1, \ldots, f_n \in L_2(\mathbb{R})$ show that

$$\left(\int_0^1 \left\| \sum_{j=1}^{n} r_j(t) f_j \right\|_2^2 dt \right)^{1/2} = \left(\sum_{j=1}^{n} \|f_j\|_2^2 \right)^{1/2}.$$

Using this, show that $(f_n, f_n^*)_{n \in A}$ is an unconditional basis in a Hilbert space H if and only if $f_n \|f_n\|^{-1}$ is a Riesz basis in H.

7.10 Suppose that $(f_n)_{n \in A}$ and $(g_m)_{m \in B}$ are unconditional bases in $L_p(\mathbb{R})$, $1 < p < \infty$. Show that $(f_n \otimes g_m)_{(n,m) \in A \times B}$ is an unconditional basis in $L_p(\mathbb{R}^2)$.

7.11 Show that 7.14 fails for $p = \infty$.

7.12 Show that the trigonometric system $\{e^{ikt}\}_{k \in \mathbb{Z}}$ is not an unconditional basis in $L_p[0, 2\pi]$ for $p \neq 2$.

8
Wavelet bases in L_p and H_1

In this chapter we will discuss properties of wavelet bases in $L_p(\mathbb{R}^d)$, $1 \leq p < \infty$, and in $H_1(\mathbb{R}^d)$ and $BMO(\mathbb{R}^d)$. To fix our attention let us assume that we have a multiresolution analysis $(V_j)_{j\in\mathbb{Z}}$ corresponding to dyadic dilations 5.2. Let us assume that the scaling function $\Phi(x)$ satisfies

$$|\Phi(x)| \leq C(1+|x|)^{-d-1}. \tag{8.1}$$

We will also assume that

$$\int_{\mathbb{R}^d} \Phi(x)\,dx = 1 \quad \text{and} \quad \sum_{\gamma\in\mathbb{Z}^d} \Phi(x-\gamma) \equiv 1. \tag{8.2}$$

This actually follows from 8.1 (see Exercise 5.1). Since we did not prove it in detail for $d > 1$, some readers may wish to treat it as an additional assumption. Note however that we know that 8.2 holds in many special cases:

- in the one-dimensional case (see Proposition 2.17)
- for a tensor product multiresolution analysis with the scaling function a tensor product of one-dimensional scaling functions
- for all examples constructed in Chapter 5.

Let us assume also that we have a wavelet set $(\Psi^s(x))_{s=1,\ldots,2^d-1}$ with wavelets Ψ^s satisfying

$$|\Psi^s(x)| \leq C(1+|x|)^{-d-1} \tag{8.3}$$

and

$$\left|\frac{\partial}{\partial x_j}\Psi^s(x)\right| \leq C(1+|x|)^{-d-1}. \tag{8.4}$$

In particular a 1-regular multiresolution analysis with a 1-regular wavelet set will do, but 8.1–8.4 will easily suffice for our purposes. Actually in 8.1, 8.3 and 8.4 we assume more decay than is actually needed in our proofs. The reader may check that all proofs in this chapter are valid if we replace the exponent $-d-1$ by an exponent $-d-\varepsilon$ for some $\varepsilon > 0$. Also we do not need the existence of the derivative of the wavelet. We need only to control the oscillation of the wavelet. The reader may check that we can replace 8.4 by the following condition

$$|\Psi^s(x) - \Psi^s(y)| \leq C|x-y| \min\left((1+|x|)^{-d-\varepsilon}, (1+|y|)^{-d-\varepsilon}\right).$$

This in particular implies that all results of this chapter are true also for piecewise-linear wavelets with exponential decay described in Chapter 1 and 3. We will not discuss more general dilations, although in some cases they can be treated by the same methods.

8.1 Projections associated with a multiresolution analysis

Our main topic in this ection is to investigate P_j, the orthogonal projections onto V_j. *We will do so under assumptions 8.1 and 8.3. In this section we do* not *use any assumption about the smoothness of the wavelets or scaling functions involved.* Explicitly the projections P_j can be written as

$$P_j f(x) = \sum_{\gamma \in \mathbb{Z}^d} \int_{\mathbb{R}^d} f(t)\overline{\Phi_{j,\gamma}(t)}\, dt\, \Phi_{j,\gamma}(x). \tag{8.5}$$

Proposition 8.1 *The projections P_j with $j \in \mathbb{Z}$ have the following properties.*

(i) *For the dilation J_r with $r \in \mathbb{Z}$ (cf. Definition 2.4 or formula 5.2) we have*

$$P_j J_r = J_r P_{j-r} \tag{8.6}$$

in particular $P_j = J_j P_0 J_{-j}$

(ii) *There exist constants $C \geq c > 0$ such that for each $j \in \mathbb{Z}$ and $1 \leq p \leq \infty$ and each sequence of scalars $(a_\gamma)_{\gamma \in \mathbb{Z}^d}$ we have*

$$\begin{aligned} c2^{jd(1/2-1/p)}\Big(\sum_{\gamma \in \mathbb{Z}^d} |a_\gamma|^p\Big)^{1/p} &\leq \Big\|\sum_{\gamma \in \mathbb{Z}^d} a_\gamma \Phi_{j,\gamma}(x)\Big\|_p \\ &\leq C2^{jd(1/2-1/p)}\Big(\sum_{\gamma \in \mathbb{Z}^d} |a_\gamma|^p\Big)^{1/p} \end{aligned} \tag{8.7}$$

(iii) *There exists a constant C such that for each $j \in \mathbb{Z}$ and $f \in L_p(\mathbb{R}^d)$, $1 \le p \le \infty$, we have*

$$\|P_j f\|_p \le C\,\|f\|_p\,. \tag{8.8}$$

Before we proceed with the proof let us establish for future use the following lemma.

Lemma 8.2 *Suppose $\phi(x)$ is a function on $\mathbb{R}^d$ such that $|\phi(x)| \le C(1+|x|)^{-d-1}$. Then for any sequence of scalars $(a_\gamma)_{\gamma\in\mathbb{Z}^d}$ and any p, $1 \le p \le \infty$, we have*

$$\Big\|\sum_{\gamma\in\mathbb{Z}^d} a_\gamma \phi(x-\gamma)\Big\|_p \le C\,\|(a_\gamma)\|_p\,.$$

Proof With q the conjugate exponent to p, i.e. $\frac{1}{p}+\frac{1}{q}=1$, let us write

$$\begin{aligned}
&\Big\|\sum_{\gamma\in\mathbb{Z}^d} a_\gamma \phi(x-\gamma)\Big\|_p^p \\
&\le \int_{\mathbb{R}^d}\Big(\sum_{\gamma\in\mathbb{Z}^d}|a_\gamma||\phi(x-\gamma)|^{1/p}\cdot|\phi(x-\gamma)|^{1/q}\Big)^p\,dx \\
&\le C\int_{\mathbb{R}^d}\Big(\sum_{\gamma\in\mathbb{Z}^d}|a_\gamma|(1+|x-\gamma|)^{-(d+1)/p}(1+|x-\gamma|)^{-(d+1)/p}\Big)^p\,dx.
\end{aligned}$$

Using Hölder's inequality for series this can be majorized by

$$C\int_{\mathbb{R}^d}\sum_{\gamma\in\mathbb{Z}^d}|a_\gamma|^p(1+|x-\gamma|)^{-d-1}\cdot\Big(\sum_{\gamma\in\mathbb{Z}^d}(1+|x-\gamma|)^{-d-1}\Big)^{p/q}\,dx.$$

Since for each $x\in\mathbb{R}^d$ we have $\sum_{\gamma\in\mathbb{Z}^d}(1+|x-\gamma|)^{-d-1}\le C$ we get

$$\begin{aligned}
\Big\|\sum_{\gamma\in\mathbb{Z}^d} a_\gamma\phi(x-\gamma)\Big\|_p^p &\le C\sum_{\gamma\in\mathbb{Z}^d}|a_\gamma|^p\int_{\mathbb{R}^d}(1+|x-\gamma|)^{-d-1}\,dx \\
&\le C\sum_{\gamma\in\mathbb{Z}^d}|a_\gamma|^p.
\end{aligned}$$

□

Proof of Proposition 8.1 The argument for (i) is just the standard change of variable argument. Since

$$J_{-j}\Big(\sum_{\gamma\in\mathbb{Z}^d} a_\gamma\Phi_{j\gamma}\Big)(x) = 2^{jd/2}\sum_{\gamma\in\mathbb{Z}^d} a_\gamma\Phi(x-\gamma)$$

we see from 6.2 that it suffices to check (ii) for $j = 0$. Then the right hand inequality follows from Lemma 8.2 and our assumption 8.1. To show the left side inequality we use elementary duality, namely given a sequence of scalars $(a_\gamma)_{\gamma\in\mathbb{Z}^d}$ we find a sequence $(b_\gamma)_{\gamma\in\mathbb{Z}^d}$ such that $\sum_{\gamma\in\mathbb{Z}^d}|b_\gamma|^q = 1$ and such that $\sum_{\gamma\in\mathbb{Z}^d} a_\gamma\bar{b}_\gamma = (\sum_{\gamma\in\mathbb{Z}^d}|a_\gamma|^p)^{1/p}$. From what we have shown we know that $\left\|\sum_{\gamma\in\mathbb{Z}^d} b_\gamma\Phi(x-\gamma)\right\|_q \leq C$ so using the orthogonality of translates and Hölder's inequality we have

$$\begin{aligned}\Big(\sum_{\gamma\in\mathbb{Z}^d}|a_\gamma|^p\Big)^{1/p} &= \sum_{\gamma\in\mathbb{Z}^d} a_\gamma\bar{b}_\gamma \\ &= \int_{\mathbb{R}^d}\Big(\sum_{\gamma\in\mathbb{Z}^d} a_\gamma\Phi(x-\gamma)\Big)\overline{\Big(\sum_{\gamma\in\mathbb{Z}^d} b_\gamma\Phi(x-\gamma)\Big)}\,dx \\ &\leq C\Big\|\sum_{\gamma\in\mathbb{Z}^d} a_\gamma\Phi(x-\gamma)\Big\|_p\end{aligned}$$

which shows the left hand inequality in (ii). Once more observe that 6.2 implies that to show 8.8 it suffices to check it for $j = 0$. From (ii) we see that we have to prove

$$\sum_{\gamma\in\mathbb{Z}^d}\left|\int_{\mathbb{R}^d} f(t)\Phi(t-\gamma)\,dt\right|^p \leq C\,\|f\|_p^p.$$

The argument for this is similar to the argument used in the proof of Lemma 8.2, namely

$$\begin{aligned}&\sum_{\gamma\in\mathbb{Z}^d}\left|\int_{\mathbb{R}^d} f(t)\Phi(t-\gamma)\,dt\right|^p \\ &\leq \sum_{\gamma\in\mathbb{Z}^d}\int_{\mathbb{R}^d}|f(t)|^p|\Phi(t-\gamma)|\,dt\cdot\Big(\int_{\mathbb{R}^d}|\Phi(t-\gamma)|\,dt\Big)^{p/q} \\ &\leq C\int_{\mathbb{R}^d}|f(t)|^p\sum_{\gamma\in\mathbb{Z}^d}|\Phi(t-\gamma)|\,dt \\ &\leq C'\int_{\mathbb{R}^d}|f(t)|^p\,dt.\end{aligned}$$

□

Note that the argument for the above proposition depends only on the orthogonality of translates and the decay estimate 8.1. Thus the same argument works for wavelets. For $s = 1,\ldots,2^d-1$ and $j\in\mathbb{Z}$ we define the orthogonal projection

$$Q_j^s f(x) = \sum_{\gamma\in\mathbb{Z}^d} \int_{\mathbb{R}^d} f(t)\overline{\Psi_{j,\gamma}^s(t)}\, dt\, \Psi_{j,\gamma}^s(x). \tag{8.9}$$

Clearly when $d = 1$ we have only one wavelet and then we drop the superscript s and consider the projections

$$Q_j f(x) = \sum_{k\in\mathbb{Z}} \int_{-\infty}^{\infty} f(t)\overline{\Psi_{jk}(t)}\, dt\, \Psi_{jk}(x). \tag{8.10}$$

The same argument as above gives:

Proposition 8.3 *Projections Q_j^s with $j \in \mathbb{Z}$ and $s = 1, 2, \ldots, 2^d - 1$ have the following properties.*

(i) *For the dilation J_r with $r \in \mathbb{Z}$ (cf. Definition 2.4 or formula 5.2) we have*

$$Q_j^s J_r = J_r Q_{j-r}^s \tag{8.11}$$

in particular $Q_j^s = J_j Q_0^s J_{-j}$

(ii) *There exist constants $C \geq c > 0$ such that for each $j \in \mathbb{Z}$ and $1 \leq p \leq \infty$ and each sequence of scalars $(a_\gamma)_{\gamma\in\mathbb{Z}^d}$ we have*

$$\begin{aligned} c2^{jd(\frac{1}{2}-\frac{1}{p})}\Big(\sum_{\gamma\in\mathbb{Z}^d} |a_\gamma|^p\Big)^{1/p} &\leq \Big\|\sum_{\gamma\in\mathbb{Z}^d} a_\gamma \Psi_{j,\gamma}^s(x)\Big\|_p \\ &\leq C2^{jd(\frac{1}{2}-\frac{1}{p})}\Big(\sum_{\gamma\in\mathbb{Z}^d} |a_\gamma|^p\Big)^{1/p} \end{aligned} \tag{8.12}$$

(iii) *There exists a constant C such that for each $j \in \mathbb{Z}$ and $f \in L_p(\mathbb{R}^d)$, $1 \leq p \leq \infty$, we have*

$$\left\|Q_j^s f\right\|_p \leq C \left\|f\right\|_p. \tag{8.13}$$

Our aim now is to show that the projections P_j approximate pointwise the identity not only in $L_2(\mathbb{R}^d)$ but also in $L_p(\mathbb{R}^d)$. To show this we first represent P_j in terms of a kernel and establish some properties of this kernel. We can rewrite 8.5 as

$$\begin{aligned} P_j f(x) &= \int_{\mathbb{R}^d} f(t) \sum_{\gamma\in\mathbb{Z}^d} \Phi_{j,\gamma}(x)\overline{\Phi_{j,\gamma}(t)}\, dt \\ &= \int_{\mathbb{R}^d} f(t) 2^{jd}\, \Phi(2^j x, 2^j t)\, dt \end{aligned} \tag{8.14}$$

where

$$\Phi(x,t) = \sum_{\gamma\in\mathbb{Z}^d} \Phi(x-\gamma)\overline{\Phi(t-\gamma)}. \tag{8.15}$$

From 8.2 we infer that for each $x \in \mathbb{R}^d$

$$\int_{\mathbb{R}^d} \Phi(x,t)\, dt = 1 \tag{8.16}$$

and from 8.1 we easily get

$$\int_{\mathbb{R}^d} |\ \Phi(x,t)|\ dt \le C. \tag{8.17}$$

From 8.1 we also see that

$$\begin{aligned} |\ \Phi(x,t)| &\le C\sum_{\gamma\in\mathbb{Z}^d} \frac{1}{(1+|x-\gamma|)^{d+1}}\frac{1}{(1+|t-\gamma|)^{d+1}} \\ &\le C\sum_{\gamma\in A}+\sum_{\gamma\in B}+\sum_{\gamma\in C}\frac{1}{(1+|x-\gamma|)^{d+1}}\frac{1}{(1+|t-\gamma|)^{d+1}} \end{aligned} \tag{8.18}$$

where the sets A, B, C depend on x, t and are defined as

$$\begin{aligned} A &= \{\gamma\in\mathbb{Z}^d\ :\ |x-\gamma| \le \tfrac{1}{2}|x-t|\} \\ B &= \{\gamma\in\mathbb{Z}^d\ :\ |t-\gamma| \le \tfrac{1}{2}|x-t|\} \\ C &= \mathbb{Z}^d\setminus(A\cup B). \end{aligned}$$

Note that for $\gamma \in A$ we have $|t-\gamma| \ge \frac{1}{2}|x-t|$, so

$$\begin{aligned} &\sum_{\gamma\in A} \frac{1}{(1+|x-\gamma|)^{d+1}}\frac{1}{(1+|t-\gamma|)^{d+1}} \\ &\quad\le \frac{1}{(1+\frac{1}{2}|x-t|)^{d+1}}\sum_{\gamma\in A}\frac{1}{(1+|x-\gamma|)^{d+1}} \\ &\quad\le \frac{C}{(1+|x-t|)^{d+1}}\sum_{\gamma\in\mathbb{Z}^d}\frac{1}{(1+|x-\gamma|)^{d+1}} \\ &\quad\le \frac{C}{(1+|x-t|)^{d+1}}. \end{aligned} \tag{8.19}$$

Analogously we obtain

$$\sum_{\gamma\in B}\frac{1}{(1+|x-\gamma|)^{d+1}}\frac{1}{(1+|t-\gamma|)^{d+1}} \le \frac{C}{(1+|x-t|)^{d+1}}. \tag{8.20}$$

To the third sum we apply Hölder's inequality to get

$$\begin{aligned}
&\sum_{\gamma\in C}\frac{1}{(1+|x-\gamma|)^{d+1}}\frac{1}{(1+|t-\gamma|)^{d+1}}\\
&\quad\le \Bigg(\sum_{\gamma\,:\,2|\gamma-x|\ge|x-t|}\frac{1}{(1+|x-\gamma|)^{2d+2}}\Bigg)^{1/2}\\
&\quad\times \Bigg(\sum_{\gamma\,:\,2|\gamma-t|\ge|x-t|}\frac{1}{(1+|t-\gamma|)^{2d+2}}\Bigg)^{1/2}\\
&\quad\le C\sum_{2|\gamma|\ge|x-t|}\frac{1}{(1+|\gamma|)^{2d+2}} \qquad (8.21)\\
&\quad\le \frac{C}{(1+|x-t|)^{d+2}}.
\end{aligned}$$

Putting 8.18–8.21 together we get the important estimate

$$|\,\Phi(x,t)|\le\frac{C}{(1+|x-t|)^{d+1}}. \tag{8.22}$$

Now we are ready to prove:

Theorem 8.4 *Suppose that $f\in L_p(\mathbb{R}^d)$ if $1\le p<\infty$, or $f\in C_0(\mathbb{R}^d)$ if $p=\infty$. Then $\|P_jf-f\|_p\to 0$ as $j\to\infty$.*

REMARK 8.1. This theorem was anticipated by Theorem 1.5.

Proof First let us show that we have uniform convergence for $f\in C_0(\mathbb{R}^d)$. Since each such f is uniformly continuous for each $\varepsilon>0$ we can find a $\delta>0$ such that $|f(x)-f(t)|<\frac{\varepsilon}{2C}$ whenever $|x-t|<\delta$, where C is the constant appearing in 8.17. Using 8.16 and 8.14 we can write

$$\begin{aligned}
f(x)-P_jf(x) &= \int_{\mathbb{R}^d}[f(x)-f(t)]2^{jd}\,\Phi(2^jx,2^jt)\,dt\\
&= \int_{|x-t|\le\delta}+\int_{|x-t|>\delta}[f(x)-f(t)]2^{jd}\,\Phi(2^jx,2^jt)\,dt\\
&=: I_1(x)+I_2(x). \qquad (8.23)
\end{aligned}$$

Clearly from 8.17 and our choice of δ we get

$$|I_1(x)|\le\frac{\varepsilon}{2C}\int_{\mathbb{R}^d}2^{jd}|\,\Phi(2^jx,2^jt)|\,dt\le\frac{\varepsilon}{2}. \tag{8.24}$$

Since f is bounded we get from 8.22

$$
\begin{aligned}
|I_2(x)| &\leq 2\,\|f\|_\infty \int_{|x-t|>\delta} 2^{jd}|\,\Phi(2^j x, 2^j t)|\,dt \\
&\leq C\,\|f\|_\infty \int_{|x-t|>\delta} \frac{2^{jd}}{(1+2^j|x-t|)^{d+1}}\,dt \\
&= C\,\|f\|_\infty \int_{|2^j x - t|>2^j\delta} \frac{1}{(1+|2^j x - t|)^{d+1}}\,dt.
\end{aligned}
$$

The integral in the last line does not depend on x and converges to 0 as $j \to \infty$. This together with 8.23 and 8.24 shows that for $f \in C_0(\mathbb{R}^d)$ the sequence $P_j f$ converges uniformly to f as $j \to \infty$. So the theorem holds for $p = \infty$.

Now let us consider the case $p = 1$. Let us take f continuous and such that $f(x) = 0$ if $|x| > R$. We write

$$
\|f - P_j f\|_1 = \int_{|x|<2R} |f(x) - P_j f(x)|\,dx + \int_{|x|>2R} |P_j f(x)|\,dx.
$$

Since we already know that $P_j f$ tends uniformly to f the first integral converges to 0 as $j \to \infty$. Let us estimate the second integral.

$$
\begin{aligned}
&\int_{|x|>2R} |P_j f(x)|\,dx \\
&\leq \int_{|x|>2R} C \int_{|t|<R} 2^{jd}|\,\Phi(2^j x, 2^j t)|\,dt\,dx \\
&\leq C \int_{|t|<R} \int_{|x|>2R} 2^{jd}(1+2^j|x-t|)^{-d-1}\,dx\,dt \\
&\leq C \int_{|t|<R} \int_{|x-t|>R} 2^{jd}(1+2^j|x-t|)^{-d-1}\,dx\,dt \\
&= C \int_{|t|<R} \int_{|u|>R} 2^{jd}(1+2^j|u|)^{-d-1}\,du\,dt \\
&\leq C \int_{|t|<R} (2^j R)^{-1}\,dt \leq C 2^{-j} R^{d-1}
\end{aligned}
$$

which clearly tends to zero as $j \to \infty$. Since continuous functions with compact support are dense in $L_p(\mathbb{R}^d)$, from 8.8 we infer that the Theorem holds also for $p = 1$.

For the case $1 < p < \infty$, note that it follows from the case $p = 1$ and from 8.8 applied for $p = \infty$ that for f continuous with compact support we have

$$\begin{aligned}\|f - P_j f\|_p &\le \left(\int_{\mathbb{R}^d} |f(x) - P_j f(x)|\, \|f - P_j f\|_\infty^{p-1}\, dx\right)^{1/p} \\ &\le \|f - P_j f\|_\infty^{1-1/p}\, \|f - P_j f\|_1^{1/p}\end{aligned}$$

which tends to 0 as $j \to \infty$ from what we have proved already. Since continuous functions with compact support are dense in $L_p(\mathbb{R}^d)$ the theorem follows from the above using 8.8. □

Now we would like to comment on the convergence of $P_j f$ as $j \to -\infty$. One would expect that $\|P_j f\|_p \to 0$ for each $f \in L_p(\mathbb{R}^d)$. Actually for $p = 1$ and for $p = \infty$ this in not the case. To see this for $p = \infty$ note that from 8.2 it immediately follows that $P_j 1 = 1$ for each $j \in \mathbb{Z}$. For $p = 1$ observe that

$$\begin{aligned}\int_{\mathbb{R}^d} P_j f(x)\, dx &= \int_{\mathbb{R}^d} \sum_{\gamma \in \mathbb{Z}^d} \int_{\mathbb{R}^d} f(t)\overline{\Phi_{j,\gamma}(t)}\, dt\, \Phi_{j,\gamma}(x)\, dx \\ &= \sum_{\gamma \in \mathbb{Z}^d} \int_{\mathbb{R}^d} f(t)\overline{\Phi_{j,\gamma}(t)}\, dt \cdot \int_{\mathbb{R}^d} \Phi_{j,\gamma}(x)\, dx \\ &= \sum_{\gamma \in \mathbb{Z}^d} \int_{\mathbb{R}^d} f(t)\overline{\Phi_{j,\gamma}(t)}\, dt \cdot 2^{-jd/2} \\ &= \int_{\mathbb{R}^d} f(t) \sum_{\gamma \in \mathbb{Z}^d} \overline{\Phi(2^j t - \gamma)}\, dt \\ &= \int_{\mathbb{R}^d} f(t)\, dt\end{aligned}$$

so $\|P_j f\|_1$ does not tend to 0 for any function $f \in L_1(\mathbb{R}^d)$ thar satisfies $\int_{\mathbb{R}^d} f(t)\, dt \neq 0$. For other p's the situation from $L_2(\mathbb{R}^d)$ prevails. We have the following proposition whose proof is a straightforward generalization of the proof of Proposition 2.14.

Proposition 8.5 *For* $1 < p < \infty$ *and for any* $f \in L_p(\mathbb{R}^d)$ *we have*

$$\|P_j f\|_p \to 0$$

as $j \to -\infty$.

Proof Since continuous functions with compact support are dense in $L_p(\mathbb{R}^d)$ we see from A1.3–III and 8.8 that it suffices to consider only

f continuous with compact support, i.e. $f(x) = 0$ for $|x| > R$. Let $B = \{x \in \mathbb{R}^d : |x| < R\}$. From 8.7 we get

$$\begin{aligned}
\|P_j f\|_p^p &\leq C 2^{jd(1/2-1/p)p} \sum_{\gamma \in \mathbb{Z}^d} \left| \int_B f(t)\overline{\Phi_{j,\gamma}(t)}\, dt \right|^p \\
&\leq C 2^{jd(p/2-1)} \sum_{\gamma \in \mathbb{Z}^d} \left(\int_B |f(t)|^p\, dt \right) \cdot \left(\int_B |\Phi_{j,\gamma}(t)|^q\, dt \right)^{p/q} \\
&\leq C \|f\|_p^p \sum_{\gamma \in \mathbb{Z}^d} \left(\int_B 2^{jd} |\Phi(2^j t - \gamma)|^q\, dt \right)^{p/q} \\
&= C \|f\|_p^p \sum_{\gamma \in \mathbb{Z}^d} \left(\int_{\gamma + 2^j B} |\Phi(u)|^q\, du \right)^{p/q}.
\end{aligned}$$

Now let us consider only j's so small that the sets $\gamma + 2^j B$ with $\gamma \in \mathbb{Z}^d$ are disjoint. Then

$$\begin{aligned}
\left(\int_{\gamma + 2^j B} |\Phi(u)|^q\, du \right)^{p/q} &\leq \left(\frac{C}{(1 + |\gamma|)^{(d+1)q}} |2^j B| \right)^{p/q} \\
&\leq C 2^{djp/q} |B|^{p/q} \frac{1}{(1 + |\gamma|)^{p(d+1)}}.
\end{aligned}$$

Thus we get

$$\begin{aligned}
\|P_j f\|_p^p &\leq C \|f\|_p^p 2^{djp/q} |B|^{p/q} \sum_{\gamma \in \mathbb{Z}^d} (1 + |\gamma|)^{-p(d+1)} \\
&\leq C \|f\|_p^p |B|^{p/q} 2^{djp/q}
\end{aligned}$$

which clearly tends to 0 when $j \to -\infty$. □

REMARK 8.2. We could have appealed to the Lebesgue dominated convergence theorem in the above proof as we did in the proof of Proposition 2.14 but because we have the estimate 8.1, it was easy to avoid it.

One useful corollary of the above considerations is:

Corollary 8.6 *Suppose $\{\Psi^s\}$ with $s = 1, 2, \ldots, 2^d - 1$ is a wavelet set satisfying 8.3 associated with a multiresolution analysis with a scaling function satisfying 8.1. Then the system $\{\Psi^s_{j,\gamma}\}$ with $j \in \mathbb{Z}$ and $\gamma \in \mathbb{Z}^d$ and s as above is linearly dense in $L_p(\mathbb{R}^d)$ for $1 < p < \infty$.*

Proof Clearly $P_{j+1} - P_j = \sum_{s=1}^{2^d-1} Q_j^s$ so this corollary follows from Theorem 8.4 and Proposition 8.5. □

8.2 Unconditional bases in L_p and H_1

In this section we do not have to assume that our wavelets are associated with a multiresolution analysis. So let us show:

Proposition 8.7 *Suppose a function $\Psi \in L_1(\mathbb{R}^d) \cap L_2(\mathbb{R}^d)$ is such that the system*

$$\left\{2^{jd/2}\Psi(2^j x - \gamma)\right\}_{j\in\mathbb{Z},\gamma\in\mathbb{Z}^d}$$

is an orthonormal system. Then $\int_{\mathbb{R}^d} \Psi(x)\,dx = 0$.

Proof Multiplying Ψ by an appropriate constant we can and will assume that $\int_{\mathbb{R}^d} \Psi(x)\,dx = \delta \geq 0$. If $\delta > 0$ we can find a cube $Q \subset \mathbb{R}^d$ with center 0, $Q \supset [-1,1]^d$ and such that for every set $A \subset \mathbb{R}^d$ with $A \supset Q$ we have $\Re \int_A \Psi(x)\,dx > \delta/2$. For each $j \in \mathbb{N}$ let us define a set of indices

$$L(j) =: \{\gamma = (\gamma_1, \dots, \gamma_d) \in \mathbb{Z}^d : -2^{j-1} \leq \gamma_s < 2^{j-1} \text{ for } s = 1, \dots, d\}$$

and a function

$$f_j =: \sum_{\gamma\in L(j)} \Psi(2^j x - \gamma).$$

One easily checks that $\|f_j\|_2 = 1$ for $j \geq 1$. Also

$$\begin{aligned}\int_{2Q} f_j &= \sum_{\gamma\in L(j)} \int_{2Q} \Psi(2^j x - \gamma)\,dx \\ &= \sum_{\gamma\in L(j)} 2^{-jd} \int_{2^{j+1}Q-\gamma} \Psi(u)\,du.\end{aligned}$$

Since $2^{j+1}Q - \gamma \supset Q$ we infer that $\Re \int_{2Q} f_j \geq \delta/2$. Since the f_j's are orthonormal, we see that for each $M > 1$

$$\Big\| \sum_{j=2}^{M} \frac{1}{\sqrt{M-1}} f_j \Big\|_2 = 1.$$

We also have

$$\Re \int_{2Q} \sum_{j=2}^{M} \frac{1}{\sqrt{M-1}} f_j \geq \sqrt{M-1}\,\frac{\delta}{2}. \tag{8.25}$$

From Hölder's inequality we get

$$\left| \int_{2Q} \sum_{j=2}^{M} \frac{1}{\sqrt{M-1}} f_j(x)\,dx \right| = \left| \int_{\mathbb{R}^d} \mathbf{1}_{2Q}(x) \cdot \sum_{j=2}^{M} \frac{1}{\sqrt{M-1}} f_j(x)\,dx \right|$$

$$\leq \|\mathbf{1}_{2Q}\|_2 \left\| \sum_{j=2}^{M} \frac{1}{\sqrt{M-1}} f_j \right\|_2 \tag{8.26}$$
$$\leq |2Q|^{1/2}.$$

Since Q is fixed and M can be arbitrarily large, comparing 8.25 with 8.26 we get $\delta = 0$. □

The aim of this section is to show that good wavelet bases (i.e. satisfying our standing assumptions 8.3 and 8.4) are unconditional bases in $L_p(\mathbb{R}^d)$ for $1 < p < \infty$ and in $H_1(\mathbb{R}^d)$. We will also investigate some relations between such bases. Our main tool and really the heart of our arguments in this section is the following technical proposition.

Proposition 8.8 *Suppose we have two wavelet sets $(\Psi^s)_{s=1,\dots,2^d-1}$ and $(\tilde{\Psi}^k)_{k=1,\dots,2^d-1}$ corresponding to dyadic dilations. Assume that the wavelets Ψ^s satisfy 8.3 and 8.4 and assume that the wavelets $\tilde{\Psi}^k$ satisfy 8.3. Then for each s and k and each ∞-atom a the series*

$$\sum_{(j,\gamma)\in\mathbb{Z}\times\mathbb{Z}^d} \langle a, \Psi^s_{j,\gamma}\rangle \tilde{\Psi}^k_{j,\gamma} \tag{8.27}$$

converges unconditionally in $H_1(\mathbb{R}^d)$ and moreover there exists a constant C such that for each $A \subset \mathbb{Z}\times\mathbb{Z}^d$ and each ∞-atom a

$$\left\| \sum_{(j,\gamma)\in A} \langle a, \Psi^s_{j,\gamma}\rangle \tilde{\Psi}^k_{j,\gamma} \right\| \leq C. \tag{8.28}$$

Proof We will prove 8.27. The estimate 8.28 will follow from the same proof. To simplify the notation we will omit the superscripts s and k. Let us take an ∞-atom $a(x)$ supported on a cube Q with center q. Fix an integer $N \in \mathbb{Z}$ such that b – the length of the side of Q – satisfies $2^{-N-1} \leq b < 2^{-N}$. Let us split the set of indices $\mathbb{Z}\times\mathbb{Z}^d$ into three sets

$$\begin{aligned} A &= \{(j,\gamma) : j \leq N\} \\ B &= \{(j,\gamma) : j > N \text{ and } 2^{-j}\gamma \in 3\diamond Q\} \\ C &= \{(j,\gamma) : j > N \text{ and } 2^{-j}\gamma \notin 3\diamond Q\}. \end{aligned}$$

In accordance with this splitting we write the series 8.27 as

$$\sum_{(j,\gamma)\in A} + \sum_{(j,\gamma)\in B} + \sum_{(j,\gamma)\in C} \langle a, \Psi_{j,\gamma}\rangle \tilde{\Psi}_{j,\gamma} =: \Sigma_A + \Sigma_B + \Sigma_C. \tag{8.29}$$

First let us find an estimate for $\left\|\tilde{\Psi}_{j,\gamma}\right\|$. From Proposition 8.7 and formula 8.3 we see that we can apply Proposition 6.20 and get $\left\|\tilde{\Psi}\right\| \leq C$. From 6.38 and 6.40 we get

$$\left\|\tilde{\Psi}_{j,\gamma}\right\| = 2^{jd/2}\left\|\tilde{\Psi}(2^j x-\gamma)\right\| = 2^{-jd/2}\left\|\tilde{\Psi}\right\| \leq C2^{-jd/2}. \tag{8.30}$$

Let us also estimate a single coefficient in the series 8.27. We have

$$\begin{aligned}
|\langle a, \Psi_{j,\gamma}\rangle| &= \left|\int_Q a(x)\overline{\Psi_{j\gamma}(x)}\,dx\right| \\
&= \left|\int_Q a(x)\overline{\left[\Psi_{j\gamma}(x)-\Psi_{j\gamma}(q)\right]}\,dx\right| \\
&\leq \|a\|_1 \sup_{x\in Q} |\Psi_{j\gamma}(x)-\Psi_{j\gamma}(q)| \\
&\leq b\sqrt{d}\sup_{x\in Q} |\text{grad } \Psi_{j\gamma}(x)|.
\end{aligned}$$

Thus from 8.4 we obtain

$$|\langle a, \Psi_{j\gamma}\rangle| \leq Cb\sqrt{d}\,2^{jd/2+j}\sup_{x\in Q}(1+|2^j x-\gamma|)^{-d-1}. \tag{8.31}$$

With the above two estimates we can show that the series Σ_A is absolutely convergent. From the triangle inequality, 8.30 and 8.31 we get

$$\begin{aligned}
\|\Sigma_A\| &\leq \sum_{(j,\gamma)\in A} |\langle a, \Psi_{j\gamma}\rangle| \left\|\tilde{\Psi}_{j\gamma}\right\| \\
&\leq C\sum_{j\leq N}\sum_{\gamma\in\mathbb{Z}^d} b\sqrt{d}2^{jd/2+j}\sup_{x\in Q}(1+|2^j x-\gamma|)^{-d-1}2^{-jd/2} \\
&\leq Cb\sqrt{d}\sum_{j\leq N}2^j\sum_{\gamma\in\mathbb{Z}^d}\sup_{x\in Q}(1+|2^j x-\gamma|)^{-d-1}. \qquad (8.32)
\end{aligned}$$

Since we consider only $j\leq N$ and $x\in Q$ we see that $2^j x - 2^j q \subset [-1,1]^d$. This implies that there exists a constant C such that

$$\sup_{x\in Q}(1+|2^j x-\gamma|)^{-d-1} \leq C(1+|2^j q-\gamma|)^{-d-1}. \tag{8.33}$$

Substituting 8.33 into the last expression of 8.32 we get

$$\begin{aligned}
\|\Sigma_A\| &\leq Cb\sqrt{d}\sum_{j\leq N}2^j\sum_{\gamma\in\mathbb{Z}^d}(1+|2^j q-\gamma|)^{-d-1} \\
&\leq Cb\sqrt{d}\sum_{j\leq N}2^j \leq Cb\sqrt{d}2^N \leq \text{const.}
\end{aligned}$$

Thus the series Σ_A is absolutely convergent.

In order to estimate Σ_C we will need another estimate for $|\langle a, \Psi_{j\gamma}\rangle|$. We use 8.3 to obtain

$$
\begin{aligned}
|\langle a, \Psi_{j\gamma}\rangle| &= \left|\int_Q a(x)\overline{\Psi_{j\gamma}(x)}\,dx\right| \\
&\leq \frac{1}{|Q|}\int_Q 2^{dj/2}(1+|2^j x-\gamma|)^{-d-1}\,dx. \qquad (8.34)
\end{aligned}
$$

Using the triangle inequality, 8.30 and 8.34 we get

$$
\begin{aligned}
\|\Sigma_C\| &\leq \sum_{(j,\gamma)\in C} |\langle a, \Psi_{j\gamma}\rangle| \left\|\tilde{\Psi}_{j\gamma}\right\| \\
&\leq C\sum_{j>N}\sum_{\gamma:\,2^{-j}\gamma\notin 3\diamond Q} 2^{dj/2}\frac{1}{|Q|}\int_Q (1+|2^j x-\gamma|)^{-d-1}\,dx\,2^{-dj/2} \\
&= C\sum_{j>N}\frac{1}{|Q|}\int_Q \sum_{\gamma:\,2^{-j}\gamma\notin 3\diamond Q}(1+|2^j x-\gamma|)^{-d-1}\,dx. \qquad (8.35)
\end{aligned}
$$

Since $x\in Q$ and $2^{-j}\gamma\notin 3\diamond Q$ we see that $|2^j x-\gamma|\geq 2^j b$ so we infer that

$$
\begin{aligned}
\sum_{\gamma:\,2^{-j}\gamma\notin 3\diamond Q}(1+|2^j x-\gamma|)^{-d-1} &\leq C\sum_{\gamma:\,|\gamma|>2^j b}|\gamma|^{-d-1} \\
&\leq C(2^j b)^{-1}. \qquad (8.36)
\end{aligned}
$$

Substituting 8.36 into the last line of 8.35 we obtain

$$
\|\Sigma_C\| \leq C\sum_{j>N}\frac{1}{|Q|}\int_Q C(2^j b)^{-1} = Cb^{-1}\sum_{j>N}2^{-j} = \text{const.}
$$

This shows that Σ_C is absolutely convergent.

In order to show that Σ_B converges unconditionally let us fix any sequence of signs $\varepsilon_{j\gamma}=\pm 1$ with $(j,\gamma)\in B$. For $M\geq N$ consider the series

$$
\Sigma_B^M = \sum_{j>M}\sum_{\gamma:\,2^{-j}\gamma\in 3\diamond Q}\varepsilon_{j\gamma}\langle a, \Psi_{j\gamma}\rangle\,\tilde{\Psi}_{j\gamma}. \qquad (8.37)
$$

Our proposition will be proved as soon as we show that there exists a constant C (valid for all signs $\varepsilon_{j\gamma}$ and all atoms a) such that

$$
\left\|\Sigma_B^M\right\| \leq C \quad \text{for all } M\geq N \qquad (8.38)
$$

and that

$$
\left\|\Sigma_B^M\right\| \to 0 \quad \text{as} \quad M\to\infty. \qquad (8.39)
$$

We will do so using Corollary 6.21. First note that for any choice of signs

$$\left\|\Sigma_B^M\right\|_2 \leq \left\|\Sigma_B\right\|_2 \leq \|a\|_2 \leq |Q|^{-1/2} \leq C2^{Nd/2}. \tag{8.40}$$

It is also clear that $\|\Sigma_B^M\|_2 \to 0$ as $M \to \infty$, so we can write

$$\left\|\Sigma_B^M\right\|_2 \leq \mu_M 2^{Nd/2} \tag{8.41}$$

for some sequence $\mu_M \to 0$.

We will prove a pointwise estimate (valid for all signs and all atoms) so that we can apply Corollary 6.21. We have

$$\begin{aligned}
\left|\Sigma_B^M(x)\right| &\leq C\sum_{j>M}\sum_{\gamma:\,2^{-j}\gamma\in 3\diamond Q} |\langle a, \Psi_{j\gamma}\rangle|\, 2^{dj/2}(1+|2^j x-\gamma|)^{-d-1} \\
&\leq \Big(\sum_{j>M}\sum_{\gamma:\,2^{-j}\gamma\in 3\diamond Q} |\langle a, \Psi_{j\gamma}\rangle|^2\Big)^{1/2} \\
&\quad\times\Big(\sum_{j>M}\sum_{\gamma:\,2^{-j}\gamma\in 3\diamond Q} 2^{jd}(1+|2^j x-\gamma|)^{-2d-2}\Big)^{1/2}.
\end{aligned}$$

Since the wavelets are orthonormal the first factor is at most $\|a\|_2 \leq |Q|^{-1/2}$ so we get

$$|\Sigma_B^M(x)| \leq |Q|^{-1}/2\Big(\sum_{j>M}\sum_{\gamma:\,2^{-j}\gamma\in 3\diamond Q} 2^{jd}(1+|2^j x-\gamma|)^{-2d-2}\Big)^{1/2}. \tag{8.42}$$

Note that for fixed $j > N$ there are at most $C2^{(j-N)d}$ points $\gamma \in \mathbb{Z}^d$ such that $2^{-j}\gamma \in 3\diamond Q$. Also for $x \notin 6\diamond Q$ and $2^{-j}\gamma \in 3\diamond Q$ we see that

$$|2^j x - \gamma| = 2^j|x - 2^{-j}\gamma| \geq C2^j|x-q|$$

(recall that q is the center of Q). This implies that

$$(1+|2^j x-\gamma|)^{-2d-2} \leq C(2^j|x-q|)^{-2d-2}.$$

Using these observations and 8.42 we get that for $x \notin 6\diamond Q$

$$\begin{aligned}
\left|\Sigma_B^M(x)\right| &\leq C|Q|^{-1/2}\Big(\sum_{j>M} 2^{(j-N)d}2^{jd}\big(2^j|x-q|\big)^{-2d-2}\Big)^{1/2} \\
&\leq C|x-q|^{-d-1}\big(|Q|^{-1}\sum_{j>M} 2^{-Nd}2^{-2j}\big)^{1/2} \\
&\leq C|x-q|^{-d-1}\big(\sum_{j>M} 2^{-2j}\big)^{1/2} \qquad (8.43)\\
&\leq C|x-q|^{-d-1}2^{-M}.
\end{aligned}$$

From 8.41 and 8.43 and Corollary 6.21 conditions 8.38 and 8.39 easily follow. This completes the proof of the proposition. □

REMARK 8.3. Note that the *only* place in the above proof where smoothness of wavelets (i.e. 8.4) is used is the estimate 8.31 which in turn is used only to estimate Σ_A. This shows that the estimates for Σ_B and Σ_C hold without any change when about the wavelets Ψ^s we assume only 8.3.

As an easy corollary from what we have proved already we obtain:

Theorem 8.9 *If a wavelet set* $(\Psi^s)_{s=1,\ldots,2^d-1}$ *associated with the dyadic dilation satisfies 8.3 and 8.4 then the system*

$$\{\Psi^s_{j\gamma}\}_{j\in\mathbb{Z},\gamma\in\mathbb{Z}^d, s=1,\ldots,2^d-1} \tag{8.44}$$

is an unconditional basis in $H_1(\mathbb{R}^d)$ *and in* $L_p(\mathbb{R}^d)$ *with* $1 < p < \infty$.

Proof Let us start with the case $H_1(\mathbb{R}^d)$. From Proposition 8.8 we infer that for each ∞-atom a the series

$$\sum_{s=1}^{2^d-1}\sum_{(j,\gamma)}\langle a, \Psi^s_{j\gamma}\rangle\,\Psi^s_{j\gamma}$$

converges in $H_1(\mathbb{R}^d)$. We also know that in $L_2(\mathbb{R}^d)$ it converges to a. This implies that this series converges in the norm of $H_1(\mathbb{R}^d)$ to a. From this and the definition of $H_1(\mathbb{R}^d)$ we immediately infer that the system 8.44 is linearly dense in $H_1(\mathbb{R}^d)$. So from 8.28 we see that this system is an unconditional basis in $H_1(\mathbb{R}^d)$. Now let us consider the case of $L_p(\mathbb{R}^d)$ with $1 < p \leq 2$. First let us make sure that the system 8.44 is linearly dense in $L_p(\mathbb{R}^d)$. If our wavelet set is associated with an appropriate multiresolution analysis this is Corollary 8.6. If we do not know this then we can use an argument outlined in Exercise 8.4. Once we know that the system 8.44 is linearly dense we need to show that there exists a constant C such that for each finite set of indices A we have

$$\Big\|\sum_A\langle f, \Psi^s_{j\gamma}\rangle\,\Psi^s_{j\gamma}\Big\|_p \leq C\,\|f\|_p. \tag{8.45}$$

It follows from 8.28 that such an inequality holds for the H_1-norm and from orthonormality of wavelets it follows for $p = 2$. So from Theorem 6.23 applied to the operator

$$P_A f = \sum_A\langle f, \Psi^s_{j\gamma}\rangle\,\Psi^s_{j\gamma}$$

we infer that 8.45 holds for $1 < p \leq 2$. This means that the system 8.44 is an unconditional basis in $L_p(\mathbb{R}^d)$ when $1 < p \leq 2$. The case $2 < p < \infty$ follows from Proposition 7.12. □

Corollary 8.10 *Suppose we have two wavelet sets* $(\Psi^s)_{s=1,\dots,2^d-1}$ *and* $(\tilde{\Psi}^s)_{s=1,\dots,2^d-1}$ *both associated with the dyadic dilation. If the set* $(\Psi^s)_{s=1,\dots,2^d-1}$ *satisfies 8.3 and 8.4 and* $(\tilde{\Psi}^s)_{s=1,\dots,2^d-1}$ *satisfies 8.3 then the map*

$$\Psi^s_{j\gamma} \mapsto \tilde{\Psi}^s_{j\gamma} \tag{8.46}$$

extends to a continuous map of $H_1(\mathbb{R}^d)$ *into itself and of* $L_p(\mathbb{R}^d)$ *into itself,* $1 < p \leq 2$. *If both wavelet sets satisfy 8.3* and *8.4 then the map 8.46 extends to an isomorphism of* $H_1(\mathbb{R}^d)$ *and of* $L_p(\mathbb{R}^d)$ *with* $1 < p < \infty$.

Proof Clearly the second claim for $1 < p \leq 2$ follows from the first applied to both wavelet sets. This claim for $2 \leq p < \infty$ follows from easy duality arguments, as in the proof of Proposition 7.12. So it suffices to show that the map defined by 8.46 is continuous. It follows from 8.28 that this map is continuous on $H_1(\mathbb{R}^d)$ and it follows from orthogonality of wavelets that it is continuous on $L_2(\mathbb{R}^d)$. From Theorem 6.23 we infer that it is continuous on $L_p(\mathbb{R}^d)$ when $1 < p \leq 2$. □

Corollary 8.11 *Let* $(\Psi^s)_{s=1,\dots,2^d-1}$ *be a wavelet set associated with the dyadic dilation and satisfying 8.3 and 8.4. Then for* $1 < p < \infty$ *there exist constants* $0 < c \leq C$ *such that for any* f *we have*

$$c\|f\|_p \leq \left\| \left(\sum_{s=1}^{2^d-1} \sum_{(j,\gamma)\in\mathbb{Z}\times\mathbb{Z}^d} |\langle f, \Psi^s_{j\gamma}\rangle|^2 |\Psi^s_{j\gamma}|^2 \right)^{1/2} \right\|_p \leq C\|f\|_p. \tag{8.47}$$

For $f \in H_1(\mathbb{R}^d)$ *we have*

$$\int_{\mathbb{R}^d} \left(\sum_{s=1}^{2^d-1} \sum_{(j,\gamma)\in\mathbb{Z}\times\mathbb{Z}^d} |\langle f, \Psi^s_{j\gamma}\rangle|^2 |\Psi^s_{j\gamma}(x)|^2 \right)^{1/2} dx \leq C\, |\!|\!| f |\!|\!| . \tag{8.48}$$

Proof Theorem 8.9 tells us that 8.47 is just a special case of Corollary 7.11. For $f \in H_1(\mathbb{R}^d)$ Theorem 8.9 and Proposition 6.14 show that the series

$$\sum_{s=1}^{2^d-1} \sum_{(j,\gamma)\in\mathbb{Z}\times\mathbb{Z}^d} \langle f, \Psi_{j\gamma}^s\rangle \Psi_{j\gamma}^s$$

converges unconditionally in $L_1(\mathbb{R}^d)$, so 8.48 follows from Corollary 7.10. □

8.3 Haar wavelets

In this section we will discuss wavelet sets on $\mathbb{R}^d$ obtained from the one-dimensional Haar wavelet using the procedure described in Example 5.1 and Proposition 5.2. Just to state the properties we are going to use let us say that we will consider a wavelet set $(h^s)_{s=1,\ldots,2^d-1}$ associated with the dyadic dilation of $\mathbb{R}^d$ such that

$$|h^s(x)| = \mathbf{1}_{[0,1]^d}(x) \quad \text{for} \quad s = 1,\ldots,2^d-1 \tag{8.49}$$

and such that

$$\text{each } h^s(x) \text{ is constant on each cube } \prod_{i=1}^{d}[\tfrac{k_i}{2}, \tfrac{k_i+1}{2}] \text{ with } k_i \in \mathbb{Z} \tag{8.50}$$

Clearly these wavelets satisfy 8.3 but do not satisfy 8.4. Our aim is to show that the Haar wavelet set generates an unconditional basis in $L_p(\mathbb{R}^d)$ for $1 < p < \infty$. Our main tool is the following proposition.

Proposition 8.12 *Suppose that $(\Psi^s)_{s=1,\ldots,2^d-1}$ is a wavelet set satisfying 8.3 and associated with dyadic dilations. For $1 < p \leq 2$ there exists a constant C_p such that for each set $V \subset \{1,\ldots,2^d-1\}\times\mathbb{Z}\times\mathbb{Z}^d$ and each $f \in L_p(\mathbb{R}^d)$ we have*

$$\Big\| \sum_{(s,j,\gamma)\in V} \langle f, h_{j\gamma}^s\rangle \Psi_{j\gamma}^s \Big\|_p \leq C_p\|f\|_p. \tag{8.51}$$

Proof With each V as in the proposition we can associate an operator

$$P_V(f) = \sum_{(s,j,\gamma)\in V} \langle f, h_{j\gamma}^s\rangle \Psi_{j\gamma}^s.$$

It is clear that P_V has norm 1 as an operator on $L_2(\mathbb{R}^d)$. To prove the proposition we will use Theorem 6.23 together with Remark 6.5. Thus we need to check 6.44. Basically we will repeat the proof of Proposition 8.8 to show that for each $s = 1, \ldots, 2^d - 1$ we have

$$\Big\| \sum_{(s,j,\gamma)\in V} \langle f, h^s_{j\gamma}\rangle \Psi^s_{j\gamma}\Big\|_1 \leq C \sum_j |\lambda_j| \tag{8.52}$$

where $f = \sum_j \lambda_j a_j$ and the a_j's are ∞-atoms supported on *dyadic* cubes. Since we are proving an estimate of the L_1-norm, to show 8.52 it suffices to show that there exists a constant C such that for every V and every $s = 1, \ldots, 2^d - 1$ and every ∞-atom a supported on a dyadic cube Q we have

$$\Big\| \sum_{(s,j,\gamma)\in V} \langle a, h^s_{j\gamma}\rangle \Psi^s_{j\gamma}\Big\|_1 \leq C. \tag{8.53}$$

Since Q is dyadic its side has length 2^{-N-1} for some integer N. As observed in Remark 8.3 we can use directly the estimates for Σ_B and Σ_C (the notation is as in the proof of Proposition 8.8) so the only thing left to prove is

$$\Big\| \sum_{(s,j,\gamma)\in V\, j\leq N} \langle a, h^s_{j\gamma}\rangle \Psi^s_{j\gamma}\Big\|_1 \leq C \tag{8.54}$$

But for $j \leq N$ each $h^s_{j\gamma}$ is supported on a dyadic cube of side $\geq 2^{-N}$. Since a is supported on a cube of side 2^{-N-1} we infer from 8.50 that each $h^s_{j\gamma}$ is constant on the support of a, so each $\langle a, h^s_{j\gamma}\rangle$ is zero. This clearly gives 8.54 with $C = 0$ and completes the proof of the Proposition.

□

Now we are ready for the first main theorem of this section.

Theorem 8.13 *Each of the Haar wavelet sets described in 8.49 and 8.50 gives an unconditional basis in $L_p(\mathbb{R}^d)$ for $1 < p < \infty$. Moreover if $(\Psi^s)_{s=1,\ldots,2^d-1}$ is any wavelet set satisfying 8.3 and 8.4 then for any p, $1 < p < \infty$, there exist constants $0 < c \leq C$ such that*

$$c\Big\|\sum_{s=1}^{2^d-1}\sum_{j\gamma} a^s_{j\gamma}\Psi^s_{j\gamma}\Big\|_p \leq \Big\|\sum_{s=1}^{2^d-1}\sum_{j\gamma} a^s_{j\gamma}h^s_{j\gamma}\Big\|_p \leq C\Big\|\sum_{s=1}^{2^d-1}\sum_{j\gamma} a^s_{j\gamma}\Psi^s_{j\gamma}\Big\|_p \tag{8.55}$$

for all sequences of scalars.

Proof It is clear, either directly from the definition or from Corollary 8.6, that $(h^s_{j\gamma})_{j\in\mathbb{Z},\gamma\in\mathbb{Z}^d,s=1,\ldots,2^d-1}$ is linearly dense in $L_p(\mathbb{R}^d)$ for $1 < p < \infty$.

Thus 8.51 applied for $\Psi^s = h^s$ shows that the Haar wavelet basis is an unconditional basis in $L_p(\mathbb{R}^d)$ for $1 < p \leq 2$. The case $2 \leq p < \infty$ follows from Proposition 7.12. If $1 < p \leq 2$ then the right hand inequality in 8.55 follows directly from Corollary 8.10 and the left hand inequality follows from 8.51. For $2 \leq p < \infty$ the formula 8.55 follows by duality from the case $1 < p \leq 2$. □

REMARK 8.4. The above results suggest the following notion of equivalence of bases. Suppose $(x_n)_{n\in A}$ and $(y_m)_{m\in B}$ are unconditional bases in a Banach space X. We say that these bases are equivalent if there exist constants $0 < c \leq C$ and a 1–1 and onto map $\pi : A \to B$ such that

$$c\Big\|\sum_{n\in A} a_n x_n\Big\| \leq \Big\|\sum_{n\in A} a_n y_{\pi(n)}\Big\| \leq C\Big\|\sum_{n\in A} a_n x_n\Big\| \tag{8.56}$$

for all finitely non-zero sequences of scalars $(a_n)_{n\in A}$. The condition 8.56 simply means that there exists an isomorphism $I : X \to X$ such that $I(x_n) = y_{\pi(n)}$.

Using this notion we can say that all unconditional bases in $L_p(\mathbb{R}^d)$, $1 < p < \infty$, given by Theorem 8.9 are equivalent among themselves and with the Haar wavelet bases discussed in this section.

Before we proceed let us introduce some notation. For $j \in \mathbb{Z}$ and $\gamma \in \mathbb{Z}^d$ let

$$\Delta_{j\gamma}(x) = |h^s_{j\gamma}(x)|.$$

It follows from 8.49 that $\Delta_{j\gamma}(x)$ oes not depend on s. It also follows from 8.49 that

$$\Delta_{j\gamma}(x) = 2^{jd/2}\mathbf{1}_{Q(j,\gamma)} \tag{8.57}$$

where $Q(j,\gamma)$ is the dyadic cube $\prod_{r=1}^d [2^{-j}\gamma_r, 2^j(\gamma_r + 1)]$ where $\gamma = (\gamma_1, \ldots, \gamma_d)$. Since this gives all dyadic cubes we may think of the family $\Delta_{j\gamma}$ as characteristic functions, normalized in $L_2(\mathbb{R}^d)$, of *all* dyadic cubes Q. Now we can formulate our next theorem.

Theorem 8.14 *Let $1 < p < \infty$. Then there exist constants $0 < c \leq C$ such that for all f*

$$c\|f\|_p \leq \Big\|\Big(\sum_{s=1}^{2^d-1}\sum_{(j\gamma)\in\mathbb{Z}\times\mathbb{Z}^d} |\langle f, h^s_{j\gamma}\rangle|^2 \Delta^2_{j\gamma}\Big)^{1/2}\Big\|_p \leq C\|f\|_p. \tag{8.58}$$

If $(\Psi^s)_{s=1,\ldots,2^d-1}$ is any wavelet set associated with dyadic dilation and satisfying 8.3 and 8.4 then

$$c\|f\|_p \leq \Big\|\Big(\sum_{s=1}^{2^d-1}\sum_{(j\gamma)\in\mathbb{Z}\times\mathbb{Z}^d} |\langle f, \Psi^s_{j\gamma}\rangle|^2 \Delta^2_{j\gamma}\Big)^{1/2}\Big\|_p \leq C\|f\|_p. \tag{8.59}$$

Proof Relation 8.58 follows directly from Theorem 8.13 and Corollary 7.11. To obtain 8.59 we apply 8.55 and Corollary 7.11. □

We can view the above Theorem 8.14 as a characterization of $L_p(\mathbb{R}^d)$ in terms of wavelet coefficients $\langle f, \Psi^s_{j,\gamma}\rangle$. Our next goal is to provide a similar characterization for the space $H_1(\mathbb{R}^d)$. Before we state our next technical proposition let us introduce some notation.

Let us fix a set $W \subset [0,1]^d$ of positive measure α and let us denote $R(x) = \mathbf{1}_W(x)$. As usual $R_{j,\gamma}(x) =: 2^{jd/2}R(2^j x - \gamma)$. Let us denote $W(j,\gamma) =: \operatorname{supp} R_{j,\gamma}$. We will also use the notation $Q(j,\gamma)$ and $\Delta_{j\gamma}(x)$ as explained before Theorem 8.14. Clearly

$$W(j,\gamma) \subset Q(j,\gamma) \quad \text{and} \quad |W(j,\gamma)| = \alpha|Q(j,\gamma)|. \tag{8.60}$$

Proposition 8.15 *Let $(\Psi^s)_{s=1}^{2^d-1}$ be any compactly supported wavelet set as described in Remark 5.1. For a family of numbers*

$$\{a(j,\gamma)\}_{j\in\mathbb{Z},\gamma\in\mathbb{Z}^d}$$

let us write

$$\varphi(x) =: \Big(\sum_{(j,\gamma)\in\mathbb{Z}\times\mathbb{Z}^d} |a(j,\gamma)|^2 R^2_{j\gamma}(x)\Big)^{1/2}$$

and suppose that $\varphi \in L_1(\mathbb{R}^d)$. Then for each $s = 1,2,\ldots,2^d-1$ the function

$$\sum_{(j,\gamma)\in\mathbb{Z}\times\mathbb{Z}^d} a(j,\gamma)\Psi^s_{j\gamma}$$

is in $H_1(\mathbb{R}^d)$ and

$$\Big\|\sum_{(j,\gamma)\in\mathbb{Z}\times\mathbb{Z}^d} a(j,\gamma)\Psi^s_{j\gamma}\Big\| \le C\|\varphi\|_1. \tag{8.61}$$

Proof Clearly in order to show that the function $\sum_{(j,\gamma)\in\mathbb{Z}\times\mathbb{Z}^d} a(j,\gamma)\Psi^s_{j\gamma}$ is in $H_1(\mathbb{R}^d)$ and to estimate its norm we must write it as a sum of atoms. This will be done by partitioning the index set $\mathbb{Z}\times\mathbb{Z}^d$ into sets $D(k,l)$ in such a way that

$$A_{kl} =: \sum_{(j,\gamma)\in D(k,l)} a(j,\gamma)\Psi^s_{j\gamma} \tag{8.62}$$

will be an appropriate multiple of a 2-atom.

For each $k \in \mathbb{Z}$ let us define

$$E_k =: \{x \in \mathbb{R}^d : \varphi(x) > 2^k\}. \tag{8.63}$$

Clearly $E_k \supset E_{k+1}$ for all $k \in \mathbb{Z}$ and we have

$$\begin{aligned}
\sum_{k=-\infty}^{\infty} 2^k |E_k| &= \sum_{k=-\infty}^{\infty} 2^k \sum_{j=k}^{\infty} |E_j \setminus E_{j+1}| \\
&\leq \sum_{k=-\infty}^{\infty} 2^k \sum_{j=k}^{\infty} 2^{-j} \int_{E_j \setminus E_{j+1}} \varphi(x)\, dx \\
&= \sum_{j=-\infty}^{\infty} \sum_{k=-\infty}^{j} 2^{k-j} \int_{E_j \setminus E_{j+1}} \varphi(x)\, dx \qquad (8.64) \\
&= 2 \sum_{j=-\infty}^{\infty} \int_{E_j \setminus E_{j+1}} \varphi(x)\, dx \\
&= 2 \int_{\mathbb{R}^d} \varphi(x)\, dx.
\end{aligned}$$

Now for each $k \in \mathbb{Z}$ let us consider a collection $\mathcal{C}_k$ of indices $(j, \gamma) \in \mathbb{Z} \times \mathbb{Z}^d$ such that the dyadic cube $Q(j, \gamma)$ satisfies

$$|E_k \cap Q(j, \gamma)| > \frac{\alpha}{2} |Q(j, \gamma)|. \tag{8.65}$$

Since the E_k's decrease we see that

$$\mathcal{C}_k \supset \mathcal{C}_{k+1} \quad \text{for all } k \in \mathbb{Z}. \tag{8.66}$$

We will denote

$$E_k^* =: \bigcup_{(j,\gamma) \in \mathcal{C}_k} Q(j, \gamma). \tag{8.67}$$

It follows from the Lebesgue differentiation theorem (Theorem 6.4) applied to the function $\mathbf{1}_{E_k}$ that for almost all $x \in E_k$ there exists a dyadic cube Q such that $|E_k \cap Q| > \frac{\alpha}{2}|Q|$, so

$$E_k^* \supset E_k \tag{8.68}$$

modulo a set of measure zero.

Let us look at the family of dyadic cubes $\{Q(j, \gamma) : (j, \gamma) \in \mathcal{C}_k\}$. We denote the maximal cubes of this family by Q_k^l. The index l runs through some unspecified index set. From elementary properties of dyadic cubes we infer that for each $k \in \mathbb{Z}$ the cubes Q_k^l are disjoint and

$$E_k^* = \bigcup_l Q_k^l. \tag{8.69}$$

From 8.69, 8.65 and 8.68 it follows that

$$\begin{aligned} |E_k^*| &= \sum_l |Q_k^l| \le \frac{2}{\alpha} \sum_l |E_k \cap Q_k^l| \\ &= \frac{2}{\alpha} \Big| E_k \cap \bigcup_l Q_k^l \Big| \qquad (8.70) \\ &= \frac{2}{\alpha} |E_k \cap E_k^*| = \frac{2}{\alpha} |E_k|. \end{aligned}$$

Let us also note that if $a(j,\gamma) \neq 0$ then the index $(j,\gamma) \in \mathcal{C}_k$ for some $k \in \mathbb{Z}$. Namely, take k such that $|a(j,\gamma)| 2^{jd/2} > 2^k$. Then on the set $W(j,\gamma)$ we have $\varphi(x) > 2^k$, so $W(j,\gamma) \subset E_k$. From 8.60 we infer that

$$|Q(j,\gamma) \cap E_k| \ge |W(j,\gamma)| = \alpha |Q(j,\gamma)|$$

so $(j,\gamma) \in \mathcal{C}_k$.

Let us denote $D_k =: \mathcal{C}_k \setminus \mathcal{C}_{k+1}$ and write

$$D(k,l) =: \left\{ (j,\gamma) \in D_k : Q(j,\gamma) \subset Q_k^l \right\}. \qquad (8.71)$$

This is the desired splitting. Since each (j,γ) such that $a(j,\gamma) \neq 0$ belongs to some $\mathcal{C}_k$ it belongs to some D_k so it belongs to some $D(k,l)$. So we get

$$\sum_{k,l} A_{kl} = \sum_{(j,\gamma) \in \mathbb{Z} \times \mathbb{Z}^d} a(j,\gamma) \Psi^s_{j\gamma}$$

where the A_{kl}'s are defined by 8.62. Note that supp $\Psi^s \subset R \diamond [0,1]^d$ for some positive constant R. This implies that supp $\Psi^s_{j\gamma} \subset R \diamond Q(j,\gamma)$. Thus the definition of $D(k,l)$ and 8.62 show that

$$\operatorname{supp} A_{kl} \subset R \diamond Q_k^l. \qquad (8.72)$$

Now let us estimate $\|A_{kl}\|_2$. Clearly

$$\|A_{kl}\|_2^2 = \sum_{(j,\gamma) \in D(k,l)} |a(j,\gamma)|^2. \qquad (8.73)$$

Note that $(j,\gamma) \in D(k,l)$ implies that $Q(j,\gamma) \subset Q_k^l$ and $Q(j,\gamma) \notin \mathcal{C}_{k+1}$, so $|E_{k+1} \cap Q(j,\gamma)| \le \frac{\alpha}{2} |Q(j,\gamma)|$. This implies that for $(j,\gamma) \in D(k,l)$ we have

$$|Q(j,\gamma) \setminus E_{k+1}| \ge \left(1 - \frac{\alpha}{2}\right) |Q(j,\gamma)|. \qquad (8.74)$$

Since

$$\varphi^2(x) \ge \sum_{(j,\gamma) \in D(k,l)} |a(j,\gamma)|^2 R^2_{j\gamma}(x)$$

we get

$$\begin{aligned}\int_{Q_k^l\setminus E_{k+1}} &\varphi^2(x)\,dx \\ \geq & \sum_{(j,\gamma)\in D(k,l)} |a(j,\gamma)|^2 \int_{Q_k^l\setminus E_{k+1}} R_{j\gamma}^2(x)\,dx \qquad (8.75)\\ = & \sum_{(j,\gamma)\in D(k,l)} |a(j,\gamma)|^2 2^{jd}|W(j,\gamma)\cap(Q_k^l\setminus E_{k+1})|.\end{aligned}$$

Since $W(j,\gamma)\subset Q(j,\gamma)\subset Q_k^l$ we see that

$$W(j,\gamma)\cap(Q_k^l\setminus E_{k+1}) = W(j,\gamma)\cap\big(Q(j,\gamma)\setminus E_{k+1}\big)$$

so from 8.74 and 8.60 we get

$$\big|W(j,\gamma)\cap(Q_k^l\setminus E_{k+1})\big| \geq \frac{\alpha}{2}|Q(j,\gamma)| = \frac{\alpha}{2}2^{-jd}. \qquad (8.76)$$

From 8.73, 8.75 and 8.76 we obtain

$$\|A_{kl}\|_2^2 \leq \frac{2}{\alpha}\int_{Q_k^l\setminus E_{k+1}} \varphi^2(x)\,dx. \qquad (8.77)$$

For $x\notin E_{k+1}$ we have $\varphi(x)\leq 2^{k+1}$ so

$$\int_{Q_k^l\setminus E_{k+1}} \varphi^2(x)\,dx \leq 2^{2(k+1)}|Q_k^l\setminus E_{k+1}| \leq 4\cdot 4^k|Q_k^l|. \qquad (8.78)$$

Note that 8.78 and 8.72 imply that $A_{kl}\in L_1(\mathbb{R}^d)$ so $\int_{\mathbb{R}^d} A_{kl}(x)\,dx = 0$. Thus A_{kl} is a multiple of a 2-atom. If we write $A_{kl} = \lambda(k,l)A'_{kl}$ where

$$\lambda(k,l) = \sqrt{|R\diamond Q_k^l|}\,\|A_{kl}\|_2 \qquad (8.79)$$

we get that A'_{kl} is a 2-atom supported on $R\diamond Q_k^l$. Thus in order to show 8.61 we must estimate $\sum_{kl}\lambda(k,l)$. From 8.79 and 8.77 together with 8.78 we get

$$\begin{aligned}\sum_{kl}\lambda(k,l) &= \sum_{kl} R^{d/2}\sqrt{|Q_k^l|}\cdot\|A_{kl}\|_2\\ &\leq R^{d/2}\sum_{kl}\sqrt{|Q_k^l|}\cdot 2\cdot 2^k\sqrt{\frac{2}{\alpha}}\sqrt{|Q_k^l|}\\ &= 2R^{d/2}\sqrt{\frac{2}{\alpha}}\sum_k 2^k\sum_l |Q_k^l|\end{aligned}$$

from 8.69 and 8.70

$$= \quad 2R^{d/2}\sqrt{\frac{2}{\alpha}}\sum_k 2^k|E_k^*|$$

$$\leq \quad 4R^{d/2}\alpha^{-1}\sqrt{\frac{2}{\alpha}}\sum_k 2^k|E_k|.$$

This estimate and 8.64 give

$$\sum_{lk}\lambda(k,l) \leq 8R^{d/2}\alpha^{-1}\sqrt{\frac{2}{\alpha}}\int_{\mathbb{R}^d}\varphi(x)\,dx$$

so 8.61 holds. □

REMARK 8.5. The use of compactly supported wavelets in the above proposition allowed us to estimate the support of A_{kl}. If we assum that our wavelets are not compactly supported but satisfy 8.3 instead, we can prove this Proposition by estimating $\|A_{kl}\|$ using proposition 6.20.

This proposition allows us to show the following theorem.

Theorem 8.16 *Suppose that* $(\Psi^s)_{s=1}^{2^d-1}$ *is a wavelet set associated with dyadic dilations and satisfying 8.3 and 8.4. The following conditions are equivalent:*

(i) $f \in H_1(\mathbb{R}^d)$

(ii) *the series* $\sum_{s=1}^{2^d-1}\sum_{(j,\gamma)\in\mathbb{Z}\times\mathbb{Z}^d}\langle f, \Psi_{j\gamma}^s\rangle\,\Psi_{j\gamma}^s$ *converges unconditionally in* $L_1(\mathbb{R}^d)$

(iii) $\left(\sum_{s=1}^{2^d-1}\sum_{(j,\gamma)\in\mathbb{Z}\times\mathbb{Z}^d}|\langle f, \Psi_{j\gamma}^s\rangle|^2|\Psi_{j\gamma}^s(x)|^2\right)^{1/2} \in L_1(\mathbb{R}^d)$

(iv) $\left(\sum_{s=1}^{2^d-1}\sum_{(j,\gamma)\in\mathbb{Z}\times\mathbb{Z}^d}|\langle f, \Psi_{j\gamma}^s\rangle|^2|\Delta_{j\gamma}(x)|^2\right)^{1/2} \in L_1(\mathbb{R}^d)$ *where* $\Delta_{j\gamma}(x)$ *are defined in 8.57.*

This theorem provides a characterization of $H_1(\mathbb{R}^d)$ in terms of wavelets. Condition (iv) describes $H_1(\mathbb{R}^d)$ in terms of wavelet coefficients, while (ii) describes $H_1(\mathbb{R}^d)$ in terms of convergence of wavelet expansions. Note also that it follows from tracing the constants in our proof or from the closed graph theorem that the integral of each of the functions appearing in (iii) and (iv) gives a norm on $H_1(\mathbb{R}^d)$ equivalent to $\|.\|$. It is a remarkable fact that very different wavelets give via (ii) or (iii) the same space $H_1(\mathbb{R}^d)$.

Proof The implications (i)$\Longrightarrow$ (ii) and (ii)$\Longrightarrow$(iii) have been already proved (cf. Corollary 8.11 and its proof).

To prove (i)$\Longrightarrow$(iv) note that we can apply Corollary 8.10 with $\tilde{\Psi}^s = h^s$ where h^s is any of the Haar wavelets described at the beginning of this section. This gives that the series

$$\sum_{s=1}^{2^d-1} \sum_{(j,\gamma)\in\mathbb{Z}\times\mathbb{Z}^d} \langle f, \Psi^s_{j\gamma}\rangle\, h^s_{j\gamma}$$

converges unconditionally in $H_1(\mathbb{R}^d)$, so also in $L_1(\mathbb{R}^d)$. Thus we can apply Corollary 7.10 to get (iv).

Now let us prove (iii)$\Longrightarrow$(i). Since we can replace each Ψ^s by an appropriate translate $\Psi^s(x-k)$ and Ψ^s is not identically zero, we can assume that each Ψ^s is not identically zero on the set $[0,1]^d$. Thus for each $s = 1, 2, \ldots, 2^d - 1$ there exists a set $W_s \subset [0,1]^d$ of measure $\geq \alpha > 0$ such that for $x \in W_s$ we have $|\Psi^s(x)| > \beta > 0$. If we write $R^s =: \mathbf{1}_{W_s}$ we infer that

$$\begin{aligned}
&\left(\sum_{s=1}^{2^d-1} \sum_{(j,\gamma)\in\mathbb{Z}\times\mathbb{Z}^d} |\langle f, \Psi^s_{j\gamma}\rangle| \cdot |\Psi^s_{j\gamma}(x)|^2\right)^{1/2} \\
&\geq \left(\sum_{s=1}^{2^d-1} \sum_{(j,\gamma)\in\mathbb{Z}\times\mathbb{Z}^d} |\langle f, \Psi^s_{j\gamma}\rangle| \cdot |R^s_{j\gamma}(x)|^2\right)^{1/2} \qquad (8.80)
\end{aligned}$$

Now let $(\tilde{\Psi}^s)_{s=1}^{2^d-1}$ be a fixed compactly supported wavelet set of class C^1. From our assumptions and 8.80 it follows that we can apply Proposition 8.15 and conclude that

$$\sum_{s=1}^{2^d-1} \sum_{(j,\gamma)\in\mathbb{Z}\times\mathbb{Z}^d} \langle f, \Psi^s_{j\gamma}\rangle\, \tilde{\Psi}^s_{j\gamma}(x) \in H_1(\mathbb{R}^d).$$

But now both wavelet sets $(\Psi^s)_{s=1}^{2^d-1}$ and $(\tilde{\Psi}^s)_{s=1}^{2^d-1}$ satisfy 8.3 and 8.4. Thus by Corollary 8.10 the map $\tilde{\Psi}^s_{j\gamma} \mapsto \Psi^s_{j\gamma}$ extends to a continuous linear map of $H_1(\mathbb{R}^d)$ into itself, which implies that

$$\sum_{s=1}^{2^d-1} \sum_{(j,\gamma)\in\mathbb{Z}\times\mathbb{Z}^d} \langle f, \Psi^s_{j\gamma}\rangle\, \Psi^s_{j\gamma}(x) \in H_1(\mathbb{R}^d).$$

The proof of (iv)$\Longrightarrow$(i) is the same; we simply take $W_s = [0,1]^d$. $\square$

8.4 Polynomial bases

In this section we want to discuss in some detail expansions of 2π-periodic functions (or equivalently functions on the circle $\mathbb{T}$) in terms of periodized wavelets. Thus we continue our discussion from Section 2.5. To fix our attention and because it leads to the most pleasing results and to the solution of an old problem we will concentrate on Meyer's wavelets from the Schwartz class S.

Let $\Phi \in S$ be a scaling function given by Proposition 3.2 with Θ of class C^∞, and let $\Psi \in S$ be the corresponding wavelet given by 3.15. We define (cf. Section 2.5)

$$\phi_{jk}(x) =: \frac{1}{\sqrt{2\pi}} \mathcal{P}\Phi_{jk}(\frac{x}{2\pi}) \tag{8.81}$$

and

$$\psi_{jk}(x) =: \frac{1}{\sqrt{2\pi}} \mathcal{P}\Psi_{jk}(\frac{x}{2\pi}) \tag{8.82}$$

where $j = 0, 1, 2, \ldots$ and $k = 0, 1, \ldots, 2^j - 1$. Note that this definition does not conform with our standard usage – neither the set $\{\psi_{jk}\}$ nor $\{\phi_{jk}\}$ consists of translations and dilations of one function. But these definitions are very natural and will be used only in this section and in some exercises. From results obtained in Sections 2.5 and 3.2 we easily see that the following hold:

the system $\psi_{jk}(x)$ with $j = 0, 1, 2 \ldots$ and $k = 0, 1, \ldots, 2^j - 1$, with the constant function $\frac{1}{\sqrt{2\pi}}$ appended, is a complete orthonormal system in $L_2(\mathbb{T})$ (8.83)

for each j the system $\{\phi_{jk}\}_{k=0}^{2^j-1}$ is an orthonormal system (8.84)

$\mathrm{span}\{\frac{1}{\sqrt{2\pi}}, \psi_{jk}\}_{j=0,k=0}^{s-1,2^j-1} = \mathrm{span}\{\phi_{sk}\}_{k=0}^{2^s-1}$ for $s = 1, 2, \ldots$ (8.85)

$$\phi_{jk}(x) = 2^{-j/2} \sum_{|s| \le 3^{-1}2^{j+1}} e^{-2\pi i s k 2^{-j}} \Theta(-\frac{2\pi s}{2^j}) e^{isx} \tag{8.86}$$

$$\psi_{jk}(x) = 2^{-j/2} \sum_{3^{-1}2^j \le |s| \le 4\cdot 3^{-1}2^j} e^{-2\pi i s k 2^{-j}} \hat{\Psi}(-\frac{2\pi s}{2^j}) e^{isx} \tag{8.87}$$

Let us note that 8.86 and 8.87 show that the functions ϕ_{jk} and ψ_{jk} are trigonometric polynomials. Those polynomials have good decay; this is made precise in the following proposition.

Proposition 8.17 *Let $t_{jk} =: \frac{2\pi k}{2^j}$. For each $l = 1, 2, \ldots$ there exists a constant C_l such that*

$$|\phi_{jk}(t)| \leq C_l 2^{j/2}(1 + 2^j|t - t_{jk}|)^{-l} \tag{8.88}$$

and

$$|\psi_{jk}(t)| \leq C_l 2^{j/2}(1 + 2^j|t - t_{jk}|)^{-l}. \tag{8.89}$$

Proof Since the proofs of both inequalities 8.88 and 8.89 are exactly the same we write it only for 8.88. Since $\Phi(x)$ is in the Schwartz class, for each l there exists a constant K_l such that

$$|\Phi(x)| \leq K_l(1 + |x|)^{-l}$$

so we obtain

$$|\Phi_{jk}(x)| \leq 2^{j/2} K_l \left(1 + 2^j|x - k/2^j|\right)^{-l}.$$

This implies that

$$\begin{aligned} |\mathcal{P}\Phi_{jk}(x)| &\leq 2^{j/2} K_l \sum_{s\in\mathbb{Z}} \left(1 + 2^j|x + s - k/2^j|\right)^{-l} \\ &\leq 2^{j/2} K_l' \left(1 + 2^j|x - k/2^j|\right)^{-l+1} \end{aligned}$$

which gives

$$|\phi_{jk}(x)| \leq K 2^{j/2} \left(1 + 2^j/2\pi|x - t_{jk}|\right)^{-l+1}.$$

This easily implies 8.88 for $l - 1$, maybe with a different constant K. Since l was arbitrary we get the claim. □

REMARK 8.6. Analogous estimates hold for the derivatives $\frac{d^s}{dx^s}\phi_{jk}$ and $\frac{d^s}{dx^s}\psi_{jk}$; namely for each $s, l = 1, 2, \ldots$ there exists a constant C_{sl} such that

$$\left|\frac{d^s}{dx^s}\phi_{jk}(x)\right| \leq C_{sl} 2^{j(s+1/2)} \left(1 + 2^j|x - t_{jk}|\right)^{-l}$$

and

$$\left|\frac{d^s}{dx^s}\psi_{jk}(x)\right| \leq C_{sl} 2^{j(s+1/2)} \left(1 + 2^j|x - t_{jk}|\right)^{-l}.$$

Now let us reorder the orthonormal system 8.83 so that it will be indexed by integers. For $n = 2^j + k$ with $j = 0, 1, 2, \ldots$ and $k = 0, 1, \ldots, 2^j - 1$ we put

$$f_n(x) =: \psi_{jk}(x).$$

This defines f_n for $n = 1, 2, \ldots$. We put $f_0(x) =: \frac{1}{\sqrt{2\pi}}$. From 8.83 we immediately see that $(f_n)_{n=0}^{\infty}$ is a complete orthonormal system in $L_2(\mathbb{T})$. Some properties of this system are summarized in the following theorem which is the main result of this section.

Theorem 8.18 *The system $(f_n)_{n=0}^{\infty}$ defined above has the following properties:*

(i) *each function f_n is a trigonometric polynomial of degree at most $\frac{4}{3}n$*
(ii) *if $k \geq 2^\mu$ then* $\operatorname{span}(f_n)_{n=0}^{k} \supset \operatorname{span}\{e^{isx} : |s| < \frac{1}{6}2^\mu\}$
(iii) *if $f \in L_p(\mathbb{T})$, $1 \leq p < \infty$, or $f \in C(\mathbb{T})$ if $p = \infty$, then the series $\sum_{n\in\mathbb{Z}} \langle f, f_n\rangle f_n$ converges to f in norm.*

Proof From 8.87 we see that $\deg\psi_{jk} \leq \frac{4}{3}2^j$, so for $n = 2^j + k$ we have $\deg f_n \leq \frac{4}{3}2^j \leq \frac{4}{3}n$. This gives (i). To show (ii) let us fix $r \in \mathbb{Z}$ and write

$$e^{irt} = \left\langle e^{irt}, \frac{1}{\sqrt{2\pi}} \right\rangle + \sum_{j=0}^{\infty} \sum_{k=0}^{2^j-1} \langle e^{irt}, \psi_{jk}\rangle \psi_{jk}.$$

From orthogonality of exponentials and 8.87 we infer that

$$e^{irt} = \left\langle e^{irt}, \frac{1}{\sqrt{2\pi}} \right\rangle + \sum_{j=0}^{\nu} \sum_{k=0}^{2^j-1} \langle e^{irt}, \psi_{jk}\rangle \psi_{jk} \tag{8.90}$$

where ν is the smallest natural number such that $\frac{1}{3}2^\nu > |r|$. This implies that $e^{irt} \in \operatorname{span}(f_n)_{n=0}^{2^{\nu+1}}$, so for each $\mu \in \mathbb{N}$

$$\operatorname{span}(f_n)_{n=0}^{2^\mu} \supset \operatorname{span}\{e^{irt} : |r| < \frac{1}{6}2^\mu\}.$$

This gives (ii). To show (iii) let us denote $S_k(f) =: \sum_{n=0}^{k} \langle f, f_n\rangle f_n$. Since the trigonometric polynomials are dense in $L_p(\mathbb{T})$, $1 \leq p < \infty$, and in $C(\mathbb{T})$, from (ii) we see that to show (iii) it suffices to prove that there exists a constant C such that

$$\|S_n(f)\|_p \leq C \|f\| \tag{8.91}$$

for all $f \in L_p(\mathbb{T})$ and $n = 0, 1, 2, \ldots$ and p, $1 \leq p \leq \infty$. For a given n let us fix $j = 0, 1, \ldots$ and $s = 0, 1, \ldots, 2^j - 1$ such that $n = 2^j + s$. Using the definition of the f_n's, 8.84 and 8.85 we can write

$$\begin{aligned} S_n(f) &= \sum_{l=0}^{j-1} \sum_{k=0}^{2^l-1} \langle f, \psi_{lk} \rangle \psi_{lk} + \sum_{k=0}^{s} \langle f, \psi_{jk} \rangle \psi_{jk} \\ &= \sum_{k=0}^{2^j-1} \langle f, \phi_{jk} \rangle \phi_{jk} + \sum_{k=0}^{s} \langle f, \psi_{jk} \rangle \psi_{jk} \\ &=: S_j^1(f) + S_n^2(f). \end{aligned}$$

We will show that $\left\| S_j^1(f) \right\|_p \leq C \left\| f \right\|_p$ for all f, n and p. The proof for S_n^2 is exactly the same and this gives 8.91. We rewrite $S_j^1(f)$ in terms of the kernel as follows:

$$S_j^1(f)(x) = \int_0^{2\pi} f(t) K_j(x,t)\, dt \tag{8.92}$$

where $K_j(x,t) = \sum_{k=0}^{2^j-1} \phi_{jk}(t)\phi_{jk}(x)$. From 8.88 we easily obtain (cf. the proof of 8.22) that

$$|K_j(x,t)| \leq C 2^j \left(1 + 2^j |x - t|\right)^{-2}.$$

From this we immediately see that there exists a C such that

$$\int_0^{2\pi} |K_j(x,t)|\, dt \leq C \quad \text{for all } x \tag{8.93}$$

and

$$\int_0^{2\pi} |K_j(x,t)|\, dx \leq C \quad \text{for all } t \tag{8.94}$$

Thus we have using 8.92–8.94, Hölder's inequality (remember $\frac{1}{p} + \frac{1}{q} = 1$) and Fubini's theorem

$$\begin{aligned} \left\| S_n^1(f) \right\|_p^p &= \int_0^{2\pi} \left| \int_0^{2\pi} f(t) K_j(x,t)\, dt \right|^p dx \\ &\leq \int_0^{2\pi} \left(\int_0^{2\pi} |f(t)|\, |K_j(x,t)|^{1/p} \cdot |K_j(x,t)|^{1/q} dt \right)^p dx \\ &\leq C^{p/q} \int_0^{2\pi} |f(t)|^p \int_0^{2\pi} |K_j(x,t)| dx\, dt \\ &\leq C^p \left\| f \right\|_p^p. \end{aligned}$$

□

REMARK 8.7. The proof of the above Theorem 8.18 is much simpler than the proof of the corresponding Theorem 8.4. This is a consequence of two facts:

- Working on the circle $\mathbb{T}$ we avoid extra complications resulting from the fact that $\mathbb{R}$ has infinite measure.
- From (ii) we know that on polynomials the series converges to the identity and we can use the Stone–Weierstrass theorem that polynomials are dense. Note however that this theorem can be relatively easily proved using methods at our disposal (cf. Exercise 8.13).

REMARK 8.8. Note that the periodic case offers some differences from the non-periodic case. Most notably we have a natural one-index ordering of periodic wavelets. Secondly our convergence results are valid also for $p = 1$ and $C(\mathbb{T})$.

Sources and comments

As already remarked, Theorem 8.4 for the Haar wavelet in one variable was proved (essentially) by M. J. Schauder [99] and similar theorems for other orthogonal systems were proved later. This means that the techniques were well developed and with the emergence of wavelets it was clear that the same methods can be applied. The situation was similar with Theorem 8.9. For the Franklin system on the interval it was proved by P. Wojtaszczyk [113] who built on the fundamental work of L. Carleson [11]. After subsequent work on orthogonal spline systems (cf. [114], [101], [102], [117]) it became clear that orthogonality and some estimates like 8.3 and 8.4 is all that matters. It also became clear that many different approaches to the proof can be taken. In this book we use atoms, following ideas from [114]. In [85] the same and much more is proved using the theory of Calderón–Zygmund operators while in Chapter 9 of [24] a one-dimensional case for L_p is presented also via Calderón–Zygmund operators. The reader interested in unconditional bases in L_p (but not in H_1) should consult the paper [45] where wavelets with no restriction on smoothness are discussed and [43] where weighted L_p spaces are considered.

Actually the one-dimensional Haar system on the interval was the first known unconditional basis in any L_p space, $p \neq 2$. This was proved by J. Marcinkiewicz [78] building on the work of Paley [90]. A simple and elementary proof yielding the best constant was given by D. Burkholder [10], cf. also [116] II.D.13. From the very beginning it was natural to compare other unconditional bases with the Haar basis. In this sense Theorems 8.13 and 8.14 are quite natural and have many predecessors. Expressions like those occuring in the middle of 7.16 or 8.58, called the square function, are widely used in various areas of analysis. Theorem 8.16 for orthogonal spline systems on the interval was proved by A.

Chang and Z. Ciesielski [14]. Our proof incorporates some essential modifications presented in [85].

In this chapter we have concentrated on norm convergence of wavelet expansions. Pointwise (or almost everywhere) convergence can also be considered. Technically this uses in a natural way the representation of operators P_j in terms of kernels (cf. 8.14) and estimates of those kernels like 8.16–8.18. Some results along these lines are presented in [58] and [111]

Besides serving as an example of how periodic wavelets behave, Section 8.4 also offers a solution of a well known problem. The following problem posed by P. L. Ulianov attracted a lot of attention starting from the early 1960s (for a short history see [117]). Suppose that $(f_n)_{n=0}^{\infty}$ is a complete orthonormal system in $L_2(\mathbb{T})$ consisting of trigonometric polynomials and such that for each $f \in C(\mathbb{T})$ the series $\sum_{n=0}^{\infty} \langle f, f_n \rangle f_n$ converges uniformly to f. What can be said about the degree of the f_n's, or more precisely how fast the sequence $\alpha_k = \max\{\deg f_n \; n = 0, 1, 2 \ldots, k\}$ does grow? Today the complete answer is known. It was shown by A. A. Privalov [95] that there exists an $\varepsilon > 0$ such that $\alpha_k \geq (\frac{1}{2} + \varepsilon)k$ for large enough k. On the other hand for every $\varepsilon > 0$ a suitable orthonormal system such that $\alpha_k < (\frac{1}{2} + \varepsilon)k$ was constructed by K. Woźniakowski in [118] and by R. A. Lorentz and A. Sahakian in [75].

Our Theorem 8.18 is a weaker result which was obtained earlier in [117], [96] and [89]. It was this last paper which pointed out the close connection of the problem with Meyer's wavelets and whose ideas we follow in Section 8.4.

It should be pointed out that all function spaces on $\mathbb{R}$ considered in this book (and many others) have natural analogues on $\mathbb{T}$, and all the results we prove in this book about wavelet expansions on $\mathbb{R}$ have natural analogues for periodized wavelets on $\mathbb{T}$.

About the exercises. The characterisation of BMO from Exercise 8.8 was first proved (for his system) by L. Carleson in [11] and later for the Franklin system on the interval in [113]. The special atom space described in Exercise 8.11 was studied in detail in [103] and the references quoted there. The characterization stated in this exercise for the Franklin system on the interval was obtained in [115].

Exercises

8.1 For $d = 2$ let us consider the dilation A given by the matrix

$$\begin{bmatrix} 1 & -1 \\ 1 & 1 \end{bmatrix}$$

and discussed in Example 5.4 on page 128. Show that any wavelet associated with this dilation and satisfying 8.3 and 8.4 gives an unconditional basis in $H_1(\mathbb{R}^2)$.

8.2 Let $Tf(x, y) = 8f(4x, 2y)$. Show that T is an isometry of $L_1(\mathbb{R}^2)$ and is continuous on $H_1(\mathbb{R}^2)$. Show also that the norms of T^n acting on $H_1(\mathbb{R}^2)$ tend to infinity as $n \to \infty$.

8.3 Suppose ϕ is a function in $L_2(\mathbb{R}^d)$ such that $\{\phi(t-\gamma)\}_{\gamma\in\mathbb{Z}^d}$ is orthonormal and $\sum_{\gamma\in\mathbb{Z}^d} |\phi(x-\gamma)|$ is bounded. Show that for each p, $1 \le p \le \infty$, there are constants $0 < c \le C$ such that

$$c\Big(\sum_{\gamma\in\mathbb{Z}^d} |a_\gamma|\Big)^{1/p} \le \Big\|\sum_{\gamma\in\mathbb{Z}^d} a_\gamma \phi(x-\gamma)\Big\|_p \le C\Big(\sum_{\gamma\in\mathbb{Z}^d} |a_\gamma|\Big)^{1/p}$$

and that

$$Pf(x) = \sum_{\gamma\in\mathbb{Z}^d} \int_{\mathbb{R}^d} f(t)\overline{\phi(t-\gamma)}\, dt\, \phi(x-\gamma)$$

is a continuous projection in $L_p(\mathbb{R}^d)$.

8.4 Show that for $1 < p < 2$ and $f \in L_1(\mathbb{R}^d) \cap L_2(\mathbb{R}^d)$ we have

$$\|f\|_p \le \|f\|_1 + \|f\|_2 \le \|\!|f|\!\| + \|f\|_2.$$

Show that this implies that if the series $\sum f_n$ with $f_n \in L_1 \cap L_2$ converges to f in the norm of L_1 and in the norm of L_2 then it converges to f in the norm of $L_p(\mathbb{R}^d)$.

8.5 Suppose that f is a continuous bounded function on $\mathbb{R}^d$. Show that $P_j f \to f$ as $j \to \infty$ almost uniformly on $\mathbb{R}^d$, where the P_j's are given by 8.5.

8.6 By considering atoms $\frac{1}{2a}(\mathbf{1}_{[-a,0]} - \mathbf{1}_{[0,a]})$ show that the Haar wavelet basis is not an unconditional basis in $H_1(\mathbb{R})$. Let P_j be the projection defined by 8.5 for the Haar wavelet. Show that there exists a constant C such that $\|\!|P_j f|\!\| \le C\,\|\!|f|\!\|$ for $f \in H_1(\mathbb{R})$ and that $\|\!|f - P_j f|\!\| \to 0$ as $j \to \infty$ for each $f \in H_1(\mathbb{R})$. Show that the same holds for Haar wavelet bases on $\mathbb{R}^d$ and $H_1(\mathbb{R}^d)$.

8.7 Let $f \in C_0(\mathbb{R}^d)$. Show that $\|P_j f\|_\infty \to 0$ as $j \to -\infty$.

8.8 Let Ψ be a wavelet on $\mathbb{R}$ satisfying 8.3 and 8.4. For pairs of integers (j,k) and (l,s) let us write $(j,s) \le (l,s)$ when

$$[2^{-j}k, 2^{-j}(k+1)] \subset [2^{-l}s, 2^{-l}(s+1)].$$

Show that $f \in BMO(\mathbb{R})$ if and only if

$$\sup_{(l,s)} \Big(2^{-l} \sum_{(j,k)\le(l,s)} |\langle f, \Psi_{j,k}\rangle|^2\Big)^{1/2} < \infty.$$

8.9 Suppose that $(f_n)_{n\in A}$ and $(g_m)_{m\in B}$ are wavelet bases on $\mathbb{R}$ corresponding to wavelets satisfying 8.3 and 8.4. Show that $(f_n \otimes g_m)_{n\in A, m\in B}$ is not an unconditional basis in $H_1(\mathbb{R}^2)$. Compare this result with Exercise 7.10.

8.10 Consider the dyadic H_1 space $\delta H_1(\mathbb{R})$ as described in Exercise 6.14. Show that the Haar wavelet is an unconditonal basis in this space.

8.11 Let $H(t)$ be the Haar wavelet on $\mathbb{R}$ defined in Definition 1.1. A *special atom* is a function

$$A(t) =: \frac{1}{a} H(at+b)$$

for some $a, b \in \mathbb{R}$ $a > 0$. Let us define a special atom space X as the space of all functions f such that $f = \sum_k \lambda_k A_k$ with $\sum_k |\lambda_k| < \infty$ and the A_k's special atoms. We define

$$\|f\|_X =: \inf\Big\{\sum_k |\lambda_k| \ : \ f = \sum_k \lambda_k A_k\Big\}.$$

Let Ψ be a wavelet on $\mathbb{R}$ satisfying 8.3 and 8.4. Show that $f \in X$ if and only if

$$\sum_{j,k\in\mathbb{Z}} 2^{-j/2} |\langle f, \Psi_{jk}\rangle| < \infty.$$

8.12 Prove Theorem 8.4 for multiresolution analyses corresponding to an arbitrary dilation matrix A assuming that the scaling function Φ satisfies 8.1 and 8.2.

8.13 Let S_j^1 be defined by 8.92. Show that $\int_0^{2\pi} K_j(x,t)\,dt = 1$ for each $x \in [0, 2\pi]$. Using arguments similar to those used in the proof of Theorem 8.4 show (without using the Stone–Weierstrass theorem) that $\lim_{j\to\infty} \big\| f - S_j^1(f)\big\|_\infty = 0$ for each $f \in C(\mathbb{T})$. Deduce from this the Stone–Weierstrass theorem.

8.14 For each $\varepsilon > 0$ construct a C^∞ function Θ satisfying 3.10–3.14 and such that supp $\Theta \subset [-\pi-\varepsilon, \pi+\varepsilon]$. Check how this improves the estimates for the degree of the corresponding f_n's.

9
Wavelets and smoothness of functions

In this chapter we will discuss connections between smoothness of functions (as measured by moduli of continuity) and properties of wavelet expansions. For the sake of simplicity we consider the one-dimensional case only.

9.1 Modulus of continuity

The main idea of modulus of continuity (present already in the notion of derivative) is to measure the difference between the function and its translate. Since there are many ways to measure the size of a function, we can have many different moduli of continuity. In this chapter we will only consider the following modulus.

DEFINITION 9.1 *The p-modulus of continuity of a function f defined on $\mathbb{R}$, $1 \leq p \leq \infty$, is the function*

$$\omega_p(f;\delta) =: \sup_{0<|h|<\delta} \|f(x) - f(x-h)\|_p \tag{9.1}$$

defined for $\delta > 0$.

Before we proceed let us note some properties of the p-modulus of continuity.

(i) For each f and p the function $\omega_p(f;\delta)$ is an increasing function of δ.

(ii) If $f \in L_p(\mathbb{R})$ for $1 \leq p < \infty$, or $f \in C_0(\mathbb{R})$, if $p = \infty$ then $\omega_p(f;\delta) \to 0$ as $\delta \to 0$. This is almost obvious when f is continuous and has compact support. Since such f's are dense in $L_p(\mathbb{R})$

and in $C_0(\mathbb{R})$ our claim follows. Also for each $\delta > 0$ we have

$$\omega_p(f;\delta) \leq 2\|f\|_p. \tag{9.2}$$

(iii) Since for any positive integer m

$$f(x) - f(x+mh) = \sum_{k=0}^{m-1} [f(x+kh) - f(x+kh+h)]$$

from 9.1 we infer that

$$\omega_p(f;m\delta) \leq m\omega_p(f;\delta). \tag{9.3}$$

This gives

$$\omega_p(f;\delta) = \omega_p(f;m\tfrac{\delta}{m}) \leq \delta \frac{\omega_p(f;\frac{\delta}{m})}{\frac{\delta}{m}} \tag{9.4}$$

so

$$\text{if } \lim_{s\to 0} s^{-1}\omega_p(f;s) = 0 \text{ then } \omega_p(f;\delta) \equiv 0. \tag{9.5}$$

The condition $\omega_p(f;\delta) \equiv 0$ clearly implies that $f(x) = f(x+h)$ a.e. for any $h \in \mathbb{R}$, so f is constant.

(iv) For translation and dilation operators we have

$$\omega(T_h f;\delta) = \omega_p(f;\delta) \tag{9.6}$$

and

$$\omega_p(J_a f;\delta) = 2^{-a/p}\omega_p(f;2^a\delta). \tag{9.7}$$

Note that it follows from 9.3 and (i) above that unless $f = \text{const.}$ we have $\omega_p(f;\delta) > 0$ for all $\delta > 0$. Also 9.3 implies that if $\omega_p(f;\delta) < \infty$ for some $\delta > 0$ then it is finite for *all* $\delta > 0$.

DEFINITION 9.2 *We will say that a function f defined on $\mathbb{R}$ has a p-modulus of continuity if $\omega_p(f;\delta) < \infty$ for some (equivalently for all) $\delta > 0$. The set of all functions having a p-modulus of continuity will be denoted by $MC_p(\mathbb{R})$.*

It is clear from Definition 9.1 and the above remarks that for each $\delta > 0$ $\omega_p(f;\delta)$ is a seminorm on $MC_p(\mathbb{R})$, i.e.

$$\omega_p(\alpha f + \beta g;\delta) \leq |\alpha|\omega_p(f;\delta) + |\beta|\omega_p(g;\delta) \tag{9.8}$$

and

$$\omega_p(f;\delta) = 0 \text{ if and only if } f = \text{const.} \tag{9.9}$$

REMARK 9.1. Note that $MC_p(\mathbb{R})$ contains many non-constant functions not in $L_p(\mathbb{R})$, cf. Exercise 9.4. Note also that the condition $\omega_\infty(f;\delta) \le C\delta^\alpha$ means simply that f satisfies the Hölder's condition with exponent α.

In view of the preceding remarks it is interesting to know how big a function in $MC_p(\mathbb{R})$ can be.

Proposition 9.3 *Suppose $f \in MC_p(\mathbb{R})$, $1 \le p \le \infty$. Then $|f|^p$ is locally integrable if $p < \infty$ and $|f|$ is locally bounded if $p = \infty$. Moreover*

$$\Big(\int_n^{n+1} |f(x)|^p\,dx\Big)^{1/p} \le C + |n|^{1-1/p}\omega_p(f;1). \tag{9.10}$$

Proof First let us prove that $|f|^p$ is locally integrable when $p < \infty$. From 9.6 we see that to show that $|f|^p$ is locally integrable it suffices to show that $\int_0^1 |f(x)|^p\,dx < \infty$. Let us fix $\alpha > 0$ such that the set $\Delta =: \{x \in [0,1] : |f(x)| < \alpha\}$ has positive measure. Since for each t with $|t| \le 1$ we have

$$\begin{aligned} \omega_p(f;1) &\ge \Big(\int_{-\infty}^{\infty} |f(x) - f(x-t)|^p\,dx\Big)^{1/p} \\ &\ge \Big(\int_\Delta |f(x) - f(x-t)|^p\,dx\Big)^{1/p} \end{aligned}$$

from the triangle inequality we infer that for each t with $|t| \le 1$ we have

$$\begin{aligned} \Big(\int_{\Delta+t} |f(x)|^p\,dx\Big)^{1/p} &= \Big(\int_\Delta |f(x-t)|^p\,dx\Big)^{1/p} \\ &\le \omega_p(f;1) + \Big(\int_\Delta |f(x)|^p\,dx\Big)^{1/p} \qquad (9.11) \\ &\le \omega_p(f;1) + \alpha|\Delta|^{1/p}. \end{aligned}$$

Now we define a function of two variables $G(x,t)$ for $x \in [-1,2]$ and $t \in [-1,1]$ by the formula

$$G(x,t) = \begin{cases} |f(x)|^p & \text{if } x \in \Delta + t \\ 0 & \text{if } x \notin \Delta + t. \end{cases}$$

The point of this definition is that for a fixed t the function $G(x,t)$ equals $|f(x)|^p$ on some translate of Δ and 0 outside this translate, while for a fixed x it takes the value $|f(x)|^p$ on the set $-\Delta + x$ which has the same measure as Δ. Using these remarks and 9.11 we get

$$\begin{aligned}
\int_{-1}^{1}\int_{-1}^{2} G(x,t)\,dx\,dt &= \int_{-1}^{1}\int_{\Delta+t} |f(x)|^p\,dx\,dt \\
&\le \int_{-1}^{1} \left(\omega_p(f;1) + \alpha|\Delta|^{1/p}\right)^p dt \qquad (9.12)\\
&= 2\left(\omega_p(f;1) + \alpha|\Delta|^{1/p}\right)^p \\
&= C' < \infty.
\end{aligned}$$

On the other hand

$$\begin{aligned}
\int_{-1}^{2}\int_{-1}^{1} G(x,t)\,dt\,dx &= \int_{-1}^{2} |f(x)|^p \left|(-\Delta + x)\cap[-1,2]\right| dx \\
&\ge \int_{0}^{1} |f(x)|^p|\Delta|\,dx \\
&= |\Delta|\int_{0}^{1} |f(x)|^p\,dx \qquad (9.13)
\end{aligned}$$

Putting together 9.12 and 9.13 and using Fubini's theorem we get the claim for $p < \infty$.

When $p = \infty$ and $f|[0,1]$ is unbounded we take $n > m$ such that the sets

$$\Delta^n =: \{x \in [0,1] \,:\, |f(x)| > n\}$$

and

$$\Delta_m =: \{x \in [0,1] \,:\, |f(x)| < m\}$$

have positive measure. There exists a number t, $|t| \le 1$, such that the set $\Delta =: (t + \Delta^n) \cap \Delta_m$ has positive measure. Clearly for $x \in \Delta$ we have $|f(x-t) - f(x)| > n - m$. Since n can be arbitrarily large (because we assume f is unbounded) we get $\omega_\infty(f;1) = \infty$.

To obtain 9.10 for $p < \infty$ we write

$$f(x+n) = f(x) + \sum_{k=1}^{n} f(x+k) - f(x+k-1)$$

so using Hölder's inequality we get

$$\begin{aligned}
&\left(\int_{n}^{n+1} |f(x)|^p\,dx\right)^{1/p} \\
&= \left(\int_{0}^{1} |f(x+n)|^p\,dx\right)^{1/p}
\end{aligned}$$

$$
\begin{aligned}
&\leq \left(\int_0^1 |f(x)|^p\,dx\right)^{1/p} \\
&\quad +\left(\int_0^1 \Big|\sum_{k=1}^{n}(f(x+k)-f(x+k-1))\Big|^p\,dx\right)^{1/p} \\
&\leq C+\left(\int_0^1 |n|^{p/q}\sum_{k=1}^{n}|f(x+k)-f(x+k-1)|^p\,dx\right)^{1/p} \\
&\leq C+|n|^{1/q}\left(\sum_{k=1}^{n}\int_0^1 |f(x+k)-f(x+k-1)|^p\,dx\right)^{1/p} \\
&\leq C+|n|^{1/q}\left(\int_{-\infty}^{\infty} |f(x)-f(x-1)|^p\,dx\right)^{1/p} \\
&\leq C+|n|^{1/q}\omega_p(f;1).
\end{aligned}
$$

The argument for $p=\infty$ is entirely analogous. □

A natural way to proceed is to consider functions with some restrictions on some modulus of continuity. One such natural restriction, mentioned earlier, is to require $\omega_\infty(f;\delta)\leq C\delta^\alpha$, which gives Hölder's condition with exponent α. As we have seen above, only a constant function satisfies this restriction for $\alpha>1$. It should be noted that any reasonable restriction on the modulus of continuity is less restrictive than say C^1_{00}. In fact for $f\in C^1_{00}(\mathbb{R})$ we easily check (see Exercise 9.5) that for each p, $1\leq p\leq\infty$, we have $\omega_p(f;\delta)\leq C\min(1,\delta)$. On the other hand it follows from 9.4 that for any non-constant function and any p, $1\leq p\leq\infty$, we have $\omega_p(f;\delta)\geq C\min(1,\delta)$ with some $C>0$. A useful way to restrict the modulus of continuity is to consider Besov norms. The definition depends on *three* parameters which is clearly awfull, and seems quite arbitrary. It happens however that Besov norms are useful as I hope to demonstrate in this chapter.

DEFINITION 9.4 *Suppose $0<\alpha\leq 1$ and p,s are such that $1\leq p,s\leq\infty$. The Besov norm $\|f\|_{p,\alpha,s}$ of the function f is defined as*

$$
\|f\|_{p,\alpha,s}=\begin{cases}\left(\int_0^\infty \left[t^{-\alpha}\omega_p(f;t)\right]^s\frac{dt}{t}\right)^{1/s} & \text{when } 1\leq s<\infty\\ \sup_{0<t<\infty} t^\alpha\omega_p(f;t) & \text{when } s=\infty.\end{cases}\tag{9.14}
$$

Actually these are only semi-norms, for they vanish on a constant function. This follows from 9.8.

Using these norms we can define two families of Besov spaces. Homogenous Besov spaces $\dot{B}^p_{\alpha,s}(\mathbb{R})$ consist of those functions f such that

$\|f\|_{p,\alpha,s} < \infty$ and inhomogenous Besov spaces $B^p_{\alpha,s}(\mathbb{R})$ consist of those functions from $L_p(\mathbb{R})$ such that $\|f\|_{p,\alpha,s} < \infty$. Thus the natural norm on $B^p_{\alpha,s}(\mathbb{R})$ is $\|f\|_p + \|f\|_{p,\alpha,s}$. A proper study of homogenous Besov spaces is beyond the scope of this book. The main reason is that in general a homogenous Besov space is not complete when equipped with the norm $\|\cdot\|_{p,\alpha,s}$. When we complete it it becomes a space of distributions not a space of functions (cf. Exercise 9.9). For this and other reasons we will discuss in this book only Besov norms *per se*, or inhomogenous Besov spaces, which are clearly spaces of functions. As the first step of this program we will conclude this section with a different expression for $\|\cdot\|_{p,\alpha,s}$. Namely we have:

Proposition 9.5 *For any α, p and s as in Definition 9.4 there exist constants $0 < c \le C$ such that for any f*

$$c\|f\|_{p,\alpha,s} \le \Big(\sum_{j\in\mathbb{Z}} 2^{\alpha j s}\omega_p(f;2^{-j})^s\Big)^{1/s} \le C\|f\|_{p,\alpha,s}. \tag{9.15}$$

Proof The expression in the middle of 9.15 is really only a discrete version of the integral defining $\|\cdot\|_{p,\alpha,s}$ (for $s < \infty$). We write

$$\begin{aligned}\int_0^\infty \left[t^{-\alpha}\omega_p(f;t)\right]^s \frac{dt}{t} &= \int_0^\infty \omega_p(f;t)^s \frac{dt}{t^{\alpha s+1}} \\ &= \sum_{j\in\mathbb{Z}} \int_{2^{-j}}^{2^{-j+1}} \omega_p(f;t)^s \frac{dt}{t^{\alpha s+1}}.\end{aligned}$$

Using the fact that $\omega_p(f;\delta)$ is increasing this can be majorized by

$$\begin{aligned}\sum_{j\in\mathbb{Z}} 2^{-j}\omega_p(f;2^{-j+1})^s 2^{j(\alpha s+1)} &= \sum_{j\in\mathbb{Z}} 2^{j\alpha s}\omega_p(f;2^{-j+1})^s \\ &= 2^{\alpha s}\sum_{j\in\mathbb{Z}} 2^{j\alpha s}\omega_p(f;2^{-j})^s\end{aligned}$$

and minorized by

$$\sum_{j\in\mathbb{Z}} 2^{-j}\omega_p(f;2^{-j})^s 2^{(j-1)(\alpha s+1)} = 2^{-\alpha s-1}\sum_{j\in\mathbb{Z}} 2^{j\alpha s}\omega_p(f;2^{-j})^s.$$

The argument when $s = \infty$ is the same except that we replace the integrals and the sums by suprema. □

9.2 Multiresolution analyses and moduli of continuity

In this section we will work with a multiresolution analysis with a C^1 scaling function $\Phi(x)$ such that

$$|\Phi(x)| \le C(1+|x|)^{-A} \tag{9.16}$$

and

$$|\Phi'(x)| \le C(1+|x|)^{-A}. \tag{9.17}$$

for some $A > 3$. It follows from 9.16 and Proposition 2.17 that

$$\int_{-\infty}^{\infty} \Phi(x)\,dx = 1. \tag{9.18}$$

REMARK 9.2. The above restriction on A can be relaxed somewhat (i.e. we can use smaller A's). In particular the constant A appearing in 9.17 can be smaller than the one appearing in 9.16. For the sake of simplicity we will use $A > 3$.

As we defined them in Section 2.1, the spaces V_j constituting the multiresolution analysis are subspaces of $L_2(\mathbb{R})$. However, with the assumption 9.16 imposed on the scaling function we see that the series $\sum_{k\in\mathbb{Z}} a_k\Phi(x-k)$ is absolutely and almost uniformly convergent for all sequences $(a_k)_{k\in\mathbb{Z}}$ satisfying $|a_k| \le C + |k|^\beta$ with $\beta < A-1$. In this chapter we will use the extended definition of V_j as the space of all functions of the form

$$f(x) = \sum_{k\in\mathbb{Z}} a_k 2^{j/2}\Phi(2^j x - k)$$

with $|a_k| \le C + |k|^\beta$ for some $\beta < A - 1$. It immediately follows from 9.17 that each such function is a C^1 function, because the series $\sum_{k\in\mathbb{Z}} a_k 2^{j/2}\Phi'(2^j x - k)$ is also absolutely and almost uniformly convergent. What we are doing here is particularly transparent for spline multiresolution analyses. A spline is a spline and we know what it is without any scaling functions. All the growth conditions imposed do not change the nature of the spline. For compactly supported $\Phi(x)$ *every* series $\sum_{k\in\mathbb{Z}} a_k\Phi(x-k)$ converges so it is also clear what we are getting. In general we have to restrict the growth of the function somehow, because we do not have an 'external' description of the V_j's.

As noted in Section 8.1, with each multiresolution analysis we can associate projections

$$P_j f(x) = \int_{-\infty}^{\infty} f(t) 2^j\ \mathbf{\Phi}(2^j t, 2^j x)\,dt \tag{9.19}$$

where $\Phi(t,x) = \sum_{k\in\mathbb{Z}} \overline{\Phi(t-k)}\Phi(x-k)$ (cf. 8.14 and 8.15 or 8.18–8.22). It follows directly from 9.16 (cf. 5.27 and 5.28) that

$$\begin{aligned} |\Phi(t,x)| &\leq C\sum_{k\in\mathbb{Z}} \frac{1}{(1+|t-k|)^A}\frac{1}{(1+|x-k|)^A} \\ &\leq C\frac{1}{(1+|t-x|)^A}. \end{aligned} \tag{9.20}$$

Also from 9.18, follows (cf. 8.16) that for each $x \in \mathbb{R}$ we have

$$\int_{-\infty}^{\infty} \Phi(x,t)\,dt = 1. \tag{9.21}$$

This implies that P_j is well defined for much broader class of functions than $L_2(\mathbb{R})$. It is well defined for all functions $f(t)$ such that

$$|f(t)| \leq C(1+|t|)^{A-2-\varepsilon}.$$

Also it follows from Proposition 9.3 that P_jf is well defined for $f \in MC_p$ provided $A > 3-\frac{1}{p}$ so our assumption about A ensures that P_jf always makes sense for $f \in MC_p$. In the rest of this chapter we will apply P_j to a function as long as it makes sense. We have seen in Theorem 8.4 that for $f \in L_p(\mathbb{R})$ the sequence P_jf approximates f in L_p norm when $j \to \infty$. We would like to see how good this approximation is. This is contained in the following proposition.

Proposition 9.6 (Jackson's inequality) *There exists a constant C such that for any $f \in MC_p$ and P_j as described above*

$$\|f - P_jf\|_p \leq C\omega_p(f;2^{-j}) \tag{9.22}$$

for all $j \in \mathbb{Z}$.

REMARK 9.3. The surprising fact about this proposition is that we do *not* assume that $f \in L_p(\mathbb{R})$.

Proof Suppose we have 9.22 for $j = 0$ with some constant C. Then from 8.6, 6.2 and 9.7 we get

$$\begin{aligned} \|f - P_jf\|_p &= 2^{-j/p}\|J_{-j}f - P_0J_{-j}f\|_p \\ &\leq C2^{-j/p}\omega_p(J_{-j}f;1) = C\omega_p(f;2^{-j}) \end{aligned}$$

so it suffices to consider $j = 0$. From 9.19 using 9.21 we can write

$$f(x) - P_0f(x) = \int_{-\infty}^{\infty} [f(x) - f(t)]\,\Phi(t,x)\,dt.$$

From 9.20 we get

$$\begin{aligned}\|f - P_0 f\|_p^p &= \int_{-\infty}^{\infty} \left| \int_{-\infty}^{\infty} [f(x) - f(t)]\, \Phi(t,x)\, dt \right|^p dx \\ &\le C \int_{-\infty}^{\infty} \left(\int_{-\infty}^{\infty} \frac{|f(x) - f(t)|\, dt}{(1 + |t - x|)^{A-1}} \right)^p dx \\ &= C \int_{-\infty}^{\infty} \left(\int_{-\infty}^{\infty} \frac{|f(x) - f(x+u)|\, du}{(1 + |u|)^{A-1}} \right)^p dx.\end{aligned}$$

Writing $A - 1 = a + b$ with $a, b \ge 0$, $ap > p + 1$ and $bq > 1$ (where as usual $p^{-1} + q^{-1} = 1$) and applying Hölder's inequality to the inside integral we get

$$\begin{aligned}&\|f - P_0 f\|_p^p \\ &\le C \int_{-\infty}^{\infty} \int_{-\infty}^{\infty} \frac{|f(x) - f(x+u)|^p}{(1 + |u|)^{ap}}\, du \left(\int_{-\infty}^{\infty} \frac{du}{(1 + |u|)^{bq}} \right)^{p/q} dx \\ &\le C \int_{-\infty}^{\infty} \frac{1}{(1 + |u|)^{ap}} \int_{-\infty}^{\infty} |f(x) - f(x+u)|^p\, dx\, du \\ &\le C \int_{-\infty}^{\infty} \frac{1}{(1 + |u|)^{ap}} \omega_p(f; |u|)^p\, du.\end{aligned}$$

We split the last integral into two parts and estimate each part separately as follows:

$$\int_{-1}^{1} \omega_p(f; |u|)^p \frac{du}{(1 + |u|)^{ap}} \le C \omega_p(f; 1)^p$$

and

$$\begin{aligned}\int_{-\infty}^{-1} + \int_{1}^{\infty} \omega_p(f; |u|)^p \frac{du}{(1 + |u|)^{ap}} &\le 2 \int_{1}^{\infty} \omega_p(f; u)^p \frac{du}{(1 + u)^{ap}} \\ &\quad \text{using 9.3} \\ &\le C \int_{1}^{\infty} u^p \omega_p(f; 1)^p \frac{du}{(1 + u)^{ap}} \\ &\le C \omega_p(f; 1)^p \int_{1}^{\infty} \frac{u^p\, du}{(1 + u)^{ap}} \\ &\le C \omega_p(f; 1)^p.\end{aligned}$$

The last inequality follows from the choice of a. The case $p = \infty$ requires the standard reinterpretation and is actually easier. □

REMARK 9.4. Note that the proof of Proposition 9.6 uses only 9.16. It uses neither 9.17 nor any other assumption about the smoothness of the scaling function Φ.

Proposition 9.7 (Bernstein's inequality) *Under the assumptions about multiresolution analyses described at the beginning of this section, for each p, $1 \le p \le \infty$, there exists a constant C such that for $f \in V_j$ we have*

$$\omega_p(f;t) \le C \min(2^j t, 1)\|f\|_p. \tag{9.23}$$

Proof Suppose we have 9.23 for $j = 0$ with some constant C. Then for $f \in V_j$ we have $J_{-j}f \in V_0$ so from 6.2 and 9.7 we get

$$\begin{aligned}\omega_p(f;t) &= 2^{-j/p}\omega_p(J_{-j}f;2^j t) \le C2^{-j/p}\min(2^j t;1)\|J_{-j}f\|_p \\ &= C\min(2^j t,1)\|f\|_p\end{aligned}$$

so it suffices to show 9.23 for $j = 0$. It also follows from 9.2 that for $j = 0$ it suffices to consider only $|t| \le 1$. Take $f \in V_0 \cap L_p(\mathbb{R})$ (if $f \notin L_p(\mathbb{R})$ there is nothing to prove). Then by 8.7 we can write $f = \sum_{k\in\mathbb{Z}} c(k)\Phi(x-k)$ with $C\|f\|_p \ge \left(\sum_{k\in\mathbb{Z}} |c(k)|^p\right)^{1/p}$, so for $|h| \le 1$ we have

$$\|f(x)-f(x+h)\|_p^p = \int_{-\infty}^{\infty}\Big|\sum_{k\in\mathbb{Z}} c(k)[\Phi(x-k)-\Phi(x-k-h)]\Big|^p\,dx \tag{9.24}$$

Since Φ is a C^1 function we infer from the mean value theorem that $|\Phi(x-k)-\Phi(x-k-h)| = |h|\,|\Phi'(x-k+\xi)|$ for some ξ between 0 and h, so from 9.17 we get (remember that $|h| \le 1$)

$$|\Phi(x-k)-\Phi(x-k-h)| \le |h|\frac{C}{(1+|x-k+\xi|)^A} \le \frac{C|h|}{(1+|x-k|)^A}.$$

Substituting this into 9.24 we get

$$\|f(x)-f(x+h)\|_p^p \le C\int_{-\infty}^{\infty}\Big(\sum_{k\in\mathbb{Z}}|c(k)|\frac{|h|}{(1+|x-k|)^A}\Big)^p\,dx.$$

From Lemma 8.2 we get

$$\|f(x)-f(x+h)\|_p^p \le C|h|^p\sum_{k\in\mathbb{Z}}|c(k)|^p$$

so $\|f(x)-f(x+h)\|_p \le C|h|\|f\|_p$ which clearly gives the claim. □

Given a function f, by $s_j(f)$ we denote the error in the best approximation of f by elements of V_j. Specifying the norm we put

$$s_j^p(f) = \inf\{\|f - g\|_p \; : \; g \in V_j\}. \tag{9.25}$$

Observe that $\|f - P_j f\|_p$ is a quantity basically equivalent to $s_j^p(f)$. Since $P_j f \in V_j$ we clearly have $s_j^p(f) \leq \|f - P_j f\|_p$. On the other hand P_j is a continuous projection from $L_p(\mathbb{R})$ onto V_j with $\|P_j\| = \|P_0\|$, so for any $g \in V_j$ we have

$$\|f - P_j f\|_p \leq \|f - g\|_p + \|P_j g - P_j f\|_p \leq (1 + \|P_0\|)\|f - g\|_p.$$

This clearly implies

$$s_j^p(f) \leq \|f - P_j f\|_p \leq (1 + \|P_0\|) s_j^p(f). \tag{9.26}$$

Now we are ready for one of the main technical propositions of this chapter, which in the framework of Besov norms ties together best approximation and multiresolution analyses and wavelets.

Proposition 9.8 *Suppose we have a multiresolution analysis with a scaling function* Φ *satisfying 9.16 and 9.17 and an associated wavelet* Ψ *also satisfying*

$$|\Psi(x)| \leq C(1 + |x|)^{-A}. \tag{9.27}$$

Let Q_j *be the projection defined by 8.10. For* $0 < \alpha < 1$ *and* $1 \leq p, s \leq \infty$ *and any function* f *the following conditions are equivalent:*

$$\left(\sum_{j \in \mathbb{Z}} \left[2^{j\alpha} s_j^p(f)\right]^s\right)^{1/s} < \infty \tag{9.28}$$

$$\left(\sum_{j \in \mathbb{Z}} \left[2^{j\alpha} \|f - P_j f\|_p\right]^s\right)^{1/s} < \infty \tag{9.29}$$

$$\left(\sum_{j \in \mathbb{Z}} \left[2^{j\alpha} \|Q_j f\|_p\right]^s\right)^{1/s} < \infty \tag{9.30}$$

$$\left(\sum_{j \in \mathbb{Z}} \left[2^{j\alpha} \left(\sum_{k \in \mathbb{Z}} 2^{j(1/2 - 1/p)p} |\langle f, \Psi_{jk} \rangle|^p\right)^{1/p}\right]^s\right)^{1/s} < \infty \tag{9.31}$$

REMARK 9.5. As usual the expression $\left(\sum_{k \in \mathbb{Z}} |a_k|^s\right)^{1/s}$ for $s = \infty$ is interpreted as $\sup_{k \in \mathbb{Z}} |a_k|$. Also it follows from inspection of the proof that there exists a constant C (which may depend on α, p, s but is independent of f) such that if one of 9.28–9.31 equals B then all the other quantities are at most CB.

It is also clear that all four quantities on the left hand sides of 9.28–9.31 are semi-norms.

Proof The equivalence 9.28$\Longleftrightarrow$9.29 follows directly from 9.26. The fact that 9.30$\Longleftrightarrow$9.31 follows directly from Proposition 8.3, in particular from 8.12. Thus really we have two distinct types of conditions in Proposition 9.8: approximation conditions and wavelet expansion conditions. Since $\|Q_j\| \le C$ independently of j and $Q_j(f - P_jf) = Q_jf$ we get $\|Q_jf\|_p \le C\|f - P_jf\|$ so 9.29$\Longrightarrow$9.30.

Thus we only have to prove 9.30$\Longrightarrow$9.29. To do this let us write $\|f - P_jf\|_p \le \sum_{r=j}^{\infty} \|Q_rf\|_p$. Using Hölder's inequality and summing geometric series we have

$$
\begin{aligned}
&\sum_{j\in\mathbb{Z}} 2^{j\alpha s}\|f - P_jf\|_p^s \\
&\quad \le \sum_{j\in\mathbb{Z}} 2^{j\alpha s}\Big(\sum_{r=j}^{\infty} \|Q_rf\|_p\Big)^s \\
&\quad\quad \text{taking } a > 0 \text{ and } a < \alpha \\
&\quad = \sum_{j\in\mathbb{Z}} 2^{j\alpha s}\Big(\sum_{r=j}^{\infty} 2^{-ar}2^{ar}\|Q_rf\|_p\Big)^s \\
&\quad \le \sum_{j\in\mathbb{Z}} 2^{j\alpha s}\Big(\sum_{r=j}^{\infty} 2^{-ars'}\Big)^{s/s'} \sum_{r=j}^{\infty} 2^{ars}\|Q_rf\|_p^s \\
&\quad \le C\sum_{j\in\mathbb{Z}} 2^{j\alpha s}2^{-jas} \sum_{r=j}^{\infty} 2^{ars}\|Q_rf\|_p^s \\
&\quad = C\sum_{r\in\mathbb{Z}} 2^{ars}\|Q_rf\|_p^s \sum_{j\le r} 2^{js(\alpha - a)} \\
&\quad \le C\sum_{r\in\mathbb{Z}} 2^{ars}2^{rs(\alpha-a)}\|Q_rf\|_p^s \\
&\quad = C\sum_{r\in\mathbb{Z}} 2^{rs\alpha}\|Q_rf\|_p^s
\end{aligned}
$$

so the implication 9.30$\Longrightarrow$9.29 is proved. This completes the proof of Proposition 9.8. □

Now we would like to connect the above conditions 9.28–9.31 with Besov norms. There is however one obstacle to this which depends on the particular multiresolution analysis involved. With the definition of the spaces V_j we are using in this chapter (see the beginning of Section

9.2) we have $V_{-\infty} =: \bigcap_j V_j \neq \{0\}$. It follows from Proposition 2.17 that always $1 \in V_{-\infty}$ but the space $V_{-\infty}$ can be bigger and clearly depends on the multiresolution analysis involved. It is clear that for $f \in V_{-\infty}$ *all four* quantities 9.28–9.31 equal zero. On the other hand if f is not constant, then $\omega_p(f;\delta)$ is not identically zero, so $\|f\|_{p,\alpha,s} > 0$. Thus to connect Besov norms with the quantities 9.28–9.31 we have to avoid this difficulty and restrict the set of functions under consideration. Before we discuss inhomogenous Besov spaces (where these difficulties disappear completely) let us formulate one technical theorem.

Theorem 9.9 *Suppose we have a multiresolution analysis with a scaling function Φ satisfying 9.16 and 9.17 and an associated wavelet Ψ also satisfying*

$$|\Psi(x)| \leq C(1+|x|)^{-A}. \tag{9.32}$$

Let Q_j be the projection defined by 8.10. For $0 < \alpha < 1$ and $1 \leq p, s \leq \infty$ and any function f such that $\|f\|_{p,\alpha,s} < \infty$ all four relations 9.28–9.31 hold.

Conversely, if f is a function such that

$$\omega_p(P_j f; 1) \to 0 \quad as\ j \to -\infty \tag{9.33}$$

and any one of 9.28–9.31 holds then $\|f\|_{p,\alpha,s} < \infty$.

Proof If $\|f\|_{p,\alpha,s} < \infty$ then from Proposition 9.5 we get that

$$\sum_{j\in\mathbb{Z}} 2^{j\alpha s}\omega_p(f;2^{-j})^s < \infty. \tag{9.34}$$

From this, using Jackson's inequality (Proposition 9.6) we immediately obtain 9.29. Thus Proposition 9.8 shows that all four conditions 9.28–9.31 hold.

To prove the converse we will show that 9.29 implies 9.34 and use Proposition 9.5. For each $j \in \mathbb{Z}$ and $\mu < j$ we can write

$$f = f - P_j f + \sum_{r=\mu}^{j-1}(P_{r+1} - P_r)f - P_\mu f$$

so using 9.8 and 9.2 we get

$$\omega_p(f;2^{-j}) \leq 2\|f - P_j f\|_p + \sum_{r=\mu}^{j-1}\omega_p((P_{r+1} - P_r)f;2^{-j}) + \omega_p(P_\mu f;2^{-j}).$$

Using 9.3 and 9.33 we can pass to the limit as $\mu \to -\infty$ and write

$$\omega_p(f;2^{-j}) \leq 2\|f - P_j f\|_p + \sum_{r=-\infty}^{j-1} \omega_p((P_{r+1} - P_r)f; 2^{-j}).$$

Since $(P_{r+1} - P_r)f \in V_{r+1}$ we can use Bernstein's inequality 9.23 to obtain

$$\omega_p(f;2^{-j}) \leq 2\|f - P_j f\|_p + C \sum_{r=-\infty}^{j-1} 2^{r-j}\|P_{r+1}f - P_r f\|_p.$$

Since $\|P_{r+1}f - P_r f\|_p \leq \|f - P_{r+1}f\|_p + \|f - P_r f\|_p$ we obtain

$$\omega_p(f;2^{-j}) \leq C \sum_{r=-\infty}^{j} 2^{r-j}\|f - P_r f\|_p. \tag{9.35}$$

The rest is just an application of Hölder's inequality and summing of geometric series. We have from 9.35

$$\begin{aligned}
&\sum_{j\in\mathbb{Z}} 2^{j\alpha s}\omega_p(f;2^{-j})^s \\
&\leq C\sum_{j\in\mathbb{Z}} 2^{j\alpha s}\Big(\sum_{r=-\infty}^{j} 2^{r-j}\|f - P_r f\|_p\Big)^s \\
&= C\sum_{j\in\mathbb{Z}} 2^{js(\alpha-1)}\Big(\sum_{r=-\infty}^{j} 2^{r}\|f - P_r f\|_p\Big)^s \\
&\quad \text{taking } a > 0,\ a+b=1,\ a < 1-\alpha \\
&= C\sum_{j\in\mathbb{Z}} 2^{js(\alpha-1)}\Big(\sum_{r=-\infty}^{j} 2^{ra}\cdot 2^{rb}\|f - P_r f\|_p\Big)^s \\
&\quad \text{using Hölder's inequality} \\
&\leq C\sum_{j\in\mathbb{Z}} 2^{js(\alpha-1)}\Big(\sum_{r=-\infty}^{j} 2^{rbs}\|f - P_r f\|_p^s\Big)\Big(\sum_{r=-\infty}^{j} 2^{ras'}\Big)^{s/s'} \\
&\leq C\sum_{j\in\mathbb{Z}} 2^{js(\alpha-1)}2^{jas}\Big(\sum_{r=-\infty}^{j} 2^{rbs}\|f - P_r f\|_p^s\Big) \\
&= C\sum_{r\in\mathbb{Z}} 2^{rbs}\|f - P_r f\|_p^s \sum_{j\geq r} 2^{js(\alpha+a-1)} \\
&\leq C\sum_{r\in\mathbb{Z}} 2^{rbs}2^{rs(\alpha+a-1)}\|f - P_r f\|_p^s
\end{aligned}$$

$$= C\sum_{r\in\mathbb{Z}} 2^{rs\alpha}\|f - P_r f\|_p^s$$

so 9.34 holds. The conclusion follows directly from Proposition 9.5. □

Now we are ready to give a characterization of inhomogenous Besov spaces.

Corollary 9.10 *Suppose we have a multiresolution analysis with a scaling function Φ satisfying 9.16 and 9.17 and an associated wavelet Ψ also satisfying*

$$|\Psi(x)| \leq C(1+|x|)^{-A}.$$

Let Q_j be the projection defined by 8.10. For $0 < \alpha < 1$ and $1 \leq p, s \leq \infty$ and any function f the following conditions are equivalent:

$$\|f\|_p + \|f\|_{p,\alpha,s} < \infty \tag{9.36}$$

$$\|f\|_p + \Big(\sum_{j\geq 0}\left[2^{j\alpha}\omega_p(f;2^{-j})\right]^s\Big)^{1/s} < \infty \tag{9.37}$$

$$\|P_0 f\|_p + \Big(\sum_{j\geq 0}\left[2^{j\alpha}s_j^p(f)\right]^s\Big)^{1/s} < \infty \tag{9.38}$$

$$\|P_0 f\|_p + \Big(\sum_{j\geq 0}\left[2^{j\alpha}\|f - P_j f\|_p\right]^s\Big)^{1/s} < \infty \tag{9.39}$$

$$\|P_0 f\|_p + \Big(\sum_{j\geq 0}\left[2^{j\alpha}\|Q_j f\|_p\right]^s\Big)^{1/s} < \infty \tag{9.40}$$

$$\Big(\sum_{k\in\mathbb{Z}} |\langle f, \Phi(x-k)\rangle|^p\Big)^{1/p}$$

$$+\Big(\sum_{j\geq 0} 2^{j(\alpha+1/2-1/p)s}\Big(\sum_{k\in\mathbb{Z}} |\langle f, \Psi_{jk}\rangle|^p\Big)^{s/p}\Big)^{1/s} < \infty \tag{9.41}$$

Proof This proof uses both Theorem 9.9 and some ideas of its proof.

9.36 $\Longleftrightarrow$ 9.37 The implication 9.36 $\Longrightarrow$ 9.37 clearly follows from Proposition 9.5. Conversely observe that if $\|f\|_p < \infty$ then from 9.2 we obtain that $\sum_{j<0}\left[2^{j\alpha}\omega_p(f,2^{-j})\right]^s < \infty$. Using this and Proposition 9.5 we see that 9.37 $\Longrightarrow$ 9.36.

9.36 $\Longrightarrow$ 9.38. If $\|f\|_p < \infty$ then 8.8 gives that $\|P_0 f\|_p < \infty$. The rest follows from Theorem 9.9.

9.38 $\Longrightarrow$ 9.39 follows directly from 9.26.

9.39 $\Longrightarrow$ 9.40. Since $\|Q_j\| \le C$ independently of j (see Proposition 8.3) and $Q_j(f - P_j f) = Q_j f$ we get $\|Q_j f\|_p \le C\|f - P_j f\|_p$ so the implication holds.

9.40 $\Longleftrightarrow$ 9.41 follows directly from Proposition 8.1, in particular 8.7, and from Proposition 8.3, in particular 8.12.

9.40 $\Longrightarrow$ 9.36. Note that from 9.40 we see that $\sup_{j\ge 0} 2^{j\alpha}\|Q_j f\|_p < \infty$ so the series $\sum_{j\ge 0} Q_j f$ is absolutely convergent in $L_p(\mathbb{R})$. This implies that

$$P_0 f + \sum_{j\ge 0} Q_j f = f \in L_p(\mathbb{R}),$$

and also that f satisfies 9.33. To see this note that for $j < 0$ from Bernstein's inequality 9.23 we obtain

$$\omega_p(P_j f; 1) \le C 2^j \|P_j f\|_p \le C 2^j \|f\|_p.$$

Since for $j < 0$ we have $Q_j f = Q_j P_0 f$ and $\|Q_j\| \le C$ for all $j \in \mathbb{Z}$ (see Proposition 8.3) we infer that $\|Q_j f\|_p \le C$ for $j < 0$. Together with 9.40 this gives

$$\sum_{j\in\mathbb{Z}} \left[2^{j\alpha}\|Q_j f\|_p\right]^s < \infty.$$

Thus we can use Theorem 9.9 to conclude that $\|f\|_{p,\alpha,s} < \infty$. This implies that 9.36 holds. □

REMARK 9.6. If in conditions 9.38–9.40 we replace $\|P_0 f\|_p$ by $\|f\|_p$ we obtain three more conditions equivalent to all the others. All this gives us nine equivalent norms on $B^p_{\alpha,s}(\mathbb{R})$. Since the norms 9.36 and 9.37 do not depend on the multiresolution analysis or wavelet involved it follows that all the other norms are equivalent for *all* multiresolution analyses satisfying our general assumptions. This is a rather surprising fact.

9.3 Compression of wavelet decompositions

In this section we want to discuss briefly and at the most elementary level the following problem of non-linear approximation:

> Suppose we have a wavelet Ψ on $\mathbb{R}$. Given a function f and $\varepsilon > 0$, find a 'small' set of wavelet coefficients A so that the function f and the sum $\sum_{(j,k)\in A} \langle f, \Psi_{jk}\rangle \Psi_{jk}$ are within ε.

Clearly we have to specify how we measure the distance between functions. The simplest and quite often the most useful way is to use the

L_2-norm. Since the wavelet basis is an orthonormal basis in $L_2(\mathbb{R})$ any f in $L_2(\mathbb{R})$ can be approximated as closely as we wish by finite sums. The real problem is how small the set A can be. Without any additional assumptions on the function f nothing sensible can be said. If $f_N = \sum_{k=1}^N \frac{1}{\sqrt{N}}\Psi(x-k)$ then $\|f_N\|_2 = 1$ and for any A with cardinality $\leq \frac{N}{2}$ we have

$$\Big\|f_N - \sum_{(j,k)\in A} \langle f_N, \Psi_{jk}\rangle \, \Psi_{jk}\Big\|_2 \geq \frac{1}{\sqrt{2}}.$$

The argument for this is quite obvious when we look at coefficients; we clearly have

$$\Big\|f_N - \sum_{(j,k)\in A} \langle f_N, \Psi_{jk}\rangle \, \Psi_{jk}\Big\|_2 \geq \Big\|\sum{}' \frac{1}{\sqrt{N}}\Psi(x-k)\Big\|_2$$

where Σ' is the sum over those k's such that $(0,k) \notin A$. Since there are at least $\frac{N}{2}$ such k's the inequality follows.

On the other hand suppose that we have $x = (x_k)_{k\in\mathbb{Z}} \in \ell_2$ such that $(\sum_{k\in\mathbb{Z}} |x_k|^p)^{1/p} \leq C$ for some $p < 2$. Let us denote $A =: \{k \in \mathbb{Z} : |x_k| < \delta\}$ and $B =: \mathbb{Z} \setminus A$. Then

$$C \geq \Big(\sum_{k\in B} |x_k|^p\Big)^{1/p} \geq (|B|\delta^p)^{1/p} = \delta |B|^{1/p}$$

so $|B| \leq C^p \delta^{-p}$. Also

$$\begin{aligned}\Big(\sum_{k\in A} |x_k|^2\Big)^{1/2} &\leq \Big(\sum_{k\in A} |x_k|^p \delta^{2-p}\Big)^{1/2} \\ &\leq \delta^{1-p/2}\Big(\sum_{k\in\mathbb{Z}} |x_k|^p\Big)^{1/2} \leq \delta^{1-p/2} C^{p/2}.\end{aligned}$$

Those remarks can be phrased in two equivalent ways:

- Suppose that $x \in \ell_2$ and $\|x\|_2 = 1$. Suppose also that for some $p < 2$ we have $\|x\|_p \leq C$. Fix a natural number N and choose the N largest coefficients of x, i.e. define a set $B \subset \mathbb{Z}$ such that $|B| = N$ and $|x_k| \geq |x_l|$ whenever $k \in B$ and $l \notin B$. Define the sequence $g = (g_k)_{k\in\mathbb{Z}}$ by the formula

$$g_k = \begin{cases} x_k & \text{if } k \in B \\ 0 & \text{if } k \notin B. \end{cases} \tag{9.42}$$

Then $\max_{k\notin B} |x_k| \leq C N^{-1/p}$ and

$$\|x - g\|_2 \leq C^{p/2} C^{1-p/2} N^{-(1/p)(1-p/2)} = C N^{1/2-1/p}.$$

- Suppose that $x \in \ell_2$ and $\|x\|_2 = 1$. Suppose also that for some $p < 2$ we have $\|x\|_p \leq C$. Fix a positive cut-off number $\varepsilon > 0$ and define the set $B \subset \mathbb{Z}$ as $B =: \{k \in \mathbb{Z} : |x_k| \geq \varepsilon\}$. Define the sequence g by 9.42. Then the number of non-zero coefficients in g (i.e. $|B|$) is at most $C^p\varepsilon^{-p}$ and $\|x - g\|_2 \leq C^{p/2}\varepsilon^{1-p/2}$.

The conclusion of the above discussion is that if a sequence is in ℓ_p for some $p < 2$ then it can be well approximated in ℓ_2. If we are interested in functions, the same argument applies if f is a function such that $\sum_{jk} | \langle f, \Psi_{jk}\rangle |^p \leq C^p$ for some $p < 2$. The language of Besov norms allows us to figure out when this happens.

Proposition 9.11 *Let α, $0 < \alpha \leq \frac{1}{2}$ be given and put $\tau = (\alpha + \frac{1}{2})^{-1}$. Then there exists a constant C such that*

$$\sum_{jk} |\langle f, \Psi_{jk}\rangle|^\tau \leq C\|f\|_{\tau,\alpha,\tau}. \tag{9.43}$$

Proof This is a direct consequence of Theorem 9.9. Substituting $p = s = \tau$ into 9.31 we get 9.43. □

REMARK 9.7. The restriction $\alpha \leq \frac{1}{2}$ is forced on us because we need $\tau \geq 1$ to use Theorem 9.8. This restriction is not really necessary because most of the content of the previous two sections can be also proved for $0 < p, s \leq \infty$.

Putting together our considerations at the beginning of this section and Proposition 9.11 we get:

Theorem 9.12 *Let α, $0 < \alpha \leq \frac{1}{2}$ be given and let $\tau = (\alpha + \frac{1}{2})^{-1}$. Suppose that $f \in L_2(\mathbb{R})$ and that $\|f\|_{\tau,\alpha,\tau} \leq C$. Then there exists a constant C' such that for every integer N we can find $A \subset \mathbb{Z} \times \mathbb{Z}$ with cardinality N such that*

$$\Big\| f - \sum_{(j,k)\in A} \langle f, \Psi_{jk}\rangle \Psi_{jk} \Big\|_2 \leq C'CN^{-\alpha}. \tag{9.44}$$

The set A can be chosen by taking N coefficients $\langle f, \Psi_{jk}\rangle$ with biggest absolute value.

Sources and comments

The modulus of continuity is a very classical and fundamental concept used throughout analysis. It is introduced and its properties are studied in most books on approximation theory or real function theory, see e.g. [105], [88], [35], [4]. Besov spaces and Besov norms were defined in full generality and investigated (naturally enough) by O. V. Besov in [5]. Today there is an extensive literature devoted to them, see e.g. [91] or [109]. Usually they are defined using higher order moduli of continuity, which allows the parameter α to be any positive number. This means that smoother functions can be handled. It results however in a much more technically complicated theory. I have chosen the simplest modulus of continuity because then the results are free from excessive technicalities.

Proposition 9.6 is one of many 'Jackson's inequalities' which deal with the precision of approximation of a function by elements from a linear subspace. Their forerunners are results of D. Jackson [53] where approximation by polynomials was considered. Proposition 9.7 is one of many 'Bernstein's inequalities' where the smoothness of functions from some special subspace is estimated by their size. Its forerunner is the classical Bernstein's inequality that for a trigonometric polynomial f of degree at most n we have $\|f'\|_\infty \leq n\|f\|_\infty$. Our proofs of Propositions 9.6 and 9.7 exploit ideas standard in approximation theory, cf. [32], [42] or [33]. Our proof of Theorem 9.8 also follows and explains a well established paradigm, although some details may be non-standard. It should be stressed that the equivalence of conditions 9.28–9.31 and their connections with Besov norms in a non-wavelet context has been known for a long time. In particular a characterization of Besov spaces on interval via orthogonal spline systems similar to condition 9.31 was given by S. Ropela [97].

Much more general Besov-type spaces are also studied (see e.g. [109]) and properties of wavelet expansions in such spaces can also be investigated, see e.g. [74]

In this chapter we have discussed so called global smoothness of functions. This means that via the Besov norms we have taken the whole function into account. We can naturally ask questions about the smoothness of functions in the neighbourhood of one point (i.e. local smoothness). A typical such question might be: Suppose that f satisfies a Hölder's condition at one point, i.e. $|f(x) - f(x_0)| \leq C|x - x_0|^\alpha$ (and f satisfies some general assumptions so that we can compute the wavelet coefficients). Can we recognize this from the wavelet coefficients? The answer is 'basically yes'. The precise results were obtained by S. Jaffard [54] and are presented also in [24] pp. 299ff.

Our treatment of compression of wavelet decomposition presented in

Section 9.3 is very elementary and straightforward, mainly because we use only the approximation in Hilbert space norm. A more sophisticated treatment with some indication of applications to statistical estimates can be found in [36]. Compression with approximation in L_p norm is treated in the paper [33]. Similar questions are discussed in [34].

About the exercises. Exercise 9.7 provides one of the standard definitions of Besov spaces while Exercise 9.8 give several instances of so called embedding theorems. These are standard facts in the theory of Besov spaces. Exercise 9.13 gives a glimpse of Besov spaces on the circle. Some results related to Exercise 9.3 are contained in [43].

Exercises

9.1 Suppose that $x \in \ell_p$, $0 < p < \infty$ and $\|x\|_p = 1$ and let $q > p$ be given. Show that there exists a $g \in \ell_q$ such that g has at most N non-zero coordinates and $\|x - g\|_q \leq N^{1/q-1/p}$.

9.2 Suppose that $f \in L_2(\mathbb{R})$ and that Ψ is a wavelet satisfying 9.16 and 9.17. Assume also that $\sum_{j,k\in\mathbb{Z}} |\langle f, \Psi_{jk}\rangle|^p < \infty$ for some $p < 2$. Show that $f \in MC_p$ and satisfies 9.33.

9.3 Let $S^n(2^j\mathbb{Z})$ be the spline multiresolution analysis described in subsection 3.3.2. Show that $\bigcap_{j\in\mathbb{Z}} S^n(2^j\mathbb{Z})$ consists of all C^{n-1} functions which are polynomials of degree at most n when restricted to $(-\infty, 0)$ and to $(0, \infty)$. Show that for $n > 1$ constants are the only such functions which are in MC_p for some p, $1 \leq p \leq \infty$. Show that the same is true for $n = 1$ and $1 \leq p < \infty$. Show that continuous functions which are linear on both $(-\infty, 0)$ and $(0, \infty)$ have an ∞-modulus of continuity. Similarly analyze the case $n = 0$.

9.4 Show that the function

$$f_\alpha(x) = \begin{cases} 0 & \text{for } x \leq 0 \\ |x|^\alpha & \text{for } x > 0 \end{cases}$$

belongs to $MC_p(\mathbb{R})$ for $p < \infty$ if and only if $\alpha < \frac{p-1}{p}$. Show that in this case $\omega_p(f_\alpha; \delta) \leq C\delta$. Let g_α be a piecewise linear function with nodes in $\mathbb{Z}$ such that $g_\alpha(n) = f_\alpha(n)$ for $n \in \mathbb{Z}$. For what p do we have $\|f_\alpha - g_\alpha\|_p < \infty$?

9.5 Let $1 \leq p \leq \infty$ and let $f \in C^1_{00}(\mathbb{R})$. Show that $\omega_p(f; \delta) \leq C \min(1, \delta)$ for some constant C.

9.6 Let $f(x) =: \mathbf{1}_{[0,\infty)}$. Show that f satisfies $\omega_p(f; \delta) \leq C\delta^\alpha$ if and only if $\alpha = 1/p$. Show also that

$$f(x) = \begin{cases} 1 & \text{if } x > 1 \\ x & \text{if } 0 \le x \le 1 \\ 0 & \text{if } x < 0 \end{cases}$$

satisfies $\omega(f;\delta) \le C\delta^\alpha$ if and only if $\alpha \ge 1/p$.

9.7 (a) Show that there exists a function $\varphi \in S(\mathbb{R})$ such that

$$\text{supp } \varphi = \{\xi \ : \ 2^{-1} \le |\xi| \le 2\}$$

$$\varphi(\xi) > 0 \ \text{ for } \ 2^{-1} < |\xi| < 2$$

$$\sum_{k\in\mathbb{Z}} \varphi(2^{-k}\xi) = 1 \ \text{ for each } \ \xi \neq 0.$$

(b) Fix any Meyer's wavelet from the Schwartz class S. Show that any of the conditions 9.28–9.31 for this wavelet is equivalent to

$$\Big(\sum_{j\in\mathbb{Z}} \big[2^{\alpha j}\|\varphi_j * f\|_p\big]^s\Big)^{1/s} < \infty. \qquad \text{(E9.1)}$$

where $\hat{\varphi}_j(x) =: \varphi(2^{-j}x)$.

9.8 Let as usual $B^p_{\alpha,s}(\mathbb{R})$ denote the inhomogenous Besov space. Show that:

(a) $B^{p_1}_{\alpha,s}(\mathbb{R}) \subset B^{p_2}_{\alpha,s}(\mathbb{R})$ for $p_1 \le p_2$
(b) $B^p_{\alpha,s_1}(\mathbb{R}) \subset B^p_{\alpha,s_2}(\mathbb{R})$ for $s_1 \le s_2$
(c) $B^p_{\alpha_1,s}(\mathbb{R}) \subset B^p_{\alpha_2,s}(\mathbb{R})$ for $\alpha_1 \le \alpha_2$
(d) $B^p_{\alpha_1,s}(\mathbb{R}) \subset B^p_{\alpha_2,t}(\mathbb{R})$ for $\alpha_1 < \alpha_2$ and arbitrary $t, s \ge 1$.

9.9 Let (a_{jk}) be a sequence of numbers such that

$$\Big(\sum_{j\in\mathbb{Z}} \Big[2^{j\alpha}\Big(\sum_{k\in\mathbb{Z}} 2^{j(1/2-1/p)p}|a_{jk}|^p\Big)^{1/p}\Big]^s\Big)^{1/s} < \infty. \qquad \text{(E9.2)}$$

Let Ψ be a wavelet satisfying 9.27.

(a) Show that

$$\sum_{j\ge 0}\sum_{k\in\mathbb{Z}} a_{jk}\Psi_{jk} \in L_p(\mathbb{R}).$$

(b) Show that for each $j \in \mathbb{Z}$

$$\Big\|\sum_{k\in\mathbb{Z}} \alpha_{jk}\Psi_{jk}\Big\|_\infty \le 2^{j/p}\Big\|\sum_{k\in\mathbb{Z}} \alpha_{jk}\Psi_{jk}\Big\|_p.$$

(c) Show that for $p < \infty$ and $\alpha < \frac{1}{p}$ the series

$$\sum_{j<0}\Big(\sum_{k\in\mathbb{Z}} a_{jk}\Psi_{jk}\Big)$$

is an absolutely convergent series of bounded functions.

(d) Suppose that $\alpha > \frac{1}{p}$. For β such that $\frac{1}{2} < \beta < \frac{1}{2} + \alpha - \frac{1}{p}$ define $a_{j0} =: 2^{-j\beta}$ if $j \le 0$ and $a_{jk} = 0$ otherwise. Show that the sequence $(a_{jk})_{j,k\in\mathbb{Z}}$ satisfies E9.2. Assuming additionally that Ψ is continuous and $\Psi(0) \neq 0$ show that the series

$$\sum_{j\in\mathbb{Z}}\sum_{k\in\mathbb{Z}} a_{jk}\Psi_{jk} = \sum_{j<0} 2^{-j\beta}\Psi_{j0}$$

does not represent a function.

9.10 Let Ψ be a C^1 wavelet satisfying 9.27 and assume also that Ψ' satisfies

$$|\Psi'(x)| \le C(1+|x|)^{-A}.$$

Let (a_{jk}) be a sequence of numbers satisfying E9.2 of Exercise 9.9. Show that for each $h \in \mathbb{R}$ the series

$$\sum_{j\in\mathbb{Z}}\sum_{k\in\mathbb{Z}} a_{jk}\big[\Psi_{jk}(x) - \Psi_{jk}(x-h)\big]$$

converges to an $L_p(\mathbb{R})$ function.

9.11 Suppose that f_n is a sequence of functions converging locally in L_p to f, i.e. for each $N > 0$ we have

$$\lim_{n\to\infty}\int_{-N}^{N} |f_n(x) - f(x)|^p\,dx = 0.$$

Show that $\liminf_{n\to\infty}\omega_p(f_n;\delta) \ge \omega_p(f;\delta)$ for each $\delta > 0$. Give an example for which $\liminf_{n\to\infty}\omega_p(f_n;\delta) > \omega_p(f;\delta)$.

9.12 Show that for $0 < \alpha < 1$ and $1 \le p, s < \infty$ any wavelet basis with wavelet satisfying 9.27 is an unconditional basis in $B^p_{\alpha,s}(\mathbb{R})$.

9.13 For a function on the circle $\mathbb{T}$ we define

$$\omega_p(f;\delta) =: \sup_{|h|<\delta} \|f(x+h) - f(x)\|_p.$$

For $0 < \alpha \le 1$ and $1 \le p, s \le \infty$ we define $B^p_{\alpha,s}(\mathbb{T})$, Besov spaces on $\mathbb{T}$, as the space of all functions on $\mathbb{T}$ such that

$$\Big(\int_0^{\pi} t^{-\alpha s-1}\omega_p^s(f;t)\,dt\Big)^{1/s} < \infty$$

when $s < \infty$, and $\sup_{0<t<\pi} t^{\alpha}\omega_p(f;t) < \infty$ when $s = \infty$.

(a) Show that if $\omega_p(f;1) < \infty$ then $f \in L_p(\mathbb{T})$.

(b) For the spaces $\tilde{V}_j =: \operatorname{span}\{\phi_{jk}\}_{k=0}^{2^j-1}$ where the ϕ_{jk} are defined in 8.81 show analogs of Jackson's and Bernstein's inequalities.

(c) Let ψ_{jk} be defined by 8.82. Show that $f \in B^p_{\alpha,s}(\mathbb{T})$ with $0 < \alpha < 1$ if and only if

$$\Big(\sum_{j\geq 0} 2^{j(\alpha+1/2-1/p)s}\Big(\sum_{k=0}^{2^j-1} |\langle f, \psi_{jk}\rangle|^p\Big)^{s/p}\Big)^{1/s} < \infty.$$

9.14 Observe that inhomogenous Besov spaces $B^2_{\alpha,2}(\mathbb{R})$ are Hilbert spaces. Show that $f \in B^2_{\alpha,2}(\mathbb{R})$ if and only if

$$\int_{-\infty}^{\infty} |\hat{f}(\xi)|^2(1+|\xi|)^{2\alpha}\,d\xi < \infty.$$

Appendix

In the first three sections of this appendix we collect several facts and notions used throughout this book. In no way is it intended to serve as a crash course on any subject. It is supposed only to refresh the reader's memory when needed. If the reader really needs to learn some of the subjects hinted at in this appendix he should consult any of the appropriate books listed in the references at the end of each section or any other appropriate textbook. The next two sections are a list of symbols which collects all symbols constantly used in the book and a list of spaces which lists some spaces of functions which are used throughout the book. In the first part of the book they are used simply as a shorthand for some properties of functions. In the second part some of them become function spaces equipped with appropriate structures.

A.1 Hilbert spaces

A.1–I. Let H be a linear space over either the real numbers $\mathbb{R}$ or the complex numbers $\mathbb{C}$. A *scalar product* on H is a function $\langle .,. \rangle$ from $H \times H$ into scalars such that

$$\begin{aligned} \langle x, y \rangle &= \overline{\langle y, x \rangle} \\ \langle \alpha x_1 + \beta x_2, y \rangle &= \alpha \langle x_1, y \rangle + \beta \langle x_2, y \rangle \\ \langle x, x \rangle &\geq 0 \\ \langle x, x \rangle &= 0 \text{ if and only if } x = 0. \end{aligned}$$

In such a situation the function $\|x\| = \sqrt{\langle x, x \rangle}$ is a norm, i.e. it satisfies

$$\begin{aligned} \|x + y\| &\leq \|x\| + \|y\| \\ \|\alpha x\| &= |\alpha| \, \|x\| \\ \|x\| &= 0 \text{ if and only if } x = 0. \end{aligned}$$

A linear space H equipped with a scalar product is a *Hilbert space* if H is complete as a metric space with the metric $d(x, y) = \|x - y\|$.

A.1–II. There are two basic examples of Hilbert spaces. For any subset $A \subset \mathbb{R}^d$, $d = 1, 2, \ldots$, in particular for the whole of $\mathbb{R}^d$ or an interval in $\mathbb{R}$, $L_2(A)$ is the space of all (equivalence classes of equal a.e.) measurable functions f such that

$$\left(\int_A |f(x)|^2 \right)^{1/2} =: \|f\|_2 < \infty.$$

The scalar product is given by

$$\langle f, g \rangle =: \int_A f(x) \overline{g(x)} \, dx.$$

If A is an interval in $\mathbb{R}$, e.g. $A = (a, b]$, then we will save brackets and write $L_2(a, b]$ instead of the more formally correct $L_2((a, b])$.

If B is a countable set then the space $\ell_2(B)$ is the space of all sequences $(a_b)_{b \in B}$ indexed by the set B such that

$$\|b\|_2 =: \left(\sum_{b \in B} |a_b|^2 \right)^{1/2} < \infty.$$

The scalar product is given by

$$\langle a, c\rangle =: \sum_{b\in B} a_b \overline{c_b}.$$

A.1–III. Two vectors x and y in a Hilbert space H are called *orthogonal* if $\langle x, y\rangle = 0$. Two subsets A and B of a Hilbert space H are orthogonal if $\langle a, b\rangle = 0$ for all $a \in A$ and $b \in B$. We denote this by $A \perp B$. A system of non-zero vectors $(x_s)_{s\in S}$ is called an orthogonal system if $\langle x_s, x_{s'}\rangle = 0$ for $s \neq s'$. If we have

$$\langle x_s, x_{s'}\rangle = \begin{cases} 0 & \text{if } s \neq s' \\ 1 & \text{if } s = s' \end{cases}$$

then the system is orthonormal. An orthonormal system $(x_s)_{s\in S}$ is an orthonormal basis in H if one of the following equivalent conditions holds:

- every $x \in H$ can be written as a convergent series $x = \sum_{s\in S} a_s x_s$ for some scalars a_s
- if $\langle x, x_s\rangle = 0$ for all $s \in S$ then $x = 0$
- for every $x \in H$ the series $\sum_{s\in S} \langle x, x_s\rangle\, x_s$ converges to x.

A.1–IV. For any number $l > 0$ the system $\left(\frac{1}{\sqrt{l}} e^{2\pi kit/l}\right)_{k\in\mathbb{Z}}$ is an orthonormal basis in the space $L_2[0, l]$. If B is a countable set, then the system $(e_b)_{b\in B}$ of sequences indexed by B where

$$e_b(a) = \begin{cases} 0 & \text{if } a \neq b \\ 1 & \text{if } a = b \end{cases}$$

is an orthonormal basis in $\ell_2(B)$. This basis is called a *unit vector basis* in $\ell_2(B)$.

A.1–V. If $(x_s)_{s\in S}$ is an orthonormal basis in a Hilbert space H then for any $x \in H$ we have

$$\|x\| = \Big(\sum_{s\in S} |\langle x, x_s\rangle|^2\Big)^{1/2}.$$

A.1–VI. If $T : H \to H$ is a continuous linear operator then T^* is a continuous linear operator defined by the relation $\langle Tx, y\rangle = \langle x, T^*y\rangle$ for all $x, y \in H$. We have $\|T\| = \|T^*\|$. An operator T is called *self-adjoint* if $T = T^*$. An operator U is called *unitary* if it is invertible and $\langle Ux, Uy\rangle = \langle x, y\rangle$. Also when $T : H_1 \to H_2$ for two different Hilbert spaces H_1 and H_2, we can define $T^* : H_2 \to H_1$ by the condition $\langle Tx, y\rangle = \langle x, T^*y\rangle$ and then also $\|T^*\| = \|T\|$.

A.1–VII. Let $T : H_1 \to H_2$ be a continuous linear operator between two Hilbert spaces. We say that T is an isomorphism if T^{-1} exists and is continuous. This is equivalent to T being onto and $\|Tx\| \geq c\|x\|$ for some positive c. If T is an isomorphism then T^* is also an isomorphism.

A.1–VIII. Suppose that $(X_s)_{s\in S}$ is a system of closed linear subspaces of H which are pairwise orthogonal. If 0 is the only vector from H which is orthogonal to all X_s then each vector $x \in H$ can be written as $x = \sum_{s\in S} x_s$ with $x_s \in X_s$. If we have two orthogonal subspaces X_1 and X_2 in a Hilbert space H, then by $X_1 \oplus X_2$ we denote the direct sum of X_1 and X_2, i.e. the subspace of H consisting of all vectors $x_1 + x_2$ with $x_i \in X_i$.

A.1–IX. For each closed linear subspace $X \subset H$ there exists a unique orthogonal projection P from H onto X which has the following properties:

- P is a continuous linear operator and $\|P\| = 1$
- $P(H) \subset X$ and $P(x) = x$ for every $x \in X$
- $\ker P \perp X$
- $P^* = P$.

If $X \subset Y$ are closed subspaces of a Hilbert space H then there exists a unique subspace $W \subset Y$ such that $Y = X \oplus W$. This subspace will sometimes be denoted as $W =: Y \ominus X$.

References. Everything mentioned above is well known and can be found in many textbooks on functional analysis or Hilbert spaces, e.g. [119], [6] or [98].

A.2 Fourier transforms

The aim of this section is to collect some facts, formulas and theorems about Fourier transforms and Fourier series.

A.2–I. For a function $f \in L_1(\mathbb{R}^d)$ we define its Fourier transform, denoted as $\hat{f}(\xi)$ or $\mathcal{F}f(\xi)$, by the formula

$$\hat{f}(\xi) =: \mathcal{F}f(\xi) =: (2\pi)^{-d/2} \int_{\mathbb{R}^d} e^{-i\langle \xi, x\rangle} f(x)\, dx.$$

Clearly $\hat{f}(\xi)$ is a bounded function on $\mathbb{R}^d$ and we have

$$\|\hat{f}\|_\infty \leq (2\pi)^{-d/2} \|f\|_1.$$

We can easily extend the definition of a Fourier transform to finite measures on $\mathbb{R}^d$. For such a measure μ we put

$$\hat{\mu}(\xi) =: (2\pi)^{-d/2} \int_{\mathbb{R}^d} e^{-i\langle \xi, x\rangle} d\mu(x).$$

With this generalization $\hat{\mu}(\xi)$ still is a bounded function on $\mathbb{R}^d$. It is also clear that $\mathcal{F}$ is a linear map.

A.2–II. If μ is a finite measure on $\mathbb{R}^d$ then $\hat{\mu}(\xi)$ is a continuous function on $\mathbb{R}^d$. If we assume that $f \in L_1(\mathbb{R}^d)$ then $\hat{f}(\xi) \in C_0(\mathbb{R}^d)$, i.e. is a continuous function vanishing at infinity.

A.2–III. If $h \in \mathbb{R}^d$ we define a translation operator (cf. Definition 2.3) by the formula $T_h f(x) = f(x - h)$. Then

$$\mathcal{F}(T_h f)(\xi) = (T_h f)^\wedge(\xi) = e^{-i\langle h, \xi\rangle} \hat{f}(\xi)$$

and

$$\mathcal{F}\big(e^{i\langle h, x\rangle} f(x)\big)(\xi) = T_h \hat{f}(\xi) = \hat{f}(\xi - h).$$

A.2–IV. **(Plancherel's theorem)** If $f, g \in L_1(\mathbb{R}^d) \cap L_2(\mathbb{R}^d)$ then

$$\langle f, g\rangle = \int_{\mathbb{R}^d} f(x)\overline{g(x)}\, dx = \int_{\mathbb{R}^d} \hat{f}(\xi)\overline{\hat{g}(\xi)}\, d\xi = \left\langle \hat{f}, \hat{g} \right\rangle.$$

This shows that the map $\mathcal{F}$ extends to the unitary operator on $L_2(\mathbb{R}^d)$. For this extension we will use the same notation as for the original transform. Further, when f is in the Schwartz class S and μ is a measure on $\mathbb{R}^d$ of bounded variation, we have

$$\int_{\mathbb{R}^d} f(x)\, d\mu(x) = \int_{\mathbb{R}^d} \hat{f}(\xi)\hat{\mu}(\xi)\, d\xi.$$

A.2–V. The inverse Fourier transform $\mathcal{F}^{-1}$ is given by

$$\mathcal{F}^{-1}f(x) =: \check{f}(x) = (2\pi)^{-d/2} \int_{\mathbb{R}^d} f(\xi)e^{i\langle x,\xi\rangle}\, dx.$$

Note that $\mathcal{F}^{-1}f(x) = \mathcal{F}g(x)$ where $g(\xi) =: f(-\xi)$. This implies that the properties of $\mathcal{F}^{-1}$ are very similar to those of $\mathcal{F}$. In particular all facts mentioned in this Appendix about $\mathcal{F}$ have their counterparts for $\mathcal{F}^{-1}$.

A.2–VI. If A is an invertible linear map $A : \mathbb{R}^d \to \mathbb{R}^d$ then

$$\mathcal{F}(f \circ A)(\xi) = f(Ax)^\wedge(\xi) = (\det A)^{-1}\hat{f}((A^{-1})^*\xi).$$

In particular if $Ax = ax$ for some $a \in \mathbb{R}$ then

$$\mathcal{F}\big(f(ax)\big)(\xi) = |a|^{-d}\hat{f}(\xi/a).$$

A.2–VII. If we put together the above observations we get the formula

$$\mathcal{F}\big(f(Ax+b)\big)(\xi) = |\det A|^{-1}e^{i\langle \xi, A^{-1}b\rangle}\hat{f}((A^{-1})^*\xi).$$

In particular when $d = 1$ we get

$$\mathcal{F}\big(f(ax+b)\big)(\xi) = |a|^{-1}e^{i(b/a)\xi}\hat{f}(\xi/a).$$

A.2–VIII. If we assume that $\frac{\partial f}{\partial x_k}$ exists and is integrable (or in $L_2(\mathbb{R}^d)$) then

$$\left(\frac{\partial f}{\partial x_k}\right)^\wedge(\xi) = i\xi_k\hat{f}(\xi).$$

If $\hat{f}(\xi)$ has an integrable partial derivative then

$$\frac{\partial \hat{f}}{\partial \xi_k}(\xi) = \mathcal{F}\big(ix_kf(x)\big)(\xi).$$

A.2– IX. In particular we have the following: if

$$|\hat{f}(\xi)| \le C(1+|\xi|)^{-N-\varepsilon}$$

for some $\varepsilon > 0$ then all partial derivatives of f of order not greater than $N-d$ are continuous and in $L_2(\mathbb{R}^d)$. In particular when f is in the Schwartz class S then $\hat{f}$ is also in S.

A.2–X. If $f, g \in L_1(\mathbb{R}^d)$ then we can define their convolution as

$$f * g(x) =: \int_{\mathbb{R}^d} f(x-y)g(y)\,dy.$$

This concept can be extended to finite measures on $\mathbb{R}^d$. If μ and ν are finite measures on $\mathbb{R}^d$ then $\mu * \nu$ is the measure defined as

$$\mu * \nu(A) = \int_{\mathbb{R}^d} \mu(A-s)d\nu(s).$$

The convolution has the following properties:

$$\begin{aligned} \|f * g\|_1 &\leq \|f\|_1 \|g\|_1 \\ f * g &= g * f \\ f * (\alpha g_1 + \beta g_2) &= \alpha f * g_1 + \beta f * g_2. \end{aligned}$$

A.2–XI. There is a close connection between convolution and the Fourier transform, namely if we define

$$\mathcal{F}_1(f)(\xi) =: (2\pi)^{d/2}\mathcal{F}(f)(\xi) = \int_{\mathbb{R}^d} f(x)e^{-i\langle \xi, x\rangle}\,dx$$

then we have

$$\mathcal{F}_1(f * g) = \mathcal{F}_1(f) \cdot \mathcal{F}_1(g).$$

Analogously we can define $\mathcal{F}_1$ for measures and with this generalization the above formula still holds.

A.2–XII. If we have an l-periodic function f on $\mathbb{R}$, locally integrable, then we can write its Fourier series

$$\sum_{k=-\infty}^{\infty} \hat{f}(k)e^{(2\pi/l)ikt}$$

where

$$\hat{f}(k) = \frac{1}{l}\int_0^l f(t)e^{-(2\pi/l)ikt}\,dt.$$

The Fourier coefficients $\left(\hat{f}(k)\right)_{k=-\infty}^{\infty}$ uniquely determine the function f. Also $\lim_{|k|\to\infty} \hat{f}(k) = 0$. If the function f is in $L_2[0,l]$ (more formally we mean that $f|[0,l] \in L_2[0,l]$) then the series converges to f in the norm of $L_2[0,l]$. Note that the notation $\hat{f}(k)$ does not reflect the period of the function. This should be clear from the context. If nothing is said we mean 2π-periodic functions.

A.2–XIII. If f is as in A1.2–XII and $f'(x)$ exists and is locally integrable then $f'(x)$ has the Fourier series

$$\sum_{k=-\infty}^{\infty} \hat{f}(k)\frac{2\pi ik}{l}e^{(2\pi/l)ikt}.$$

It implies that if $f'(x)$ is locally integrable then

$$|\hat{f}(k)| \leq \frac{C}{1+|k|}.$$

Conversely, if $|\hat{f}(k)| \leq C(1+|k|)^{s+\varepsilon}$ for some $\varepsilon > 0$ then $f^{(s-1)}(x)$ exists and is continuous.

A.2–XIV. **(Poisson summation formula)** For $f \in L_1(\mathbb{R})$ we define a function $\mathcal{P}(f)(x) = \sum_{n\in\mathbb{Z}} f(x+n)$. This is clearly a 1-periodic function on $\mathbb{R}$ and $\mathcal{P}(f)|[0,1] \in L_1[0,1]$ so $\mathcal{P}f$ is locally integrable. The Fourier coefficients of $\mathcal{P}f$ are connected with the Fourier transform of f as follows:

$$\widehat{\mathcal{P}f}(k) = \sqrt{2\pi}\hat{f}(-2\pi k).$$

A.2–XV. A function f on $\mathbb{R}^d$ is said to be $l\mathbb{Z}^d$-periodic if $f(x) = f(x+l\gamma)$ for each $\gamma \in \mathbb{Z}^d$. If such a function is locally integrable then we can write its Fourier series

$$\sum_{\gamma\in\mathbb{Z}^d} \hat{f}(\gamma)e^{(2\pi/l)i\langle\gamma,x\rangle}$$

where

$$\hat{f}(\gamma) =: \left(\frac{1}{l}\right)^d \int_{[0,1]^d} f(t)e^{-(2\pi/l)i\langle\gamma,t\rangle}\,dt.$$

The sequence of Fourier coefficients $\left(\hat{f}(\gamma)\right)_{\gamma\in\mathbb{Z}^d}$ uniquely determines the function f. Also $\lim_{|\gamma|\to\infty} \hat{f}(\gamma) = 0$. If the function f is in $L_2[0,1]^d$ (more formally we mean that $f|[0,1]^d \in L_2[0,1]^d$) then the series converges in $L_2[0,1]$ norm.

A.2–XVI. If f is as in A1.2–XV and $\frac{\partial f}{\partial x_j}$ exists and is locally integrable then $\frac{\partial f}{\partial x_j}$ has the Fourier series

$$\sum_{\gamma\in\mathbb{Z}^d} \hat{f}(\gamma)\frac{2\pi}{l}i\gamma_j e^{(2\pi/l)i\langle\gamma,x\rangle}$$

where $\gamma = (\gamma_1,\ldots,\gamma_d)$. This implies that if each $\frac{\partial f}{\partial x_j}$ is locally integrable then $|\hat{f}(\gamma)| \leq C(1+|\gamma|)^{-1}$. Conversely, if $|\hat{f}(\gamma)| \leq C(1+|\gamma|)^{s+\varepsilon}$ for some $\varepsilon > 0$, then $f \in C^{s-d}(\mathbb{R}^d)$.

A.2–XVII. **(Poisson summation formula)** For $f \in L_1(\mathbb{R}^d)$ we can define a function $\mathcal{P}f(x) = \sum_{\gamma\in\mathbb{Z}^d} f(x+\gamma)$. This is clearly a $\mathbb{Z}^d$-periodic function on $\mathbb{R}^d$ and $\mathcal{P}f \mid [0,1]^d \in L_1[0,1]^d$, so $\mathcal{P}f$ is locally integrable. The Fourier coefficients of $\mathcal{P}f$ are connected with the Fourier transform of f as follows:

$$\widehat{\mathcal{P}f}(\gamma) = \left(\sqrt{2\pi}\right)^d \hat{f}(-2\pi\gamma).$$

References. Everything mentioned above is well known and can be found in many textbooks, e.g. [37], [57],[98], [105] or [108].

A.3 Banach spaces

A.3–I. Let X be a linear space over the real numbers $\mathbb{R}$ or complex numbers $\mathbb{C}$. A norm on X is a function $x \mapsto \|x\|$ from X into the non-negative reals such that

$$\begin{aligned} \|x+y\| &\leq \|x\| + \|y\| \\ \|\alpha x\| &= |\alpha| \cdot \|x\| \text{ for all numbers } \alpha \\ \|x\| &= 0 \text{ if and only if } x = 0. \end{aligned}$$

A linear space equipped with a norm is called a normed space. Every norm gives a metric on X by the formula $d(x, y) =: \|x - y\|$. If X is complete when equipped with this metric we call $(X, \|\,.\,\|)$ a Banach space.

A.3–II. If X and Y are two Banach spaces and $T : X \to Y$ is a linear operator, then by $\|T\|$ we denote

$$\sup\{\|Tx\| : \|x\| \leq 1\}.$$

An operator T is continuous if and only if $\|T\| < \infty$. For this reason we use the phrase 'T is bounded' as synonymous with 'T is continuous'. Clearly we have $\|Tx\| \leq \|T\| \cdot \|x\|$.

A.3–III. It is clear that if T is a linear map (into some Banach space) defined on a dense linear subspace Z of a Banach space X and if

$$\sup\{\|Tx\| : x \in Z \text{ and } \|x\| \leq 1\} < \infty$$

then T extends uniquely to a continuous linear operator from X into Y. Also if a sequence of operators $T_n : X \to X$ is such that $\sup \|T_n\| < \infty$ and $T_n(z) \to z$ for each $z \in Z$, then $T_n(x) \to x$ for every $x \in X$. Analogously if $T_n(z) \to 0$ for each $z \in Z$, then $T_n(x) \to 0$ for all $x \in X$.

A.3–IV. An operator $T : X \to Y$ is an isomorphic embedding if there are constants $0 < c \leq C$ such that for all $x \in X$ we have $c\|x\| \leq \|Tx\| \leq C\|x\|$. T is an isomorphism if additionally it is *onto* Y. An operator $P : X \to X$ is a projection if $P^2 = P$. Two norms $\|\,.\,\|^1$ and $\|\,.\,\|^2$ on a linear space X are called equivalent if there are constants $0 < c \leq C$ such that $c\|x\|^1 \leq \|x\|^2 \leq C\|x\|^1$ for all $x \in X$. Equivalent norms define the same topological structure on X, i.e. the identity acts as an isomorphism between $(X, \|\,.\,\|^1)$ and $(X, \|\,.\,\|^2)$. An operator T is continuous with respect to the norm $\|\,.\,\|^1$ if and only if it is continuous with respect to the norm $\|\,.\,\|^2$.

A.3–V. A linear map from a Banach space X into the scalars is called a linear functional on X. Since the scalars form a Banach space with absolute value as a norm, the above facts A1.3–II and A1.3–III hold in particular for functionals. The space of all continuous (linear) functionals on X is denoted by X^* and is called the dual space of X. It is a Banach space when equipped with the norm of a functional. The Hahn–Banach theorem asserts in particular that for each closed, linear subspace $Y \subset X$ and each $x \in X \setminus Y$ there exists a linear functional $x^* \in X^*$ such that $x^*(y) = 0$ for all $y \in Y$ but $x^*(x) = 1$. It also asserts that for each $x \in X$ there exists a functional $x^* \in X^*$ such that $x^*(x) = \|x^*\| \cdot \|x\|$.

A.3–VI. Let us consider the space $L_p(A)$ for some subset $A \subset \mathbb{R}^d$. For $1 \leq p < \infty$ the dual space $L_p(A)^*$ equals $L_q(A)$ where as usual $\frac{1}{p} + \frac{1}{q} = 1$. This means that for every continuous linear functional $\varphi \in L_p(A)^*$ there exists a unique function $g_\varphi \in L_q(A)$ such that

$$\varphi(f) = \int_A f(x) g_\varphi(x)\, dx$$

and the norm of the functional φ equals $\|g_\varphi\|_q$. Conversely, each $g \in L_q(A)$ gives by the above formula a functional on $L_p(A)$. The same holds for sequence spaces ℓ_p. Note that the classical Hölder's inequality can be stated as $|\varphi(f)| \leq \|f\|_p \|g_\varphi\|_q$.

A.3–VII. If $T : X \to Y$ is a continuous linear operator between Banach spaces then its adjoint operator $T^* : Y^* \to X^*$ is defined as $T^*(y^*)(x) =: y^*(Tx)$. It is a continuous linear operator and $\|T\| = \|T^*\|$. T maps X *onto* Y if and only if T^* is an isomorphic embedding of Y^* into X^*. P is a projection in X if and only if P^* is a projection in X^*.

References. All the above facts can be found in (almost) every textbook on functional analysis, e.g. [98] or [6].

A.4 List of symbols

$\mathbb{R}$	The set of real numbers, the real line
$\mathbb{Z}$	The set of integers
$\mathbb{N}$	The set of natural numbers
$\mathbb{C}$	The set of complex numbers
$\mathbb{T}$	The unit circle in the complex plane $\mathbb{C}$, i.e. the set of all complex numbers of absolute value 1
$\mathcal{F}$	The Fourier transform of a function or a measure normalized to be a unitary map
$\mathcal{F}_1$	$(2\pi)^{d/2}\mathcal{F}$
$\langle .,. \rangle$	The scalar product, either in the abstract Hilbert space or in $L_2(\mathbb{R}^d)$, i.e. $\int f(x)\overline{g(x)}\,dx$. Also in $\mathbb{R}^d$ or $\mathbb{C}^d$, i.e. $\langle x, y\rangle = \sum_{i=1}^d x_i \overline{y_i}$.
$\vert . \vert$	Either absolute value of a number or the euclidean norm of a vector in $\mathbb{R}^d$ or Lebesgue measure of a subset of $\mathbb{R}^d$ or cardinality of a discrete set
$\mathbf{1}_A$	The indicator (or characteristic) function of the set A $$\mathbf{1}_A(x) = \begin{cases} 1 & \text{if } x \in A \\ 0 & \text{if } x \notin A \end{cases}$$
$\vert\vert\vert f \vert\vert\vert$	The H_1 norm of the function f
$\Vert.\Vert_p$	The norm of the function in $L_p(A)$ or the norm of a sequence in ℓ_p, $1 \le p \le \infty$
$\Vert.\Vert_\infty$	The sup-norm; the supremum of absolute value of a function or a sequence
$\Vert.\Vert_{*,p}$	The norm in BMO_p defined in Definition 6.10

$\|\cdot\|_*$ The generic bounded mean oscillation norm; see comments after Corollary 6.17

$\|\cdot\|_{p,\alpha,s}$ The Besov norm as defined in Definition 9.4

$=:$ The equality which defines one side

$\Im z$ The imaginary part of a complex number z

$\Re z$ The real part of a complex number z

supp f The support of a function f

Supp f The smallest closed *interval* containing the support of f

$c \diamond Q$ For a cube Q this denotes the cube with the same center whose sides are c times the sides of Q

f_Q The mean value of the function f on a cube Q, i.e.

$$f_Q =: \frac{1}{|Q|} \int_Q f(x)\, dx$$

$\mathcal{P}f$ The periodization of a function f on $\mathbb{R}^d$, i.e.

$$\mathcal{P}f(x) =: \sum_{\gamma \in \mathbb{Z}^d} f(x+\gamma)$$

$F_{jk}(x)$ For a function F on $\mathbb{R}$ this is $2^{j/2}F(2^j x - k)$ where $j, k \in \mathbb{Z}$

$F_{j\gamma}(x)$ For a function F on $\mathbb{R}^d$ and a dilation matrix A (to be understood from the context) this is $|\det A|^{j/2}F(A^j x - \gamma)$ where $j \in \mathbb{Z}$ and $\gamma \in \mathbb{Z}^d$. In particular for the dilation matrix $Ax = 2x$ this is $2^{jd/2}F(2^j x - \gamma)$.

A.5 List of spaces

$L_p(\mathbb{R}^d)$ If A is any measurable subset of $\mathbb{R}^d$, in particular the whole of $\mathbb{R}^d$, then $L_p(A)$ with $1 \le p \le \infty$ is the space of all (classes of equal a.e.) measurable functions such that when $1 \le p < \infty$

$$\|f\|_p =: \left(\int_A |f(x)|^p \, dx \right)^{1/p} < \infty.$$

If $p = \infty$ then $L_\infty(A)$ is the space of all (classes of equal a.e.) essentially bounded measurable functions on A. We use only $A = \mathbb{R}^d$ or A an interval in $\mathbb{R}$ or a cube in $\mathbb{R}^d$.

$C(\mathbb{R}^d)$ The space of all continuous functions on $\mathbb{R}^d$

$C_0(\mathbb{R}^d)$ The space of all continuous functions on $\mathbb{R}^d$ such that the limit $\lim_{|x|\to\infty} f(x)$ exists and equals 0

$C_{00}(\mathbb{R}^d)$ The space of all continuous functions on $\mathbb{R}^d$ with compact support, i.e. functions $f \in C(\mathbb{R}^d)$ such that $f(x) = 0$ if $|x| > R$ for some R (depending on the function)

$C^k(\mathbb{R}^d)$ For $k = 1, 2, \ldots$ this denotes the space of all functions on $\mathbb{R}^d$ which have partial derivatives up to the order k continuous. For $k = 0$ we mean $C^0(\mathbb{R}^d) = C(\mathbb{R}^d)$ and for $k = -1$ we mean by $C^{-1}(\mathbb{R}^d)$ the space of all measurable functions on $\mathbb{R}^d$

$C^\infty(\mathbb{R}^d)$ The space of all infinitely differentiable functions on $\mathbb{R}^d$, i.e. $C^\infty(\mathbb{R}^d) = \bigcap_{k=1}^\infty C^k(\mathbb{R}^d)$

$C_{00}^\infty(\mathbb{R}^d)$ The space of all C^∞ functions with compact support, i.e. $C_{00}^\infty(\mathbb{R}^d) = C^\infty(\mathbb{R}^d) \cap C_{00}(\mathbb{R}^d)$

$C_{00}^k(\mathbb{R}^d)$ The space of all C^k functions with compact support, i.e. $C^k(\mathbb{R}^d) \cap C_{00}(\mathbb{R}^d)$

$S(\mathbb{R}^d)$ The Schwartz class, i.e. the space of all functions f belonging to $C^\infty(\mathbb{R}^d)$ such that for each multiindex $\alpha =$

$(\alpha_1, \ldots, \alpha_d)$ with $\alpha_i \geq 0$ and each $k = 0, 1, \ldots$ there exists a constant $C = C(f, \alpha, k)$ such that

$$|\partial^\alpha f(x)| \leq C(1 + |x|)^{-k}.$$

$H_1^p(\mathbb{R}^d)$ The Hardy space obtained using p-atoms, see Definition 6.12

$H_1(\mathbb{R}^d)$ The Hardy space; see Theorem 6.18

$BMO_p(\mathbb{R}^d)$ The space of functions with bounded p-mean oscillation, see Definition 6.10

$BMO(\mathbb{R}^d)$ The space of functions of bounded mean oscillation, see Corollary 6.17 and comments after it

$MC_p(\mathbb{R})$ The space of functions having p-modulus of continuity, see Definition 9.2

$B^p_{\alpha,s}(\mathbb{R})$ An inhomogenous Besov space, see comments after Definition 9.4 on page 220

$\dot{B}^p_{\alpha,s}(\mathbb{R})$ A homogenous Besov space, see comments after Definition 9.4 on page 220

REMARK The reader may have observed that all the above symbols denoting spaces of functions consist of two parts: the first indicating the nature of the functions and the second ($\mathbb{R}$ or $\mathbb{R}^d$ or A in the above list) indicating the set on which the functions are defined. Quite often we will omit the part indicating on which set the functions are defined. Then the set should be clear from the context or nonessential. Thus e.g. the phrase 'C^k function' simply means 'function all of whose partial derivatives up to order k are continuous'. This is a standard usage with which the reader is most likely already familiar. In any case it should not cause any problems.

Bibliography

[1] P. Auscher, *Solution of two problems on wavelets*, J. Geometric Analysis 5(2) (1995) pp. 181–236

[2] G. Battle, *A block spin construction of ondelettes. Part I: Lemarie functions*, Comm. Math. Phys. 110 (1987) pp. 601–615

[3] G. Battle, *Phase Space Localization for Ondelettes*, J. Math. Phys. 30–10 (1989) pp. 2195–2196

[4] C. Bennet & R. Sharpley, **Interpolation of operators**, Academic Press 1988

[5] O. V. Besov, *On a family of function spaces: embedding and extension theorems*, Doklady AN SSSR 126 (1959) pp. 1163–1165 (in Russian)

[6] B. Bollobas, **Linear Analysis**, Cambridge University Press, Cambridge 1990

[7] A. Bonami, F. Soria & G. Weiss, *Band limited wavelets*, J. Geometric Analysis, 3(6) (1993) pp. 543–578

[8] M. Bownik, *Wavelety i podziały samopodobne* $\mathbb{R}^n$, Master's thesis at Warsaw University, 1995 (in Polish)

[9] Ch. Brandt & G. Gelbrich, *Classification of self-affine lattice tilings*, J. London Math. Soc. (2) 50 (1994) pp. 581–593

[10] D. Burkholder, *A proof of Pelczyński conjecture for the Haar system*, Studia Math. 91 (1988) pp. 79–83

[11] L. Carleson, *An explicit unconditional basis in* H_1, Bull. des Sci. Math. 104 (1980) pp. 405–416

[12] P. G. Casazza & O. Christensen, *Frames containing a Riesz basis and preservation of this property under perturbations*, preprint

[13] P. G. Casazza & O. Christensen, *Hilbert space frames containing a Riesz basis and Banach spaces which have no subspace isomorphic to* c_0, preprint

[14] Sun-Yung A. Chang & Z. Ciesielski, *Spline characterisation of* H^1, Studia Math. 75 (1983) pp. 183–192

[15] C. K. Chui, **An Introduction to wavelets**, Academic Press 1992

[16] Z. Ciesielski, *Properties of the orthonormal Franklin system II*, Studia Math. 27 (1966) pp. 289-323

[17] Z. Ciesielski, *The* $C(I)$ *norms of the orthogonal projections onto subspaces of polygonals*, Trudy Math. Steklov Institut AN SSSR, 134 (1975) pp. 366–369

[18] Z. Ciesielski, *Spline bases in spaces of analytic functions*, in **Canadian Math. Soc. Conference Proceedings 3; Second Edmonton Conference on Approximation Theory** Amer. Math. Soc., Providence R.I. 1983 pp. 81–112

[19] Z. Ciesielski & J. Domsta, *Construction of an orthonormal basis in* $C^m(I^d)$ *and* $W_p^m(I^d)$, Studia Math. 41 (1972) pp. 211–224

[20] A. Cohen, *Ondelettes, analyses multiresolutions et filters miroir en quadrature*, Annales Inst. H. Poincaré, Anal. non-linéaire 7 (1990) pp. 439–459

[21] A. Cohen & I. Daubechies, *Non-separable bidimensional wavelet bases*, Revista Mat. Iberoamericana, (1) (1993) pp. 51–137

[22] R. R. Coifman & G. Weiss, *Extensions of Hardy spaces and their use in analysis*, Bull. Amer. Math. Soc. 83 (1977) pp. 569–645

[23] X. Dai, D. R. Larson & D. Speegle, *Wavelet sets in* $\mathbb{R}^n$, preprint

[24] I. Daubechies, **Ten lectures on wavelets**, Society for Industrial and Applied Math., Philadelphia 1992

[25] I. Daubechies, *Orthonormal bases of compactly supported wavelets*, Comm. Pure Appl. Math 41 (1988) pp. 909–996

[26] I. Daubechies & Ying Huang, *How does truncation of the mask affect a refinable function*, Constructive Approx. 11 (1995) pp. 365–380

[27] G. David, **Wavelets and singular integrals on curves and surfaces**, Lecture Notes in Math. 1465, Springer-Verlag 1991

[28] S. Demko, *Inverses of band matrices and local convergence of spline projections*, SIAM J. Numer. Anal. 14(4) (1977) pp. 616–619

[29] S. Demko, W. F. Moss & Ph. W. Smith, *Decay rates for inverses of Band Matrices*, Math. of Computation, 43(168) (1984) pp. 491–499

[30] G. Derfeld, N. Dyn & D. Levin, *Generalised refinement equation and subdivision processes*, J. Approx. Th. 80(2) (1995) pp. 272–297

[31] G. de Rham, *Sur un example de fonction continue sans derivée*, Enseign. Math. 3 (1957) pp. 71–72

[32] R. A. DeVore & B. J. Lucier, *Wavelets*, Acta Numerica 1991, pp. 1–56

[33] R. A. DeVore, B. Jawerth & V. Popov, Compression of wavelet decompositions, Amer. J. Math. 114 (1992) pp. 737–785

[34] R. A. DeVore, P. Petrushev & X. M. Yu, *Non-linear wavelet approximation in the space* $C(\mathbb{R}^d)$, in **Progress in Approximation Theory** A. A. Gonchar & E. B. Saff eds., Springer-Verlag 1992, pp. 261–283

[35] Z. Ditzian & V. Totik, **Moduli of Smoothness**, Springer-Verlag, Berlin 1987

[36] D. L. Donoho, *Unconditional bases are optimal bases for data compression and for statistical estimation*, Applied and Computational Harmonic Analysis, 1 (1993) pp. 100–115

[37] H. Dym & H. P. McKean, **Fourier series and integrals**, Academic Press 1972

[38] K. J. Falconer, **The geometry of fractal sets**, Cambridge University Press, Cambridge 1985

[39] C. Fefferman & E. Stein, H_p *spaces of several variables*, Acta Math. 129 (1972) pp. 137–193

[40] G. B. Folland & E. Stein, **Hardy Spaces on homogenous groups**, Princeton University Press, Princeton, NJ 1982

[41] Ph. Franklin, *A set of continuous orthogonal functions*, Math. Annalen 100 (1928) pp. 522–29

[42] M. Frazier, B. Jawerth & G. Weiss, **Littlewood–Paley theory and the study of function spaces**, CBMS Regional Conference Series in Math. 79, Amer. Math. Soc. 1991

[43] J. Garcia-Cuerva & J. Kazarian, *Spline wavelet bases in weighted L_p spaces $1 < p < \infty$*, Proc. Amer. Math. Soc. 123(2) (1995) pp. 433-439

[44] J. Garcia-Cuerva & J. L. Rubio de Francia, **Weighted norm inequalities and related topics**, Math. Studies 116, North–Holland, Amsterdam 1985

[45] G. Gripenberg, *Wavelet bases in $L_p(\mathbb{R})$*, Studia Math. 106(2) (1993) pp. 175–187

[46] G. Gripenberg, *A necessary and sufficient condition for the existence of a father wavelet*, Studia Math. 114(3) (1995) pp. 207–226

[47] K.-H. Gröchenig, *Analyse multiéchelle et bases d'ondelettes*, C. R. Acad. Sci. Paris 305 Série I (1987) pp. 13–17

[48] K. Gröchenig, *Orthogonality criteria for compactly supported scaling functions*, Appl. Computational Harmonic Analysis 1 (1994) pp. 242–245

[49] K. Gröchenig & A. Haas, *Self–similar lattice tilings*, J. Fourier Analysis 1 (1994) pp. 131–170

[50] K. Gröchenig & W. R. Madych, *Multiresolution analysis, Haar bases and self–similar tilings of $\mathbb{R}^n$*, IEEE Trans. Inform. Theory, 38(2), (March 1992) pp. 556–568

[51] Young-Hwa Ha, Hyeonbae Kang, Jungseob Lee & Jin Keun Seo, *Unimodular wavelets for L^2 and the Hardy space H^2*, Michigan Math. J. 41 (1994) pp. 345–361

[52] A. Haar, *Zur theorie der orthogonalen Funktionensysteme*, Math. Annalen 69 (1910) pp. 331–371

[53] D. Jackson, **The theory of approximation**, Amer. Math. Soc. Colloquium Publ. XI, New York 1930

[54] S. Jaffard, *Exposants de Hölder en des points donnés et coefficients d'ondelettes*, C. R. Acad, Sci, Paris 308 Serie 1 (1989) pp. 79–81

[55] F. John & L. Nierenberg, *On functions of bounded mean oscillation*, Comm. Pure Appl. Math. 14 (1961) pp. 415–426

[56] B. S. Kashin & A. A. Saakian, **Orthogonal series**, Nauka, Moscow 1984 (in Russian), also English translation Amer. Math. Soc. 1989

[57] Y. Katznelson, **An introduction to harmonic analysis**, John Wiley & Sons, New York 1968

[58] S. E. Kelly, M. A. Kon & L. A. Raphael, *Pointwise convergence of wavelet expansions*, Bull. Amer. Math. Soc. 30 (1994) pp. 87–94

[59] J. C. Lagarias & Y. Wang, *Integral self–affine tiles in $\mathbb{R}^n$ I. Standard and non-standard digits sets*, J. London Math. Soc. (to appear)

[60] J. C. Lagarias & Y. Wang, *Integral self–affine tiles in $\mathbb{R}^n$ II. Lattice tilings* preprint

[61] J. C. Lagarias & Y. Wang, *Haar bases for $\mathbb{R}^n$ and algebraic number theory*, J. Number Theory (to appear)

[62] J. C. Lagarias & Y. Wang, *Haar-type orthonormal wavelet bases in $\mathbb{R}^2$*, J. Fourier Anal. Appl. 2(1) (1995) pp. 1–14

[63] Ka-Sing Lau & Jianrong Wang, *Characterisation of L_p-solutions for two-scale dilation equation*, SIAM J. Math. Anal. 26(4) (1995) pp.

1018–1046

[64] W. Lawton, *Necessary and sufficient conditions for constructing orthonormal wavelet bases*, J. Math. Phys. 32 (1991) pp. 57–61

[65] W. Lawton, *Tight frames of compactly supported wavelets*, J. Math. Phys. 31 (1990) pp. 1898–1901

[66] P. G. Lemarie, *Ondelettes à Localisation Exponentielle*, Journ. de Math. Pures et Appl. 67 (1988) pp. 227–236

[67] P. G. Lemarie, *Base d'ondelettes sur les groupes de Lie stratifiés*, Bull. Soc. Math. France, 117 (1989) pp. 211–232

[68] P. G. Lemarie, *Some remarks on wavelet theory and interpolation*, preprint Orsay 91–13

[69] P. G. Lemarie-Rieusset, *Projecteurs invariants, matrices de dilatation, ondelettes de dimension n et analyses multi–resolution*, Revista Mat. Iberoamericana 10(2) (1994) pp. 283–347

[70] P. G. Lemarie-Rieusset, *Distributions dont tous les coefficients d'ondelettes sont nuls*, C. R. Acad. Sci. Paris, Série I 318(12) (1994) pp. 1038–86

[71] P. G. Lemarie & G. Malgouyres, *Support des fonctions de base dans une analyse multi-resolution*, C. R. Acad. Sci. Paris 313 (1991) pp. 377–380

[72] J. Lindenstrauss & L. Tzafriri, **Classical Banach spaces I**, Springer-Verlag, Berlin 1977

[73] J. Lindenstrauss & L. Tzafriri, **Classical Banach spaces II**, Springer-Verlag, Berlin 1979

[74] J. Lippus, *Wavelet coefficients of functions of generalised Lipshitz classes*, in Proc. Workshop 'Sampling Theory and Applications SampTA95' Jurmula, Latvia pp. 167–172

[75] R. A. Lorentz & A. Sahakian, *Orthonormal trigonometric Schauder bases of optimal degree for $C(K)$*, J. Fourier Analysis and Appl. 1(1) (1994) pp. 103–112

[76] W. R. Madych, *Some elementary properties of multiresolution analyses of $L_2(\mathbb{R}^d)$*, in **Wavelets – a tutorial in theory and applications**, C.H. Chui ed., Academic Press 1992, pp. 259–294

[77] S. Mallat, *Multiresolution approximation and wavelet orthonormal bases of $L_2(\mathbb{R})$*, Trans. Amer. Math. Soc. 315 (1989) pp. 69–88

[78] J. Marcinkiewicz, *Quelques theorèmes sur les séries orthogonales*, Ann. Soc. Polon. Math. 16 (1937) pp. 84–96

[79] J. Marcinkiewicz, *Sur l'interpolation d'operations*, C. R. Acad. Sci. Paris 208 (1939) pp. 1272–1273

[80] P. R. Massopust, **Fractal functions, fractal surfaces and wavelets**, Academic Press, New York 1994

[81] Y. Meyer, *Principe d'incertitude, bases hilbertiennes et algebres d'operateur*, Seminaire Bourbaki 38 no. 662 (1985–6)

[82] Y. Meyer, *Constructions de bases orthonormées d'ondelettes*, Revista Mat. Iberoamer. 4(1) (1988) pp. 31-39

[83] Y. Meyer, *Wavelets and operators*, in **Analysis in Urbana 1** ed. E. R. Berkson *et al.*, London Math Soc. Lecture Notes Series 137, Cambridge University Press, Cambridge 1989, pp. 256–365

[84] Y. Meyer, **Ondelettes et opérateurs**, Hermann, Paris 1990

[85] Y. Meyer, **Wavelets and operators**, Cambridge University Press, Cambridge 1992; English translation of the first volume of [84]

[86] Y. Meyer, *Wavelets: their past and their future* in **Progress in wavelet analysis and applications** ed. Y. Meyer, S Roques, Editions Frontières, Gif-sur-Yvette 1993 pp. 9-18

[87] J. Morlet & A. Grossman, *Wavelets, ten years ago* in **Progress in wavelet analysis and applications** ed. Y. Meyer, S Roques, Editions Frontières, Gif-sur-Yvette 1993 pp. 3-7

[88] I. P. Natanson, **Constructive function theory**, Moscov 1949 (in Russian)

[89] D. Offin & K. Oskolkov, *A note on Orthonormal Polynomial Bases and Wavelets*, Constructive Approx. 9 (1993) pp. 319–325

[90] R. E. A. C. Paley, *A remarkable series of orthogonal functions I*, Proc. London Math. Soc. 34 (1932) pp. 241–264

[91] J. Peetre, **New thoughts on Besov spaces**, Duke University Math. Series, Duke University Press, Durham, NC 1976

[92] S. Pitter, J. Schneid & Ch. W. Ueberhuber, **Wavelet Literature Survey** Technical University Vienna, Wien, Austria 1993

[93] G. Plonka & M. Tasche, *A unified approach to periodic wavelets*, in **Wavelets, theory algorithms and applications** ed. C. K. Chui, L. Montefusco, L. Puccio; Academic Press, San Diego 1994 pp. 137–151

[94] D. Pollen, *Daubechies' scaling function on [0,3]*, in **Wavelets – a tutorial in theory and applications**, C.K. Chui, ed. Academic Press, Boston 1992, pp. 3–13

[95] A. A. Privalov, *On the growth of degrees of polynomial basis and approximation of trigonometric projectors*, Mat. Zametki 42(2) (1987) pp. 207–214 (in Russian)

[96] A. A. Privalov, *On an orthonormal trigonometric basis*, Mat. Sbornik 182(3) (1991) pp. 384–394

[97] S. Ropela, *Spline bases in Besov spaces*, Bull. Acad. Pol. Sci. serie math., astr., phys. 24(5) (1976) pp. 319–325

[98] W. Rudin, **Functional Analysis**, McGraw-Hill, New York 1973

[99] M. J. Schauder, *Einige Eingenschaft der Haarschen Orthogonalsystems*, Math. Zeit. 28 (1928) pp. 317–320

[100] L. Schumaker, **Spline functions; basic theory**, J. Wiley & Sons, New York 1981

[101] P. Sjölin & J.-O. Strömberg, *Spline systems as bases in Hardy spaces*, Israel J. Math. 45 (1983) pp. 147–156

[102] P. Sjölin & J.-O. Strömberg, *Basis properties of Hardy spaces*, Arkiv för mat. 21 (1983) pp. 111–125

[103] G. Soares de Souza, *Spaces formed by special atoms I*, Rocky Mnts. J. of Math. 14 (1984) pp. 423–41

[104] E. Stein, *Harmonic Analysis: real-variable methods, orthogonality and oscillatory integrals*, Princeton Math. series 43, Princeton University Press, Princeton, NJ 1993

[105] E. Stein & G. Weiss, **Introduction to Fourier Analysis on Euclidean spaces**, Princeton University Press, Princeton, NJ 1971

[106] R. S. Strichartz, *Wavelets and self–similar tilings*, Constr. Approx. 9 (1993) pp. 327–347

[107] J.-O. Strömberg, *A modified Franklin system and higher order spline systems on $\mathbb{R}^n$ as unconditional bases for Hardy spaces*, in **Conference in Harmonic Analysis in Honor of A. Zygmund** vol. II, Wadsworth, Belmont 1983, pp. 475–493

[108] A. Torchinsky, **Real-variable methods in harmonic analysis**, Academic Press, New York 1986

[109] H. Triebel, **Theory of function spaces**, Birkhäuser-Verlag, Basel 1983

[110] H. Volkmer, *Asymptotic regularity of compactly supported wavelets*, SIAM J. Math. Anal. 26(4) (1995) pp. 1075–1087

[111] G. G. Walters, *Pointwise convergence of wavelet expansions*, J. Approx. Theory, 80(1) (1995) pp. 108–118

[112] R. O. Wells jr., *Parametrizing smooth compactly supported wavelets*, Trans. Amer. Math. Soc. 338(2) (1993) pp. 919–931

[113] P. Wojtaszczyk, *The Franklin system is an unconditional basis in* H_1, Arkiv för matematik, 20 N.2 (1980) pp. 293–300

[114] P. Wojtaszczyk, H_p*–spaces*, $p \leq 1$ *and spline systems*, Studia Math. 77 (1984) pp. 289–320

[115] P. Wojtaszczyk, *Two remarks on the Franklin system*, Proc. Edinburgh Math. Soc. 29 (1986) pp. 329–333

[116] P. Wojtaszczyk, **Banach spaces for analysts**, Cambridge Studies in Advanced Math. 25, Cambridge University Press, Cambridge 1991

[117] P. Wojtaszczyk & K. Woźniakowski, *Orthonormal polynomial bases in function spaces*, Israel J. Math. 75 (1991) pp. 167–191

[118] K. Woźniakowski, *Orthonormal polynomial basis in* $C(\Pi)$ *with optimal growth of degrees*, Israel J. Math. (to appear)

[119] N. J. Young, **Introduction to Hilbert spaces**, Cambridge University Press, Cambridge 1988

Index